BENTLEY
FACTORY CARS
1919–1931

Christmas Greetings
1927

BENTLEY
FACTORY CARS
1919-1931

Michael Hay

First published in Great Britain in 1993 by Osprey
an imprint of Reed Consumer Books Limited
Michelin House, 81 Fulham Road
London SW3 6RB
and Auckland, Melbourne, Singapore and Toronto

A catalogue record for this book is available from the British Library.

ISBN 1-85532-240-4

Editor Shaun Barrington
Page design Gwyn Lewis
Typeset by Tradespools Ltd, Frome, Somerset
Printed by BAS Printers Limited, Over Wallop, Hampshire
Jacket photography: Richard Bird.

Contents

Acknowledgements

INEVITABLY, IN PRODUCING a book of this type, contributions and assistance come from many quarters.

Perhaps first should be the late Michael Frostick's *Bentley Cricklewood to Crewe*, in many respects the forerunner of this book.

However, the task would not even have been started without the Bentley Drivers Club's archives, and the author is grateful for having been allowed ready access to these. Principal among the archives are Bentley Motors' own press cuttings books, which were presented to the Club by Rolls-Royce many years ago. Bentley retained a press cuttings agency, Romeike & Curtis, the cuttings then being pasted into large books kept at Hanover Court and later at Pollen House. These proved an invaluable source of reference. Also of particular use were Woolf Barnato's similar press cuttings books, Nobby Clarke's writings, both handwritten and typed, which add a distinctive element to the narrative, and personal correspondence and papers of WO's. My particular thanks go to Barbara Fell and Bill Port at the BDC offices in Long Crendon.

Marjorie Burgess, F.T. Burgess' daughter, provided early correspondence between WO and Burgess in 1920 while the latter was still in Coventry, along with a copy of the 1928 Annual Report, documents which shed invaluable light onto certain aspects of the Old Company's activities, as well as a wealth of photographs. Mrs Valerie Owen-Hughes and Mrs Willow Prizeman, H.M. Bentley's daughters, produced many fascinating photographs and some insights into the workings of H.M. Similarly, David Northey very kindly produced many of WO's photos and papers.

The Public Records Office at Kew very kindly provided a full set of photocopies of all the records relating to Lecoq and Fernie, Bentley and Bentley, the multifarious Bentley Motors Companies and the Aerolite Piston Co. Those for Amherst Villiers Superchargers Ltd. and Baromans have unfortunately been destroyed. The Public Records Office at Chancery Lane similarly provided copies concerning Villiers v Bentley Motors in 1929, The London Life Association v Bentley Motors and Woolf Barnato in 1931, and Bentley Motors (1931) Ltd v Lagonda and W.O. Bentley in 1945. The British Newspaper Library at Colindale provided access to *The Times* and various motoring magazines of the period.

The Rolls-Royce Ltd. experimental file on Bentley Motors, preserved at the Rolls-Royce Enthusiasts' Club headquarters at Paulerspury, gave an invaluable insight into the Bentley car from the viewpoint of a contemporary manufacturing concern.

Ian Lloyd's *Rolls-Royce – the Years of Endeavour* provided much of the financial data for the appendices. The sales figures come from WO's own notebook, preserved by the BDC.

Michael Evans of the Rolls-Royce Heritage Trust made available various papers relating to the periods 1930–32 and 1944–45, which added greatly to the accuracy of chapters dealing with the events of those years.

The usual published works (WO's three books, Hillsteads' two, *The Other Bentley Boys* et al) have all been used as reference sources to correlate details and piece together the story.

Those parts of the book dealing with the Works have been greatly enhanced by recollections from past employees, notably Frank Ayto, Jack Baker,

Albert Deavin, Rivers Fletcher, Ernest Hanson, Walter Hassan, George Hawkins, Albert Holdsworth, W.G. Plant, Clarence Rainbow, Billy Rockell and Mrs Dora Steele. To all of these the author owes a particular debt of gratitude.

The author would also like to acknowledge various assistance from the late Jim Bentley, Mrs Jim Bentley, Mrs Arthur Watson, Mrs Diana Barnato-Walker, Mrs Janet Towers, Victor Barclay of Jack Barclay Ltd., John Fasal, Ray Roberts and Dr John Porter. Norman Lilley skilfully made copies from many faded photographs.

Finally, my thanks go to Tim Houlding, Hugh Young and Tom Clarke, who read the manuscript and made numerous improvements.

Photographic Acknowledgements

Patents material is reproduced courtesy of the Controller, Her Majesty's Stationery Office. The aerial view of the Works in 1925 is reproduced courtesy of Hunting Aerofilms Ltd. Some drawings are reproduced courtesy of *Autocar & Motor* magazine. My thanks for the loan of, and permission to reproduce the remaining illustrations, go to: Chas K Bowers & Sons Ltd., Dr T D Houlding, W D S Lake esq., Mrs V Owen-Hughes, Mrs W Prizeman, Mrs J Towers, Mrs J Bentley, N Portway esq., Miss M Burgess, A F Rivers Fletcher esq., W Hassan esq., C Rainbow esq., P Streete esq., J R Andrews esq., J Weedon esq., J Pike esq., R Fox esq., J Wallace esq., D L P Humfrey esq., J Mackinnon esq., A Cocks esq., J Fasal esq; and especial thanks to the Bentley Drivers Club.

Introduction

WHEN *Thoroughbred & Classic Cars* reviewed Michael Frostick's book *Bentley Cricklewood to Crewe* (Osprey Publishing) in February 1981, the review concluded that "the definitive Bentley history has yet to be written". This book attempts to fill that gap, but only concentrates on the Cricklewood era, up to and including the Rolls-Royce takeover. While the ground has indeed been well trodden in the past, many of the accounts have been written by the participants, and are inevitably coloured by their viewpoints. Michael Frostick put it very succinctly: "there is...an understandable wish, when elderly gentlemen write their life stories, to gloss over the rough with the smooth." Indeed, if those accounts are placed together, certain of the discrepancies become difficult to overlook. The story is distributed between the published works, none of which can really be said to be comprehensive. This book largely arose as an attempt to bring together all the disparate accounts and reconcile them, while adding material from other sources in order to create a coherent narrative that is probably as near to the truth as is possible, given that over sixty years have passed since the events described ended.

The basic framework of the book is a chronological history, covering events prior to 1919 as necessary to give the requisite background, and ending in late 1931 when Bentley Motors ceased to exist and Bentley Motors (1931) Ltd had taken over, acting as a subsidiary of Rolls-Royce. The scope of the book covers the personalities, the commercial and financial aspects, the engineering and development work, and, in overview, the motor racing. The latter activity is inseparable from the history of the Old Company, and has such an impact on all the other activities that it cannot be left out. In essence, it is intended to cover all the aspects of Bentley Motors except the production cars themselves, which are dealt with in the author's *Bentley – The Vintage Years*. Some of the technical and development sections have been split from the main text, as they tend to break up the story and are perhaps too detailed. The reader may skip over these or read them separately to choice. The author makes no pretence to being an automobile engineer, and these sections may appear simplistic to those more versed in motor car engineering. There is no Technical section on the 4 Litre, because the technical aspects of this model are integral to considerations of Company policy in 1931.

A number of myths and legends have grown up around the Cricklewood Bentley, and there are those who might be critical of attempts to dispel these. However, this approach does a great disservice to all those involved in Bentley Motors. It ignores the fact that they were human, and subject to human failings and weaknesses. The story of any company is that of the individuals involved, of their efforts to achieve success in a chosen field, or perhaps, as in this case, of their efforts to stave off or deal with calamity.

The reputation of both WO himself and his motor car were forged in the white heat of international competition, and nothing can diminish that legacy. It is to be hoped that a fuller understanding of the difficulties against which WO and his devoted followers struggled to produce a world-beating motor car will serve to enhance this reputation.

Michael Hay

1 DFP Days

THE STORY OF the halcyon years of Bentley Motors is inextricably linked with that of one man, Walter Owen Bentley, or just "WO" as he has long been known. WO was born on 16th September 1888, "at the usual disagreeably early hour of the morning", the ninth child of well-to-do parents living in a large house in Avenue Road, near Regent's Park, North London and was known from an early age as "the bun". It seems almost an impertinence to write about someone else's early life, except insofar as is necessary, and in WO's case, he has written his own story in *WO – An Autobiography* and *The Cars in My Life*, later more or less amalgamated and published in WO's own words as "The complete WO" entitled *My Cars and My Life*. Suffice to say that WO developed a love for steam engines and railways at an early age. The interesting aspect of WO's account of his time at Lambrook (his prep school) and Clifton College are the insights into his character, and the facets that were to result in future successes. WO was single-minded about the subjects he was interested in, particularly physics and chemistry, and in these subjects he needed to see the logical development of argument and to understand each step before advancing to the next. A lack of understanding at any stage resulted in retracing the steps until he was happy. This dogged persistence and the need to understand crops up again later on, particularly during the aero-engine period and the design of EXP 1 in 1919/1920.

From Clifton, WO went to the Great Northern Railway Co. plant in Doncaster as a premium apprentice, the two separated by a brief period at King's College in London studying theoretical engineering, an experience that was clearly not to the young Bentley's liking. It was at Doncaster that he learnt a great deal of practical engineering, and an innate feel for the subject in general. WO was evidently deeply enamoured of the railway engine, and determined to make the grade. "To be even thirty seconds late was an unforgiveable crime. The Doncaster regime really was a tough one and you had to be a devoted disciple to survive it. The first session was from 6 to 8.15 am, when there was a break for breakfast; then a four hour stretch to one o'clock, and the final one from 2 to 5.30. Including half Saturdays it was just short of a sixty hour week, and there was no slacking, no knocking off for cups of tea and gossips. It was hard going every day and the timekeeping and discipline were to military standards." (*My Life and My Cars*.) For this, the premium apprentices paid their way in, at a cost of £75 for the five years. During this period WO bought himself a bicycle, and the means of independent transport this afforded him was a source of considerable pleasure. Bicycles were soon followed by motorcycles which became faster and more powerful at each trade-in and were increasingly used for trials, both speed trials and long distance observed trials. In all these WO achieved a not inconsiderable amount of success, often with his brothers Horace Milner (who will later be referred to as HM) and Arthur competing as well. Although WO's first love was the railway, and it would persist throughout his life, it was clear that the industry would not offer an adequate position for an ambitious young man with fairly well-developed tastes.

The railway did, though, furnish WO with a real gift: "It is not often that childhood dreams mature into reality; it is rarer still for there to be no disappointment or disillusion when they do. I realise

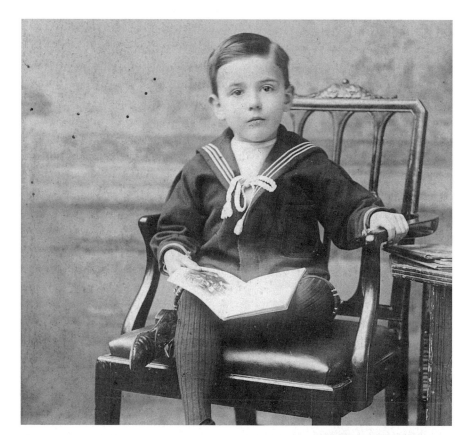

RIGHT **A young WO, aged three. 'The Bun', looking very solemn in sailor suit and shorts.**

BELOW **The Bentley brothers all together. Captioned on the back is the line 'sadder as age creeps on'! From left to right, William, Hardy, Leonard, Arthur, Horace (HM), Walter (WO).**

ABOVE LEFT **The Bentley family at the back of No. 78, Avenue Road. Emily and Alfred Bentley are in the centre, WO top right with HM beneath him.**

LEFT **Motorcycles – WO on his 3½ hp Speed Model Rex, the machine he bought in 1909 and started racing on. This is probably the bike that WO crashed in the 1909 Tourist Trophy.**

BELOW **From WO's own negative, the start of a race.**

With apologies for the print, WO's Indian – his favourite bike, a machine possessed of an 'astonishing performance'. WO rode an Indian in the 1910 TT, retiring because of tyre problems. He sold his second Indian soon after, part-exchanging it for a Riley and transferring his allegiances from two wheels to four.

now how lucky I was that the sensation of being on the footplate of a Great Northern Atlantic, heading an express north out of London, was more thrilling and wonderful than I had ever thought it could have been. I was fascinated by the feeling of power as we pulled out of King's Cross, up the steep gradients and tunnels of north London, up the steady grind for another eleven miles to Potters Bar, and by the sudden irresistible surge of acceleration when the track levelled off and fell away. There is nothing I know to compare with the sensation of rushing through the night without lights and with that soothing mechanical rhythm beating away continuously, even leading to a dangerous tendency to surrender to the power quivering beneath the steel floor. And then the signals flash into view, your absolute guide and master, and from time to time the lights of a town, the searing white flash of a station – and back into darkness.'' (*My Life and My Cars*.)

It must have been somewhat galling to reach the end of a long, hard training only to find that the arrival left so much to be desired. WO was by then well aware of the potential of the internal combustion engine, both on land and in the air, not least from his own motorcycling activities. As WO said, ''Internal combustion had to be taken seriously'', and his next step was a letter to E.M.P. Boileau, a friend who was on the staff of *The Autocar* magazine. This contact was successful, and in 1910 WO started at the National Motor Cab Company in Hammersmith, with the responsibility for the

maintenance of a fleet of 250 two cylinder Unic taxis. These bright red cabs were all landaulettes, with ample passenger and luggage space. A good natured form of undercover warfare was conducted between WO and the Cab Company on one side and the drivers on the other. Every ploy by the drivers to fiddle the company had to be found out and met by counter measures, all without words being exchanged as good cabbies were hard to come by and would have been eagerly snapped up by their rivals. WO obviously enjoyed this, even at one point disguising a private detective as a cabby to find out where some of the drivers were having a minute hole drilled in the glass of the meters so they could be wound back with a pin! While it is unlikely that WO's salary was very substantial, this did not reduce his social life. ''There were Henry Wood Promenade Concerts at Queen's Hall, and of course music-halls at the Palace and Empire, with late suppers at the Piccadilly to the music of de Groot's orchestra to round off an evening.'' 'Sometimes – but only as a last resort – I would hail one of our cabs in the West End late at night, and then there would be no peace the next day. 'Bentley gallivanting round the bright lights again last night – coo, *e's a masher, Bentley is!*'' (*My Life and My Cars*.) WO was to stay for only two years, joining Lecoq and Fernie in 1912, and that event marks the beginning of the Bentley story.

Lecoq and Fernie were established in December 1911 as an agency for three makes of French car, La Licorne, Buchet and DFP, (Doriot, Flandrin and Parant). The firm was based in a converted stables at the back of a block of service flats in Hanover Street, described as Hanover Court Yard, Hanover Street, in their letterhead. This company was established to take over and carry on the business of G.A. Lecoq Ltd, with Major (retd.) Francis Hood Fernie and George Arthur Lecoq as the principal shareholders. Between them they held 1000 £1 shares, two fully paid up and 998 ''for considerations other than cash''. The secretary of the concern was G.P.H. de Freville. Lecoq was the chairman and managing director of a firm of trunk and suitcase manufacturers, Vuitton Trunks. Without any attempts to push sales, the cars sold at a slow rate but Lecoq and Fernie themselves seemed content to let the business slide into a state of decline. When their advert appeared for a director to join the firm in 1912 the business was more or less moribund, Lecoq being too busy making suitcases and Fernie proving to be a blustering, red faced, ineffectual individual, only good at pushing people around. Initially the family thought L.H.

Bentley, another of the Bentley brothers, would join, as he had lost interest in farming and had some money to invest. Lecoq and Fernie were more interested in HM, a chartered accountant, but in the end it was WO who joined the firm in January of that year, being elected as a director on the 15th. Lecoq and Fernie had already disposed of the franchises for Buchet and La Licorne, retaining that for the DFP.

Although the firm seemed dead, WO knew that there was a lot of money to be made out of agencies for good continental cars. In his opinion the DFP was a well-made sporty car, with first class steering and brakes and that indefinable quality that makes certain cars a pleasure to drive. What was needed was some enthusiasm and publicity to generate sales, but such an occurrence was unlikely while Lecoq and Fernie themselves remained in the business. WO discussed the situation with HM, who suggested that they should buy out the other two. WO was reluctant to sink into the general feeling of

WO with the DFP, circa 1912/14.

despair at Hanover Street, and in a short time proposals were under way for Lecoq and Fernie to sell their remaining shares to HM. WO intimates that was what Lecoq and Fernie had hoped would happen, and HM was elected as a director on 14th February 1912. Fortunately there was money in the Bentley family. WO's grandfather, Thomas Waterhouse, had left his daughter, Emily Bentley, £60,000 – the interest being paid to her (from which source WO's allowance emanated), with the principal being divided among the children on her death. It is conceivable that without this source of funds there would have been no Bentley Motors.

In May 1912 a new company Bentley and Bentley Ltd was formed, a special resolution being passed in June to effect the change in name. Of the share capital of £1000, £500 each was held by WO and HM. Apart from the premises at Hanover Street, occupied by HM, a secretary and G.P.H. de Freville, the Sales Manager, they also had a service station in New Street Mews, rented from the coachbuilders, Easters, and used by WO as a base for his racing activities. Shortly afterwards they obtained Leroux as the chief and only mechanic from the Paris factory. Leroux performed every function from preparing and servicing customers cars to building, tuning and acting as riding mechanic in WO's racing machines.

Three models of DFP were offered, the 10–12 HP, the 12–15 HP and the 16–20 HP. Prices of these ranged from £275 to £555. The customer was offered a wide range of bodywork and could, presumably, have had any style he desired built on his particular chassis. Of these three, the 10–12 was a pleasant but unremarkable car, the 16–20 heavy and sluggish, but the 2 litre 12–15 HP was a much livelier car with its 70x130 mm long stroke engine. It was the tuning potential of the latter that interested WO, who had Leroux easing things up and increasing the compression ratio of his own car with a view to an assault on the Aston Hill-Climb on the 8th June, 1912. Right from the start WO decided that racing was the best means of attracting publicity for minimal expense, and HM was soon converted to this opinion. It seems remarkable that WO should have set the fastest time of the day for Class 2 in a time that was a record for a 2 litre car at Aston, first time out without even knowing the hill, but that is what he did. DFPs won five other hill-climbs in that first year, and Class 2 at Aston was to be a DFP benefit until 1914.

In the closing months of 1912, Harrisons of

ABOVE **High-grade six light limousine coachwork on a DFP, with very modern lines for a pre-Great War car.**

LEFT **Another DFP saloon, photographed in Hanover Square. Hanover Street stretches away in the background.**

TOP RIGHT **The record-breaking 12/15 DFP at Brooklands in November 1913, with WO at the wheel. This was taken at the time of the record attempts of 15 November, when the DFP set six new Class B records including the 1 Hour at 82.15 mph.**

CENTRE RIGHT **Further development of the 12/15 DFP, circa February 1914. In that month the DFP broke six Class B records ranging from the half-mile to ten miles, at speeds of up to 89.7 mph.**

RIGHT **The DFP in final form at Brooklands, with further streamlining to the nose cone. WO is on the left, Leroux is working on the engine.**

ABOVE **Tuck at the wheel of the Humber, WO's rival for Class B records in 1913/14.**

LEFT **AHMJ Ward, director of Bentley and Bentley, the Aerolite Piston Co., and Bentley Motors. Ward was married to Edith, WO's sister.**

Drummond Street fitted a light 2 seat shell body with a pointed tail to the 12–15, and in this form on November 9th, they set the 10 lap record at Brooklands at 66.78 mph. In this car WO entered into a series of ding-dong battles with Tuck, Humber's chief tester, at hill climbs and at Brooklands. The DFP broke Tuck's records only for the Humber to take them back again. WO journeyed to the factory at Courbevoie, on the outskirts of Paris, to meet M Doriot, less to discuss ways of making his car go faster than to ask whether the resulting modifications could be incorporated into cars for the UK market. Doriot was already rather surprised at the speeds WO and Leroux had extracted from his family car, but it was the toy on his desk that most interested WO. This was a piston cast in aluminium by DFP's foundry, and it set WO's mind working. He later contacted DFP and asked them to have a set of pistons cast, in aluminium, a proposal met with grave doubt by the parent company, but the foundry were interested and worked out a formula with 12% copper to 88% aluminium and the pistons were duly fitted to WO's 12–15. Previous attempts to extract more power had failed

THIS PAGE **WO at the wheel of the 12/40 Speed Model DFP, prepared for the 1914 Tourist Trophy race. These photos look to have been taken just after it was finished, in Hanover Square, judging by the railings. Note the detachable rear portion of the scuttle, made to hinge in the same fashion as the bonnet.**

THIS PAGE **The DFP pit, with Leroux (bottom left) and WO with reversed cap and goggles.**

because of piston failure, ring breakage with steel pistons or cracking with cast iron. Initial results with the new pistons were good, and after further increases in compression ratio it was found that higher power outputs were possible without detrimental effects on reliability. These changes were incorporated in a new 12–40 Speed Model for the British market, and the marked increase in speed allowed WO to beat Tuck once more at Brooklands. While the DFP story is not strictly relevant to the story of Bentley Motors, it is nevertheless a very important aspect to the creation of the Bentley cars and shows WO's abilities as a development engineer, to take a chassis and extract the maximum from it. The 10 lap record was raised from 66.78 mph to 77.93 mph on 3rd September 1913 and then to 81.98 mph on 15th November of the same year. In February 1914 the DFP set the flying half mile record at 89.7 mph, certainly an impressive figure for a 2 litre car.

A.H.M.J. Ward joined the board of Bentley and Bentley on 1st May 1914. Ward was an electrical engineer by profession and was married to Edith,

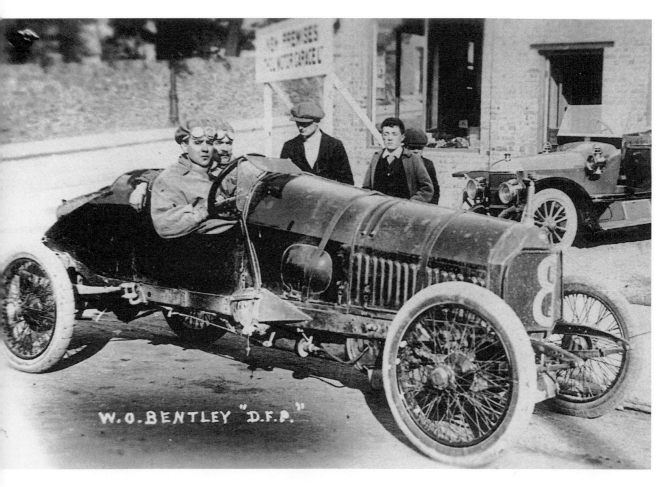

W.O.BENTLEY "D.F.P."

one of the Bentley sisters. As Ward later held virtually a majority share in Bentley and Bentley (exactly half of the Company) it is strange that he was never mentioned in any of WO's books. A.F.C. Hillstead, Bentley and Bentley's salesman, however, had a lot to say about Ward in *50 Years with Motor Cars*, and it is clear he played a very important role in the early days.

Although WO has long been credited as the first to use aluminium pistons, as with most ideas it seems to have occurred to more than one person at a similar time but in isolation from each other. According to a letter in *The Vintage & Thoroughbred Car* for November 1954, Corbin in France had been making aluminium pistons since 1910, the letter suggesting that it was their pistons that were used by WO in the 1914 TT. Corbin's pistons were apparently also used by Chenard-Walcker and Panhard-Levassor. The 1088cc Violet-Bogey entered in the 1913 Amiens GP for cyclecars was also reported as using aluminium pistons. It is interesting that shortly after the start of the war, Hillstead (who was unfit for military service) applied

WO and Leroux in the DFP on the Isle of Man, the DFP looking rather well-used. Doriot's son can be seen in the centre. Note the streamlining to the handbrake/gear lever and the bulge in the body for the driver's feet.

for a job as a salesman at Bentley and Bentley, and was shown round by Ward. Ward said that HM and WO had already joined up, and he (Ward) was principally concerned at the time in promoting a small company called The Aerolite Piston Company, which had been founded "just before the war" to market aluminium pistons, about which venture more will be said later.

In a fit of confidence, the Bentley brothers entered the little DFP for the 1914 Tourist Trophy race on the Isle of Man, held on the 10th and 11th June. The race was run over 600 miles on a winding course with steep climbs. The regulations allowed a maximum engine size of 3310 cc, and a minimum weight of $21\frac{1}{2}$ cwt. In the event, quite apart from an engine capacity roughly two-thirds of the other

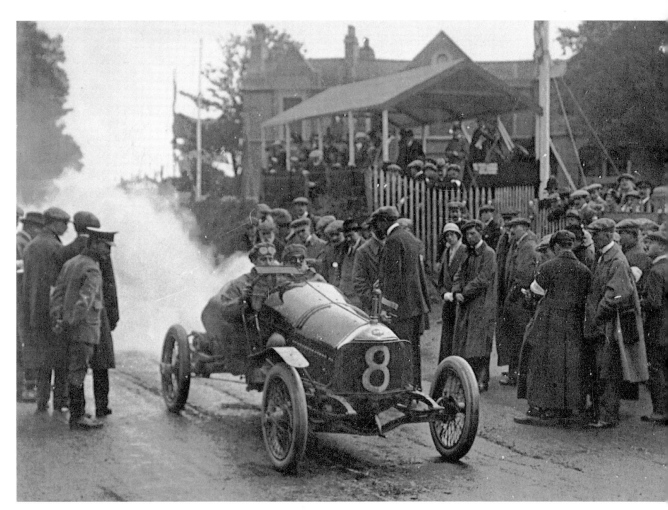

cars, the DFP was also some 2 cwt heavier. WO already knew the circuit from his three previous entries in the motor cycle TT races. On the day the DFP (sometimes referred to as "Deserves First Place") performed impeccably to finish 6th, albeit last, at an average of 48.38 mph, witnessed by M Doriot himself. They even had to use the axles off the larger 16–22 model to meet the regulations! Of the 22 starters the DFP entry probably attracted more attention from the press in attendance than the winning Sunbeam.

Another notable entrant in that year was a team of three TT Humbers, led by F.T. Burgess who was responsible for the design. None of the over-head camshaft cars, based on Henri principles, finished, but Burgess and the Henri design philosophy derived from the 1912 Grand Prix

LEFT **The DFP at the start of the race, with a mudguard lashed-up over the offside front wheel; the wet and miserable conditions can be clearly seen. This card was sent to WO by Bertie Kensington Moir in 1933. Moir wrote on the back: 'Dear WO, Thought you'd like this found it in a shop here. Am looking after a 2.3 Bugatti for today's race. Moir'.**

FAR LEFT, BELOW **Half-way through the TT, the DFP passes the score-board to complete another lap. Seven of the 22 starters have already retired.**

BELOW **The DFP on the circuit.**

W.O. BENTLEY.

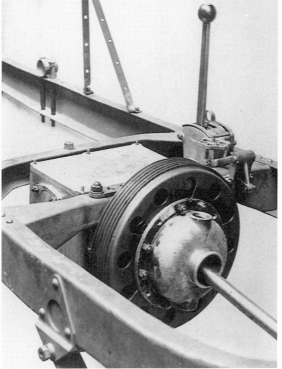

ABOVE **The 1914 TT Humber. This car is significant not just because of the Burgess connection, but because of the influence that the design of this car had on the 3 Litre Bentley. The front axle, in particular, is virtually identical to the unbraked Bentley axle apart from the dip in the centre of the beam.**

ABOVE RIGHT **The back axle bears a remarkable resemblance to the 3 Litre Bentley, and the frame configuration and the rear chassis cross-member are also very similar.**

LEFT **Midships view of the Humber chassis. Again, the gearbox bears a striking resemblance to the A type Bentley box, although the transmission brake is a feature that was not used by Bentleys.**

RIGHT **Close-up of the Humber front axle. The general configuration of this assembly and the front of the frame, with the ball-joint to the top steering arm and forked ends to the track-rod arms, was closely followed on the 3 Litre Bentley.**

Peugeots were to have a significant impact on Bentley Motors in later years. With four valves per cylinder the engines produced about 100 bhp at 3200 rpm, a bore and stroke of 82 x 156 mm giving a capacity of 3295cc. Two Straker-Squires were entered and again, that team was to have subsequent connections with Bentley Motors. Straker-Squire 1, which finished 4th at an average of 52.75 mph was driven by R.S. Witchell, later to be Works Manager at Bentley's Cricklewood factory. The second car was driven by Frank Clement, later head of the Bentley Experimental/Racing Shops and a prodigious driver of racing Bentleys. Clement retired on the second day with a broken piston.

LEFT **The TT Humbers outside the Humber Works. FT Burgess to the left of No.2, the car he drove in the race, with his mechanic Arthur Saunders on the right. Saunders later became head of the racing shop at Bentleys.**

BELOW LEFT **FT and Mrs Burgess with the three TT Humber cars in 1914.**

BELOW **The 1914 TT brought together many of the future Bentley staff. One of the Straker-Squires was driven by RS Witchell, seen here, later works manager at Bentleys.**

Following the TT success, Bentley and Bentley persuaded DFP to produce a 12–40 Tourist Trophy Speed Model using their aluminium pistons and wire wheels, but world events were to overtake these plans. On the 1st August, less than two months after the Tourist Trophy Race, war was declared following the shooting at Sarajevo. By the 5th of that month, Great Britain had been drawn into the conflict.

WO had married Leonie Withers on New Year's Day 1914 at Hampstead Parish Church, but had not taken a proper honeymoon until August only to have it cut short. HM joined up with the Army and WO applied to join one of the new armoured car divisions, going to Derby on the 9th October 1914 for the purpose, only to be told to go home and wait until things had sorted themselves out a little. It became increasingly obvious to WO that he could contribute more to the war effort with his aluminium pistons, a secret which as far as he knew nobody else was aware of. This was not as easy as it might seem, as widespread publication of the idea would give it to the other side as well. In due course he was signed up by Commander Briggs, and was attached to the Royal Naval Air Service's Technical Department as a member of the Royal Naval Volunteer Reserve in June 1915. After convincing Rolls-Royce and Sunbeams of the merits of the aluminium piston, WO was sent by Briggs to Gwynnes at Chiswick to sort out the Clerget rotary engine. Sunbeam's Louis Coatalen commented "Bentley, you're one of the best salesmen I've ever met" - an interesting insight. Within weeks of the start of the war, Leroux was killed in Flanders.

The other TT Straker-Squire was driven by Frank Clement, here seen at Brooklands. Clement became head of the experimental department at Bentleys in 1921, and drove in virtually every major race that Bentley Motors entered.

2 Aero Engines

IN THE FIRST WORLD War period, three basic patterns of aero-engines were made – rotaries, radials, and in-lines. The in-line pattern is simply an engine with the cylinders in line, vee or w formation, driving the propellor off the end of the crankshaft. The other two have the cylinders arranged in a ring, the connecting rods converging onto a very short crankshaft. In the case of a radial, the cylinders remain stationary and the propeller is bolted to the rotating crankshaft. In the case of the rotary, the crankshaft remains stationary and the propeller is bolted to the revolving cylinder assembly. Many of the fighter aircraft of the era had rotaries, because of their high power/weight ratio and the absence of vulnerable (to small arms fire) cooling systems. The rotary is also comparatively short, an important feature in such aircraft as the Sopwith Camel, which was only 18′ 9″ long. It was with the French-designed rotary Clerget that WO was to become involved.

The problems with the Clerget's reliability centred around the obturator ring. Rotary engines are air-cooled, the result being that the leading side of the cylinder is substantially better cooled than the trailing side. In practice this leads to a non-circular cylinder bore because of differential thermal expansion. Clerget's solution to this was a thin light alloy washer, the obturator ring, intended to take up this distortion. In practice this ring tended to break up leading to immediate piston seizure, resulting in an engine life of approximately 15 hours and as neither the RNAS nor the RFC used parachutes, the result of an engine failure was often fatal. Something needed to be done, and reading between the lines it seems that Briggs, recognising WO's abilities as a development engineer, sent him

to Gwynnes to do rather more than merely convert them to the use of aluminium pistons as had been the case with Rolls-Royce and Sunbeams.

WO obviously felt a considerable responsibility for the lives of the men flying behind these engines and on several occasions by-passed procedures to solve problems or effect improvements. A problem with a batch of incorrectly tempered, brittle oil pump springs was sorted out by WO by having a batch made up properly and then taking them over to France himself on a destroyer in a suitcase. On another occasion, an engine on test was found to give slightly more power than normal, a phenomenon eventually traced to a small hole in the induction pipe. Soon after, WO and a batch of small drills were on their way to France to show the squadron fitters where to drill the induction pipes of their engines.

It wasn't long before WO wanted to design his own engine, but things were far from a bed of roses at Gwynnes. It was gradually made clear that they would rather drop WO and his ideas than their Clerget, then being worked on by another designer imported from France for the purpose. WO made several attempts to improve the obturator ring itself, but more importantly wanted to attack the heat distribution problem by using an aluminium cylinder with a cast iron liner shrunk in.

In March 1916 WO outlined his ideas to Briggs and volunteered to produce the designs for a new engine. Special permission was given by Commodore Murray Sueter, head of the Admiralty Air Department, for WO to do the design work at home, as the Technical Department had no facilities. Gwynnes themselves were not much in favour of WO's new design, for political rather than for

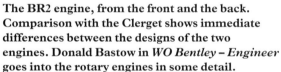

The BR2 engine, from the front and the back. Comparison with the Clerget shows immediate differences between the designs of the two engines. Donald Bastow in _WO Bentley – Engineer_ goes into the rotary engines in some detail.

technical reasons. WO had to circumvent them by making up a single cylinder to his own design attached to a Clerget crankcase to prove his point. On the strength of the results he obtained a transfer to Humber at Coventry who welcomed him with open arms and work proceeded rapidly on the Admiralty Rotary engine, later known as the Bentley Rotary and then the BRl after the design of the BR2 was completed.

Superficially, the new engine looked like the Clerget with which WO was well familiar, a state of affairs that has led to some acrimony in more recent years, but there is no doubt but that the Bentley Rotary was a significant improvement on the Clerget in certain areas. In the circumstances it would have been madness for WO to have created a completely new design when there were sound aspects of a proven engine available, and his decision to take the best of what was available and redesign the rest to the same standard was without doubt the correct decision. The design philosophy was well explained by WO in _An Illustrated History of the Bentley Car_: "There was nothing revolutionary or comic about

those rotaries of mine, apart from no [obturator] ring and the wide use of alloys. This was no time for experimentation. The Clerget valve gear worked very well, and was duplicated in the BRs, with the obvious advantage of interchangeability of spares, a tremendously important factor in the field." In practice the BR1 differed from the Clerget principally in the cylinder design and the method of attaching the cylinders to the crankcase. The resulting engine proved reliable and powerful. The BRl was received with delight by the squadrons and by the early part of 1918 WO was working on the BR2 rotary, a bigger and more powerful design. In fact, the BR2 was conceived first, but WO was instructed by Briggs to design the BR1 before the bigger engine. The new Clerget was known as the "LS", or long stroke, rated at 140 bhp to 130 bhp of the standard Clerget. The BR1 was rated at 150 bhp.

WO's accounts of his trips to the squadrons are detailed in _W.O. – An Autobiography_, including his first encounter with Petty Officer R.A. "Nobby" Clarke in a canal near an aerodrome when they were being strafed by none other than Baron Manfred von Richthofen. Many years later Nobby wrote an account of his war, and of his time at Bentleys (Nobby became Bentley Motors first shop floor employee in 1919, and remained with Bentley Motors (1931) Ltd until he retired in 1957); the basis of an autobiography that never saw the light of

The Sopwith Camel, probably the best British fighter of the Great War. The Camel soon gained a reputation as a difficult plane to fly, and it killed many novice pilots. This was largely because of the weight concentration in the nose and the gyroscopic effect of the rotary engine, which meant the Camel could turn to the right incredibly quickly, but only comparatively slowly to the left.

day. The following extract refers to his encounters with WO and the introduction of the first Bentley rotary-engined Sopwith Camel to Nobby's squadron:

"Around the mess table, mechanics discussed the merits and demerits of the new machine from the limited information at their disposal. Said one "Alum Pistons" "Alum Cylinders" they want their "Ruddy heads" seeing to – why another rotary engine – haven't we enough trouble with the present one – don't all "rotaries" suffer from the same disease.

"What about a good water-cooled stationary engine, such as the new engine in the FE2D or the Hispano Suiza in the SPAD, hadn't we any engineers who could design a good Aero engine, admittedly the RAF engine on the RE8 had its shortcomings, but it was air cooled, no water jackets to be shot away. Anyhow, he for one disliked these rotary contraptions. Look how much better off the night bomber men were, with their stationary engines – no top overhaul at eight hours and complete overhaul at sixteen hours, all hours of the night – simply change engines when they were unserviceable, and send the "old uns" back to base – no more hangar night work, with one eye on the job, and the other to see if the "Ack Ack" gun's crew had been "alerted" to await the sneak Hun raid from the sea, at low level.

"Then the conversation drifted to the plane's constructional features: dihedral bottom plane, flat top-plane, twin Vickers guns, foolproof interrupter gear, wonderful climb, ceiling 18,000 feet, all these reveries were disturbed by the "croaking" of the squadron Dismal Jonah – "that's what you think, old boy", enumerated this great man after cadging his second "Tot" from one and an issue of "Cooper Nail" from another – "I can't see," said he, with the air of an expert, "what good this departure from standard practice will do, except to make the crate unstable and vicious. GET INTO A SPIN, with it mate and unless you've got height, you've had it"–and so the merits and demerits of the new machine, were debated far into the night, enlivened at intervals by shouts of "Pipe down – put a sock in it" and other service "epithets" designed to close down old arguments in the service of those days.

"Morning broke with an H.O.P. [offensive patrol] by "A Flight" and the Buzz that Flight Commander "B" was going to collect the first Camel allocated to the Squadron.

"During that morning, a small, humped-back machine, appeared low over the aerodrome, circled over, hopped over the Dromes, landed, taxied up to the hangars, the pilot switching off his engine and jumping out. It was "S" of "B Flight", he was immediately surrounded by a crowd of curious pilots and mechanics, anxious to have a look, see and to hear first hand, the comments on its behaviour and qualities from one who, to everyone assembled there, was recognised as an "ACE" in every sense of the word, and whose assessment of the new machine would be as near "factual" as a short test flight would be capable of disclosing – "YES" it seemed good, the engine most certainly was excellent, but from one or two things that had happened in the short test flight – it was a machine with which no liberties should be taken – how right did "S's" prophesy prove to be.

"The day wore on, first one pilot, then another flight-tested the machine until quite a few had given it a "flip." "YES" the general comment was "good." "Climb, boy oh boy" could the B---d climb, but NO LIBERTIES – anyhow, for a start, and what an engine, smooth, plenty of power above 10,000 feet, excellent for formation flying.

"And so, around the end of May 1917, The Bentley Aero Engine started to become a byword amongst men who knew, and men who had cause to know a good thing when they saw, flew or maintained it. The change-over from Pups to Camels was effected smoothly and the day arrived for the first all Camel Flight H.O.P. when much to the surprise of Hun Fighters in the area, they were outclimbed, out manoeuvred, out fought, and the first Victory fell to "S" who claimed a "cert" East of Roulers 4.6.17. Which date pinpoints the date of the Marque's entry into the arena – grim as it possibly was.

"Soon after this a quiet spoken unknown engineer officer [WO] arrived on a tour of inspection at the aerodrome, and accompanied by the C.O. made a round of the hangers – introduced to the senior personnel in turn, somehow or other he seemed to ask only "relevant" questions, whether it was a pilot or a mechanic, if to a mechanic, the question was obviously engendered by an engineering knowledge, far and away more intimate than the usual run of questions emanating from the ordinary type of visitor. He seemed in no hurry for a safer place, in point of fact a little enemy action, seemed only to add zest to his visit.

"The pilots on their return from patrol interested him with their reports on the behaviour of the engines in air fighting, but his questions on the particular "teething troubles" being experienced, always seemed to us to be engendered by actual experience, and were questions which we alone with our knowledge of service troubles could answer. He had one peculiar "trait" in his character however, which many people, years later were to experience.

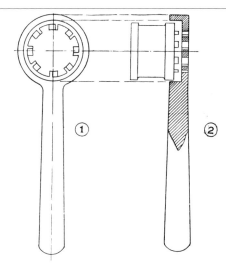

SPANNER FOR CYLINDER TIE ROD NUTS, B.R. 1 ENGINE.

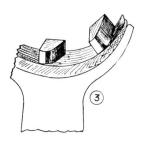

Drawing No. A. 5. Traced by W. G. F. Approved, W. Sempill for D.A.S. R.N. Air Department, N. 12 Section, Hotel Cecil.

A.T.O. 444.

444.—Special Spanner for B.R. 1 Cylinder Tie Rod Nuts.
(Air 134400.—4.1.1918.)

Considerable troubles are being experienced on the above-mentioned type of engine due to cylinder liners and barrels cracking. Although there are other reasons which account for this trouble, the main one is undoubtedly due to the fact that mechanics are in the habit of tightening up the nuts on the tie rods too tightly, with the result that when the engine becomes hot the cylinder liner or barrel, or both, crack, as expansion is restricted owing to the cylinder head being too tight when the engine was cold.

Owing to the fact that these engines will not always be looked after by skilled mechanics, it becomes necessary to arrange that damage cannot occur to the engine due to the above cause.

The obvious method of doing this is by providing some suitable spanner. A ratchet type of spanner could be designed, but it would be complicated, liable to get out of order, and difficult to manufacture.

The attached drawing A. 5 shows how the teeth of the spanner illustrated, which is of a simple and robust type, are rounded off. It can be readily appreciated that it will not allow of excessive tightening of the holding-down nuts, as the spanner will slip over the castellations.

The removal of the nut is perfectly easy, even should it be rusted or the threaded portion of the tie rod burred over, as the teeth are only rounded off on the working face when tightening up the nuts.

Arrangements have been made for a tool of this description to be incorporated in future B.R. 1 tool kits, and for a separate supply to be arranged for, for use with engines now in service.

A similar tool will be supplied in the B.R. 2 tool kit.

An example of work resulting from service reports on the BR engines.

Inevitably a question would be answered to the best of our ability, and then would follow the next question "WHY?" I wonder how many other people, famous and otherwise, have been asked in the same quiet voice, many years later, the same question "WHY?," and for the moment have been "clean bowled" for an answer, or have stood stock still, fidgeting from one leg to the other, in their efforts to be evasive.

"Another curious characteristic of his visits was the inexplicable fact, that rarely did he make a visit of inspection un-expected, the Hun always seemed only too pleased to put on a show for his benefit whenever he arrived on the scene, in consequence, many discussions, in darkened hangers were abruptly interrupted. A typical example of this is a diary entry given as "Stood to" 23 hrs ACK ACK busy in neighbourhood, Hun made an abortive attempt to "bomb us" out of it – near miss – stick of bombs 400 yards away in the dunes.

"During one of these periodical visits, we were suffering from an epidemic of broken oil pump plunger springs, and his method of dealing with this trouble shook us to the foundations, inured as we were to months elapsing before any action would be taken. Through the usual channels, he received information that we were having trouble with these pump springs – unannounced he arrived from England – How? that is his secret, held an inquest on the spot, and in his characteristically quiet manner said, Beg, Borrow, Steal all the springs you can from the surrounding depots, but change them frequently to prevent fatigue, and off he went with samples of the defective springs with the laconic "I'll be back in a few days". Much to our surprise four or five days later he returned, his pockets full of coppered springs – small parcels of them in his valise – everywhere, enough to keep us going even with breakages, for many months. That story, typical of the man, is his story, how he thumbed a lift back to England in a returning bomber [actually a destroyer], cadged the spring makers, to push through a quantity of redesigned springs without the usual Ministry Forms and Instructions, pocketed the springs and cadged a lift in a bomber [in fact a destroyer again] back to France without the "High Ups" being aware that anything out of the ordinary was happening.

And so, during the critical first days on the line, he "popped up" at all times, unheralded, unannounced, saw what he wanted to see, talked to the people using and servicing the A.R.1. [Admiralty Rotary] as it was then known, and disappeared as quickly and mysteriously as he came."

Those who have ever read any of Nobby's letters will recognise the style! It is interesting to see WO's methodical and meticulous approach to a problem, his desire to get to the root of a problem, and a refusal to accept any second hand information. How much easier it would have been to have stayed at Chiswick and read the service reports. Incidentally, although Nobby doesn't mention it, he spent some time at Humber's works in May, 1917, doing an instructional course on the BR rotaries. Nobby served with No. 4 Squadron RNAS (later No. 204 Squadron, RAF), and it was Flight Sub-Lieutenant Chadwick of No. 4 Squadron who scored the first victory with a BR1 powered Sopwith Camel on 25th June, 1917 (however, Nobby claimed the first was on 4th June – perhaps the earlier "cert" was not officially confirmed).

Humber's Chief Designer was Frederick Tasker Burgess, widely known as "Monkey" Burgess because of his facial expression. He and WO were already on familiar terms. Before the war Burgess had been responsible for the TT Humbers and had driven one of his own in the 1914 TT Race in which WO had competed with the DFP. The two men got on well, and it was obvious that the subject of the cars they would produce after the war would have been discussed. Burgess was thinking along the lines of a small car made up of lots of pressings, but was soon taken by WO's concept of a fast, sporting car. Without a doubt, they must have started roughing out the concept for their post-war creation and talking through many of the aspects of it. WO could hardly have failed to be impressed by Burgess remarkable drawing abilities. Burgess later revealed that he had given up a "damn good job to join this outfit" (Bentleys) and along with almost all the early participants, he put his money where his mouth was in the shape of a number of shares. Sadly, Burgess died relatively young in 1929, after a long period of illness.

The events of the Great War brought together a number of the subsequent players in the Bentley scheme of things, principally Sammy Davis, who was none other than the Admiralty Inspector for aero engines! The BR2 was running its fifty hour Air Ministry acceptance test just two weeks after the prototype had first been run. WO said in *The Cars in My Life* "certainly they [BR1 and BR2] gave me more satisfaction than any of the car engines we later produced." The success of them also seems to have suprised WO, and increased his confidence in his abilities as a designer, going so far as to say himself that had the rotaries not been successful there would not have been a Bentley car.

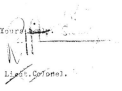

```
                                          Headquarters,
                                          7th Brigade, R.A.
                                          18, Marine Parade
                                          DOVER.

Dear Bentley,                            15th April, 1918,

        Attached are copies of a number of reports received
from different sources in connection with B.R.1.engines. I though
that possibly they might be of interest to you to see how they
have turned out.       A great deal of this is absolute balls and
from the way it is written you will know what I mean.

        As I am now in charge of "Q" Section, and have been
for sometime, at the R.N.A.S. Headquarters (now 7th Brigade R.A.)
all equipment has come under me and your engines have never let
us down and we have never had any complaints.    A few things,like
the old troubles cracked cylinders and rocker arms, appear
from time to time but when one takes into consideration the
extremely hard conditions under which these engines have to
work, I consider them negligible.    No engine has ever stood
up to the work on the Front as this engine has and everybody
who has had anything to do with it, loves it and trusts it
every time.

        All this correspondence is sent to you unofficially
but as I know you are full out for the Pilots on the Line, that
is the reason why I am sending it.

        You will probably know a number of the Officers
who have made these remarks and will be able to judge and
follow all their reports.

        This correspondence may be of little use to you
in the design of the B.R.2, if not it will be interesting.

                         Yours       ,

                         Lieut.Colonel.

Captain Bentley,
38, Netherall Gardens,
Hampstead,
    LONDON.N.W.
```

This letter stands as eloquent testimony both to WO and to the BR rotary engines. It was written by Lt. Col. Richard Marker, then attached to 7th Brigade RAF Dover.

The BR2 was almost too late for the war, and was fitted to Sopwith Camels and Snipes in late 1918. An order for 30,000 engines was placed, but after the Armistice of 11th November, it was cancelled. WO was advised to apply to the Royal Commission for Awards for recompense for his design work. The Royal Comission was set up in March 1919, and it was announced on the 13th that the Commission was to be chaired by Mr. Justice Sargant. Other members of the Committee were H.J. Mackinder MP, R. Young MP, Prof. the Hon. R.J. Strutt, Sir J.J. Dobie, W. Temple Franks, G.L. Barstow and A.C. Cole. The Commission was to deal with awards up to £50,000, with greater awards to be dealt with by the Treasury direct. In due course the proceedings went through and WO was awarded £8,000. Incidentally, although WO implied that he was probably the longest serving junior officer because he was never officially demobilised, he was demobbed "in or about May 1919" having received a gratuity of £1,000 in January.

The business of WO's claim was somewhat unpleasant. The following report is reproduced from *The Daily Telegraph* for 27th January 1920:

"The hearing of the claim by Captain Bentley in respect of a new aeroplane engine which he invented was continued yesterday before the Royal Commission on Awards to Inventors, presided over by Mr. Justice Sargant. The first hearing of the claim was heard in December, when it was stated that the engine, in which aluminium was largely used, had proved very successful, and a large number of aeroplanes had been fitted with it.

"Mr. Carthew appeared for the applicant, and Sir Gordon Hewart, K.C., Attorney-General, represented the Treasury.

"Commodore Murray Sueter, R.N., said Captain Bentley was employed under the Admiralty and was not employed in designing engines. The officer under whom Captain Bentley worked [Briggs] came to the witness [Sueter] and asked him if Captain Bentley could design an engine, using aluminium largely for the pistons and for the cylinders. He gave permission. "We were out for new engines all the time," he said. "This was creative work, and was a little outside his ordinary duties."

"The Attorney General, opening the case for the Treasury, said a large sum was involved in this claim respecting the BR1 and BR2 engines. The applicant alleged that not only were these engines very useful, but they were so unusual in character that it was fair to say they contributed more than any other factor to the British air supremacy at the front. The claim amounted to £107,000. Not one witness would say that the invention contributed more than any other factor to our supremacy in the air. These engines were of very substantial utility, the utility indicated by the extent to which they were used; but when one was asked to go further and say that these engines exhibited these superlative merits, then they joined issue, and he would call evidence to show that any claim to be a determining factor in British air supremacy was wholly illusory. The very engines which were here in question were the subject of another claim for a sum of about £150,000 by Messrs Gwynnes, the owners of the Clerget engine.

"The Air Department recognised the valuable work and great zeal of Captain Bentley, and he (the Attorney General) would not for one moment suggest that there should not be granted to him a reasonable sum, but if Government departments were faced with claims of this magnitude they would have to protect themselves by stringent agreements with assistants such as those in many engineering firms, where all inventions were the property of the firm. That result would be a misfortune for the country and for the officers employed by the Government, and it was because this case would appear to be typical of the sort of claim which might lead to that undesirable result that it was being so opposed.

"Mr. Trevor Watson, for the Treasury, said he proposed first to call evidence on the allegation that the BR engine was the first main factor which established the British air supremacy in France.

"Captain C. Fairbarn, the head of the Historical Records Branch of the Air Ministry, was called for the Crown. He said he had had considerable experience as a pilot, having flown about 1,000 hours up to 1917. "I think British air supremacy was established in the beginning of the war, but we temporarily lost it, and regained it at the end of the spring of 1917," said witness.

"After technical evidence had been called, Mr. Trevor Watson, addressing the Commission, said Captain Bentley, a young man who had spent slightly more than two years on this particular problem which he was fortunate in having presented to him by the naval authorities, was asking the Commission for an award which would secure him for the rest of his life an annuity on a slightly larger scale than the salary of a Cabinet Minister. Was this Commission satisfied that his work could not have been done by any of half a dozen people in this country?

"Mr. Hogg, for the applicant, submitted that Captain Bentley had found out something which no one else knew. We got a new and infinitely better engine, which was made the standard engine for our 1919 campaign, if that had been necessary. It was less valuable for ground strafing, and could fly higher than others. It required less spares and less overhauling, and it was mean of the Treasury to turn round and say, "Any competent designer could have found that out."

"The chairman announced that the Commission would consider the award."

In March 1920 WO received a letter from the Air Ministry reading "to forward herewith draft for £5,600 in payment to Captain W.O. Bentley of the award of £8,000 recommended by the Royal Commission on Awards to Inventors less income tax of £2,400". Douglas Hogg, QC, WO's counsel, was so horrified by what he considered to be a niggardly

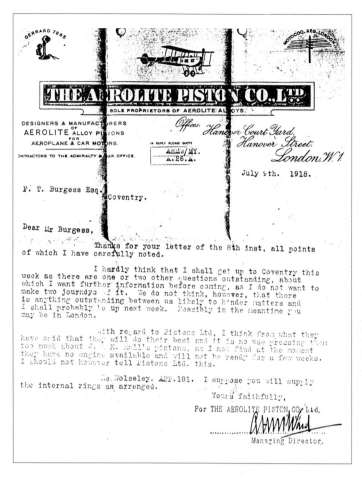

A letter from Ward to Burgess, on Aerolite paper, giving the Bentley and Bentley address of Hanover Court Yard, Hanover Street.

pittance that he promptly halved his fee. The first report of the Commission listed WO's award of £8,000, along with the award to Gwynnes of £110,375.

The Inland Revenue's attempt to tax the award became a test case, but the action was withdrawn and WO received the full £8,000. However, the difficulties with the money deducted for tax were not resolved until early in 1922. Needless to say that £8,000 probably disappeared very quickly! As WO noted in his autobiography, the Great War had not yet finished with him. Leonie, his wife, died in the Spanish flu epidemic that swept across Europe just after the war.

As far as the DFP business went, Ward kept the showrooms open, largely on the strength of the piston business – among others, they had substantial contracts to supply pistons for Bentley Rotaries.

137, Victoria Street,
London, S.W.1.

10th December, 1919.

Messrs. Bentley and Bentley, Ltd.,
Hanover Court,
London, W.1.

Dear Sirs,

You will remember I purchased a 1914 model 12/15 h.p. D.F.P. car from you in July, 1918, and sold it in October of this year. I would like to inform you that in the intervening period I have driven this car 30,000 miles and besides never having had a mechanical breakdown as you know I have never had to call upon you for replacement.

I found the petrol consumption of this car remarkable, inasmuch as I carried out a careful test over Cornish roads with three up and luggage and the result worked out at well over 30 miles to the gallon.

In view of the fact that this car was not one of your "speed" models but simply a touring type the speed I have obtained with a 70 m/m bore engine appears to me to be very good indeed. With a scientifically correct speedometer I have touched 58 miles per hour with one passenger and luggage.

The car could be driven at speed without effort to driver or engine continuously over suitable roads. This car was the sixth make I have owned and I am now of the opinion that I had no idea of the real pleasures of motoring until you supplied me with this D.F.P. As you know I am about to take delivery of a 1920 "Speed" model 12/40 h.p. D.F.P. from you and from the run I have had on the chassis I have every hope that this car will prove superior to the 1914 model.

I might mention that this is the first testimonial I have ever written and I only write this because I think it is so richly deserved.

Yours faithfully,

M. V. ROBERTS.

I ... whatsoever in your firm
... one of your cars.

An encomium from a satisfied customer, the proud owner both of a DFP and a "scientifically correct" speedometer.

WO was involved in the business, from a technical rather than commercial aspect. Hillstead also dealt with the pistons and DFP spares, as there were no new cars after 1915. The pistons themselves were cast at Rowland Hill's works in Coventry. Another venture was a sub-agency for the Emerson Farm Tractor. This was designed to be started on petrol and then run on paraffin, but despite demonstrations at Farningham, Sussex, Hampshire, and the Isle of Wight, sales were disappointing and the agency was soon sold. As the war drew towards a close, there was an increasing turn-over of second hand DFPs, the income from which was to pave the way for the Bentley car.

The business of The Aerolite Piston Company is of some interest. WO does not mention the company at all, only Hillstead having anything to say about it in *Fifty Years with Motor Cars*. Firstly, Hillstead tells us "I was told [by Ward] that just prior to the War a small subsidiary company had been formed called The Aerolite Piston Company, for the sole purpose of marketing the aluminium pistons which WO had pioneered, and that several useful contracts for aero engines were pending". Later he says "the mainstay of the firm [Bentley and Bentley] at that time was, of course, The Aerolite Piston Company, and several contracts for aero engine pistons had been secured, including one for W.O. Bentley's BR1 rotary". HM, we are told, "had a staff job at the Air Ministry and was resplendent in the usual red trimmings". Turning now to the Public Records Office files, the story goes as follows. The Aerolite Piston Company was not founded before the war – it was incorporated on the 29th June, 1916. The company was established solely to take over the piston business of Bentley and Bentley Ltd., under an agreement dated 13th July 1916. This agreement ascribed all Bentley and Bentley's piston contracts and all related activities to The Aerolite Piston Co., in exchange for 497 £1 shares in that company. The other three shares were held by Ward, James Tolley, of whom nothing is known, and Alfred Bentley, WO's father.

What is perhaps significant is that neither HM nor WO appear as directors of The Aerolite Piston Company–indeed, A.H.M.J. Ward was the sole director. WO and HM were not elected as directors until the 23rd April 1919, with the appointments made as from 1st January 1919. The Aerolite Piston Company moved to 36 North Audley Street on 9th September 1920, along with Bentley and Bentley. WO and HM resigned from Aerolite in November 1922. It would seem that The Aerolite Piston Company was set up with Ward as the sole director, with Bentley and Bentley as shareholders only, to distance HM, and particularly WO, from the piston business. It would seem at best unethical that WO should be a director of a company with major contracts to make pistons for his own design of engine in the middle of a war. HM was responsible for the movement of new aircraft to France, so again he was involved with aircraft supply in some capacity.

There is a final twist to the Aerolite story. The Aerolite Piston Company went into liquidation on the 19th January 1925, the liquidator being E.G. Davies, later the liquidator of the second Bentley Motors company. However, the Aerolite name passed to The Light Production Co., who in 1930 were marketing the Aerolite Mk. IV compensating piston. The Light Production Co. was Martin Valentine Roberts' firm, Roberts having become a director of Bentley Motors in July 1919!

3 The Formation of Bentley Motors

THE FIRST BENTLEY Motors effectively consisted of just three people – WO himself, Burgess and Harold Varley, based in an upstairs office at 16 Conduit Street, London, starting on the 20th January 1919. Burgess has already been mentioned with reference to Humber cars, and was an extremely gifted, intuitive design engineer. Varley served his apprenticeship with an engineering company, Lloyd & Plaister, before working for Vauxhall. Varley might not have had Burgess' flair, but he was a vital element in the scheme of things and probably had a stronger theoretical grasp than Burgess. Burgess was still living in Coventry, working from there until October 1920, when he moved to Finchley. These three evolved WO's concept to produce a fast touring car into a production proposition, designing every component from scratch.

Early in his life, WO was to be found dismantling bicycles to the considerable concern of his mother who was sure that all the bits would never go back together again. Despite some problems, the re-assembly was always successfully achieved. While serving an apprenticeship at British Aerospace, the author was told that the best design engineers were often not the most academically gifted, but those who from an early age had messed about dismantling and reassembling bicycles, motor bikes, engines and the like, because apart from having an innate feel for machinery, such people acquire an insight into problems and how other people have tackled them in the past, in a manner that can never be taught in a classroom. WO was a remarkable product of this school of engineering.

His feel for machinery developed from the bicycle to the motor bike, with a different aspect added by railway engines (if an example of the application of this form of experience to a design problem is needed, compare the three-throw camshaft drive of the six cylinder cars and the strut gear used to brace the chassis to locomotive practice). This experience was further compounded by the Unic taxi, WO's private cars, and the DFP, before blooming with the design of the BR1 rotary.

The BR1 and BR2 were still linear descendants of the Clerget (a statement that WO would understand and agree with), but the 3 Litre was the first time that WO had been in a position to sit down with a clean sheet of paper to design a car. It is inevitable that the result would be an amalgamation of WO's experiences and the solutions he had seen used by other people in the past. Nobby's "Why?" results in WO, Burgess and Varley's analysis of other designs to take the best aspects of all that was available and to question why previous designers had adopted certain solutions. From all this analysis the trio produced a car with the characteristics that WO desired without any radical ideas that would have caused insurmountable development problems to a firm that had no development facilities. It is important to realise that the motor industry in 1919 was different from even ten years later in the late twenties when firms such as Timken produced off-the-shelf axles and gearboxes for car manufacturers – in 1919 the only available components were such things as steering wheels, electrical equipment, carburettors and road wheels.

The two cars that provided the most study material were the 1912 Peugeot, Henri's brainchild, and the 1914 Grand Prix Mercedes. WO was well familiar with the latter, having helped to remove an example from Mercedes' showrooms at Long Acre in London in 1915. He was present when the engine was stripped and examined by Rolls-Royce at Derby, and kept a record of the results. The engine was run on test at Derby in January 1916, giving an output of 100 bhp at 2500 rpm, and 99 bhp at 3000 rpm. No results were taken above this speed, so presumably 3000 rpm represents the safe maximum. With a stroke of 165 mm, 3000 rpm represents a very high piston speed and piston acceleration. The three plugs per cylinder were fired by two Bosch magnetos, one double and one single spark.

Examination of drawings of the Mercedes engine shows that the 3 Litre Bentley engine incorporated many of the basic principles of the Mercedes unit, in terms of the use of bevel gears to drive the overhead camshaft (and initially the oil pump as well in EXP 1 and 2) and the use of four valves per cylinder driven off a single overhead camshaft. The Mercedes also used long studs through the crankcase to secure the main bearings, but unlike the Bentley engine these did not double as studs for the cylinder block. In general terms, the detail design of the Bentley engine is superior to that of the Mercedes and would have been cheaper to manufacture. WO was keen on the twin overhead camshafts used by Zuccarelli and Henri, which also rendered possible the fitting of one of the spark plugs in the roof of the combustion chamber. The Bentley and Mercedes units have their plugs fitted on each side, the Bentley having two plugs per cylinder to the three of the Mercedes. However, the noise that would have been generated by chain drive was unacceptable to WO.

Of the three, (WO, Burgess and Varley) there is little doubt but that the concept and overview of the 3 Litre was WO's – he was, in fact the "Ideatore". Frank Ayto, who worked in the Drawing Office at Oxgate Lane from 1921 to 1931 and then worked closely with WO at Lagonda for many years, knew WO well. In all those years, he says that he cannot recall seeing WO actually draw anything. WO came up with the original idea for a design, and it was then turned into drawings by the draughtsmen who worked for him. But WO was around the Drawing Office the whole time, observing, commenting, and guiding. Sometimes he would take a pencil, and perhaps add a line or two to a drawing to express his point more clearly. Ayto recalls WO as a shrewd engineer with very good ideas who knew what he wanted and would stick to his guns to get it. WO would listen, but would over-rule others if he felt he was right, making sure that he got what he wanted in the end. Frank Ayto also says that in all the years he worked with WO, he

cannot remember him ever being wrong. There is no doubt but that WO was a man of great integrity and this is reflected in his designs.

Burgess' proposed post-war car was a very different kettle of fish ("a pretty prosaic little affair all made up of little pressings" as Sammy Davis put it), an idea he seems to have dropped completely in favour of WO's. Burgess obviously had a high opinion of both WO and his proposed car. But the Company's intent was always to produce a bread and butter car, once manufacture of the 3 Litre was well-established. Burgess' role in the design team was to take WO's ideas and turn them into feasible design schemes. Burgess was particularly responsible for the back axle, gearbox and much of the chassis design. WO himself was principally interested in the engine, and the roadholding and performance of the complete car. WO commented on Burgess' remarkable drawing abilities, that he could turn an idea into a paper sketch as fast as WO could express it. Burgess did not go in for much in the form of stress calculations to determine the size of components – his eye and feel tended to result in generously-sized components, a practice which also means over-engineered components – a considerable factor in the remarkable longevity of the Vintage Bentley. It is also very easy to overlook the impact of the pocket calculator – particularly in the aircraft industry, where such calculations had to be done, teams of mathematicians each checking the other's work were often employed.

Varley had a greater theoretical grasp than Burgess and seems to have been responsible for turning Burgess' schemes into the production drawings from which the first components were manufactured. Unfortunately, working relations between Burgess and Varley were not all they could have been and Varley left in 1924. Both had volatile temperaments. WO remarked that he could manage either of them separately, but not together! In truth, all three were talented design engineers with strong characters and it was Varley's misfortune to be the one who eventually had to leave. It was not known that Burgess was a very sick man at the time. Although he was still with the Company when he died of a brain tumour in November 1929, he had been off work for long periods before then.

The first announcement of the new car was contained in *The Autocar* of 6th March 1919, followed up by a descriptive article in their issue of 17th May. This article stated that the drawings had been finished and the machine was under construction! The 4 cylinder engine of 2994 cc (sic) was planned to give 65 bhp at 3200 rpm, and the accompanying chassis drawing showed the single magneto/water pump set up, dry sump and other features characteristic of EXP 1, with a steering wheel described as being "of racing type with a narrow rim of large diameter". The estimated chassis weight of 13 cwt was inevitably considerably exceeded, as was the chassis price, "in the neighbourhood of £750" including running boards, front mudguards, five lamps, five wheels and tyres, tool-kit, a rev counter and speedometer. Some of the latter items were also quietly discarded in due course. The whole was preceded by Gordon Crosby's painting of a Bentley at speed, similar in outline to EXP 1 and with (almost right first time) the classic Bentley radiator. It is perhaps significant that the copies of this article that were reprinted for the Company as hand-outs had the chassis price blacked out!

Bentley Motors in its first creation was registered on 18th January 1919, the day the Certification of Incorporation was signed by the Registrar of Joint Stock Companies two days before WO, Burgess and Varley started work on the 20th. The registered office was 16 Conduit Street. The company was registered by William Robert Caesar, a solicitor working for Druces & Attlee of 10, Billeter Square, EC3. The Articles of Association gave the share capital as £13,000, but this was changed to £20,000 before the document was even registered. The articles read that "The objects for which the Company is established are:- To purchase or otherwise acquire from Bentley and Bentley Ltd the designs and drawings for an improved Motor Car chassis and certain rights in association therewith belonging to Bentley and Bentley Ltd..." for the sum of £7,000. The schedule of drawings consisted of 6 items:

A. General arrangement of Chassis
B. Front view of Chassis
C. Side elevation of engine
D. Cylinder head showing arrangement of valves
E. General arrangement of cylinder
F. Crankshaft

For which, under the terms of the agreement, Bentley Motors paid Bentley and Bentley Ltd. £7,000. This latter sum of course, was just paper – in effect the Bentley brothers paying themselves for the work they had done, while retaining significant shareholdings. WO was managing director, HM also a director and A.H.M.J. Ward the third and last director of the new firm. All three were directors of Bentley and Bentley Ltd, Ward also being a director of The Aerolite Piston Company. WO was "Captain, RAF" and HM "Lt. Colonel, RAF",

WO in 1919 – this photograph was used by *The Autocar* in the first announcement of 'A New British Sporting Car' in their issue of 8th March 1919.

and Ward was secretary of all three concerns. Of the 7000 £1 shares issued to represent the money paid to Bentley and Bentley Ltd., 2334 were issued to HM and 2333 each to WO and Ward. Two clauses of the lengthy agreement between the two parties are of particular interest:

– "And whereas the said Walter Owen Bentley is the inventor of certain improvements in relation to Motor Car Chassis, which are incorporated in the said designs and drawings on some or one of them and Bentley and Bentley Limited by virtue of agreements between them and the said Walter Owen Bentley entitled to the full benefit of all such inventions and of any further inventions or improvements or such inventions which the said Walter Owen Bentley may make but no provisional protection or Letters Patent have yet been granted in respect thereof.

– "The said Walter Owen Bentley and the New Company shall execute an Agreement (the engrossment whereof has been approved by the parties hereto and is intended to bear even date herewith [28/2/19]) providing for the said Walter Owen Bentley entering the service of the New Company for the purposes and period and upon the terms contained in such Agreement".

The basic terms of this agreement were that Bentley Motors would employ WO as Chief Engineer for life, with a minimum royalty payment of £2,000 per annum for the use of his patents. WO was barred from either leaving Bentley Motors and designing for a rival using his patents, or to compete with Bentley Motors while still working for them, by using his patents. It is clear that WO was a valuable asset. WO resigned his directorship of Bentley and Bentley on 31st October 1920, but he had long before ceased to play any part in the company.

Throughout 1919 WO, Burgess and Varley laboured away while HM and Hillstead earned the money at Bentley and Bentley by trading second-hand DFPs and importing as many new chassis as the Courbevoie factory would let them have. At one point William Bentley (another of the Bentley brothers) and Hillstead even ran a ferry service,

HM, the other half of Bentley and Bentley. His role in the early years was crucial to the success of the Bentley Company. HM's hobby was woodworking, and he was a skilled furniture maker.

picking up bare chassis from the works near Paris and driving them back to London! Using Hillstead's experience, they also dealt in other makes of car secondhand, to WO's disgust when they blocked his entrance to Hanover Court Yard.

The post-war DFP was inferior to the pre-war 12–15 hp in many aspects – a deterioration that eventually led to the closure of the parent company. By 1916 Ward had taken up a third of all the shares in Bentley and Bentley and by October 1921 owned half of it before it went into liquidation in December 1922. De Freville had left Bentley and Bentley to set up Ware and de Freville Ltd, a firm for which Hillstead worked for a while before Ward offered him the job at Hanover Street. De Freville had also set up a company connected with pistons – Aluminium Pistons Ltd – and was later involved in setting up the Alvis Car Company.

Hillstead's accounts of the early days at Hanover Street are very interesting. Ward told him "don't forget there are three of us to please, and not all temperaments are the same." Hillstead described the brothers Bentley thus: "HM was of average height, well-built, neither dark nor fair, cheerful, ready to crack a joke and possessed of a decided twinkle in his brown, myopic eyes; WO was short, stocky, dark, inclining to ocular fierceness, deliberate, monosyllabic and decidedly dour." And probably at that time under a great deal of pressure with technical problems with pistons and all the work on the BR rotaries on hand. It seemed for a time that there was no real job for Ward at Bentley Motors and he decided to keep Bentley and Bentley. Ward offered Hillstead a job, but the writing was on the wall for the DFP with no new design work underway, so Hillstead declined. Ward was back on the board of Bentley Motors by May, 1920.

The post-war DFP 12/40 chassis. Note the disc wheels, cantilever rear springs and redesigned radiator. Hillstead was very scathing about this car.

4 The Formation of the Second Bentley Motors

WITH THE CONCEPT for the new car sketched out and evolving towards a production proposition, attention needed to be focussed on the mechanisms needed for manufacturing. There was no need to use the first company as anything more than a holding company, as all the business activities could be taken care of and staff employed by Bentley and Bentley. Restructuring the Company left all the options open, and it seems likely from the manner in which the first Company was indeed only used as a holding company that HM had planned the restructuring in advance. Certainly it left them entirely free to wind up the January, 1919 Bentley Motors and establish a new one on a securer financial footing.

In April/May 1919 the oft-recalled dinner at Verreys in Regent Street was held, attended by WO, HM, Hardy Bentley the lawyer, A.F.C. Hillstead, F.T. Burgess and potential investors in the form of General Whittington and Colonel Wolfe Barry (introduced by HM) and Martin Valentine Roberts (introduced by Hillstead). Roberts was already the owner of a 12–15 DFP, with which he had expressed great satisfaction, and had bought a new 12–40 in December 1919.

The terms of the new company were discussed and a sum of money was guaranteed, to be paid up in £1 shares. For perhaps the first and last time in his life, WO spoke passionately in public, something he always disliked! Hardy Bentley was instructed to draw up the documentation for the new company, this being the second company formed after the liquidation of the first Bentley Motors on 10th July 1919. We know (from Hillstead) that Hardy Bentley was instructed to draw up the documentation and indeed that he registered the new company on 8th July 1919 (in *Cricklewood to Crewe* Michael Frostick questions the sequence of events and timing, but in fact the events described here and the timing fit all the published accounts and correspond with the Public Records Office files).

By this time they had evidence of some design work to show to potential investors. They would also have had more confidence in what they were doing, and for the second company a sum of £100,000 for capitalisation was agreed upon. This was subject to a 10/- call (so only £50,000 might be raised initially) and as the first £30,000 of shares were issued to WO and HM for considerations other than cash (ie all the existing design work), the financial footing of the Company was distinctly shaky. By June 1920, the Company had just £18,750 in cash. With a chassis price of £750 as initially projected and published in *The Autocar*, that sum represented the retail price of just 25 chassis – to do most of the design and all the development work, build works, buy machines, employ men, advertise etc, etc – all the activities that need to be performed before a penny is seen to return on the first sale. It is no wonder they were always desperately short of money. In the early days, Hillstead and HM spent most of their time selling the Company, not cars, while HM and brother Hardy worked tirelessly to try and get the Company on its feet, shielding WO from these pressures so that he could concentrate on the engineering.

It is interesting to look in detail at the agreement signed between Bentley Motors in liquidation (ie the January, 1919 company) and the new Company in formation. This agreement took a similar form to that between Bentley and Bentley Ltd and

The third of the Bentley brothers to be involved in the old company, Hardy, the lawyer, here seen in the 1930s.

the first Bentley Motors. In this case though, £30,000 was allotted made up of £26,770 for cash, stock in trade, drawings and designs, and £3,230 for the residue of premises agreed to be sold. (This presumably represented 16 Conduit Street and the DFP showrooms/offices in Hanover Street.) The schedule covering design work was exactly the same as that listed earlier. In other words, WO, HM and Ward sold WO, HM and Roberts for the best part of £30,000 that which they had paid £7,000 for six months earlier. On the face of it, it all looks very dubious, but there were no shareholders concerned, and, according to *Topham's Company Law* for 1919, as long as all the directors were aware of what was going on this practice was entirely legal. It seems that the January, 1919 company was little more than a necessary first move, as the whole operation was kept fairly low key and no effort seems to have been made to sell shares. Certainly, had there been any shareholders it would have made matters much more difficult.

The new Bentley Motors was incorporated on the 10th July 1919, with WO, HM and Martin Valentine Roberts as directors. Roberts' own firm, The Light Production Co., based at 60–66 Rochester Row, SW1, was a well-equipped engineering works, and Roberts soon lost interest when it became apparent that no work would be coming his way from Bentley Motors. Roberts' disinterest spread to Wolfe Barry and Whittington, HM having to juggle with income from other investors to pay them off. Although Hillstead was accused of "writing with a chip on his shoulder" by one reviewer, his accounts are far more accurate and consistent with documented records (eg, the company returns held by the Public Records Office) than anybody else's. It does seem, however, from reading between the lines that Hillstead probably found WO difficult and it is quite conceivable that the two men had little time for each other. It does not take much of a stretch of the imagination to think just how difficult those early years must have been, with the constant uncertainty – "the Sword of Dammycockles has been hanging over me", as Hillstead put it in 1925! Shortly after the creation of the new firm A.H.M.J. Ward rejoined the Board. The qualification to be a director was the holding of at least £2,000 in shares, so Roberts and Ward were putting up a lot of money.

Concurrently with this financial activity was a need to expend money on premises, staff, machinery and running prototypes. As the drawings were produced they were put out to tender for manufacture and as the components became available WO approached Nobby Clarke to put together the engine in his own inimitable way. Jimmy Enstone, the Company Secretary, wrote to Nobby on the 9th July 1919, to arrange an appointment for Nobby to see WO. WO said that Nobby was more or less expecting the invitation and the only entry in Nobby's diary for 1919 records that he went to Bentley Motors at Conduit Street on Wednesday 18th July – if only Nobby had kept his diary, what a story it would have told! We do, though, have Nobby's account of his joining the Company:

"The hectic war days of the spring and summer of 1918 drew to a close – first rumours of a French Mutiny, then a tightening up of our own discipline, then the advance of Autumn 1918, over the old Battlefields at the rear of stagnation until 11 am 11th November 1918, when I witnessed virtually the last shots and casualties of the war in that area on the old Napoleonic battlefield of Oudenards between our Artillery and French whippet tanks, and the retreating Huns, having again, with my typical luck, got tangled up in a battle for a sunken road between the whippets and the retreating Huns, complete with Crossley Tender etc as usual. I well remember this scene, the Jerry rifle pits dug into our side of the road bank, the casualties and debris of war strewn about, the "Froggy" rifle pits on the other bank and the rattle of the whippet tanks as they dived down one bank and up the other side across our roads. Needless to say we got out of that hole "pretty damn quick." And then silence, the inevitable rumours, the advance and then demobilisation after a winter of hopes and fears culminating in my saying goodbye to the RAF after $4\frac{1}{2}$ years, at Waddington in Lincolnshire, after just escaping going to South Russia to aid the White Russians. Civvy Street – job, and then I thought of the number of times that I had met WO during the war, thought all sorts of things and in the end wrote to my old CO. He wrote to me by return, gave me WO's address and plucking up my courage, I wrote to the great man, was called to Conduit Street for an interview, and so in a day in June [sic] 1919, an association that was to last nearly 40 years commenced, and Bentley Motors more or less became my life from that day on, so I consider that from now on, the title of the story should be titled Bentley Motors and all That, the latter with a capital *T*.

"WO was still the same WO that I met on the Western Front, the same man that when we were short of oil pump springs, borrowed transport and got some coppered springs over to us to keep us going, stood with us on the top of the Nissen huts at T ... watching the Handley Pages at C ... being bombed out with low altitude bombing, until the Jerry dropped a brick behind us, and we, either slid off or were blown off the hut, on to our beam ends on to the ground below, imperturbable WO in one of his understatements "That was close", when everyone else considered it was too damned close to be comfortable. From this interview, I gathered that WO was building a motor car, that parts were being at that time manufactured all over the country and that on completion the car was to create a type unknown in this country, a docile yet virile sports car capable of fulfilling a long felt want, a car for the open road and yet docile enough to be used in town, as distinct from the pre-World War 1 racing cars, with their attendant difficulties when used for touring purposes. Soon after this, I was to make the acquaintance of FTB [Frederick Tasker Burgess] of immortal memory, RCG, [Clive Gallop] FCC [Frank Clement] and RSW, [Dick Witchell] to name a few, not forgetting of course ACH, [A.F.C. Hillstead] HMB, [HM, of course!] HFV,

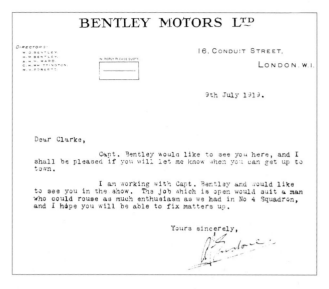

Enstone, the Company Secretary, writes to Nobby Clarke to invite him to join the Company.

[Harry Varley] all of whom played a distinguished part in putting into practice the ideas of the master mind of the whole show. RCG schooled in the arts and wiles of racing practice at a famous Continental firm [Peugeot] and an early RFC pilot, was a mine of information on things that happen to motors that go fast, FCC and RSW both products of the English School of Engineering and pre-war racing drivers, coupled with the redoubtable FTB of pre-war fame, formed a magnificent foundation upon which, as subsequent events proved, the fame of Bentley Motors was founded."

Colonel Clive Gallop was a fighter pilot during the Great War, but more importantly had worked at Peugeot pre-war on the Grand Prix engines and was employed by WO principally to work on the overhead valve gear. Gallop did not last long, soon going to Higham to work for Zborowski, but crops up again at the 1926 Le Mans and at Welwyn with the Birkin Team.

Working in an upstairs room of Easter's coach-building works in New Street Mews, Nobby lovingly assembled that first engine with the minimum of equipment – a "gut buster" drill and a huge grinding wheel human powered by young Leslie Pennal, who had worked downstairs from Nobby with DFPs in the Bentley and Bentley service station for some time, until Nobby realised his potential and brought him upstairs to work on the first Bentley. Overcoming considerable problems with the prototype overhead camshaft gear, the engine was at last ready for its test run on the bench, with

WO, Clive Gallop, HM, Hillstead and various investors in attendance. The initial anticipation turned to frustration as the engine refused to start and Nobby carefully checked over the timing. Finally WO called for some benzole, and EXP 1 at last burst into song for the first time – with the straight cut gears whining and exhaust emerging straight from the manifold! Complaints from the matron of the nearby nursing home that they were disturbing the sleep of a dying man led to an early shut off – as one wag remarked, "A happy sound to die to – the exhaust roar of the first 3 Litre Bentley engine!"

However, there was much to be done before EXP 1, the first car, could take to the road. The chassis was assembled downstairs in the Easter works at New Street Mews, and it is easy to hypothesise that quantities of TT Humber parts were used. Burgess would probably have had access to spares at Coventry. It is significant that Burgess remained in Coventry until October 1920 and that there was much to-ing and fro-ing between there and London until the latter date when Burgess moved to 16 Ravensdale Avenue, North Finchley. With Burgess' close links with Humber and the known quantity of TT Humber spares available the hypothesis becomes more likely.

The chassis of EXP 1 and 2 with their distinctive down-turn to the front dumb-irons bore a remarkable resemblance to the TT Humber chassis, and it is significant that when C.G. Brocklebank rebuilt his 1913 GP Peugeot he used the old chassis of EXP 1 – and the TT Humber chassis was largely copied from the GP Peugeot. Unfortunately the evidence was destroyed when Captain Toop went over the top of the banking at Brooklands in 1924, sadly ending his own life as well as that of his motor car. The Peugeot was a 1913 GP car, with dimensions of 8'9" wheelbase and 4'6" track. EXP 1 and 2 had wheelbases of 9'4" and track of 4'8", which correspond remarkably with the 1913 Coupé de L'Auto Peugeot's dimensions of 9'4" and 4'8½", so perhaps Burgess not only copied the Peugeot design for the 1914 TT Humber but also the chassis design for the 3 Litre Bentley! Again, when C.D. Wallbank bought his TT Humber in 1925 it was sold as a Peugeot, Wallbank not realising this until he took it to Brooklands, where Cobb and Hawkes pointed out that it was, in fact, a Humber. Wally Hassan, mechanic at Bentley Motors from 1919 to 1931, observed in 1978 that the early 3 Litre Bentley chassis were identical to the TT Humber.

The chassis frames of EXP 1 and 2 differed from the later cars in that their wheelbase of 9'4" was

ABOVE **HM and Clive Gallop. Gallop joined Bentley Motors in the Summer of 1919, and was heavily involved in the work on EXP 1 and 2.**

RIGHT **EXP 1 engine, mid/late 1919. These photos show the details of that first engine. The sump is the first dry-sump pattern, with the double-length oil pump with scavenger pump. The crankcase, flywheel and conical front casting are very similar to the production engine, but with extra breathers/ oil returns to the crankcase casting. The starter motor is by CAV. The cylinder block is very similar to the production unit, but there are no drain tubes between the end pairs of cylinders. Oil drained back from the overhead gear from the holes at the top of the block on the offside, through external pipes to the crankcase breathers. The magneto turret with ML G4 magneto and water pump driven off the other end of the cross-shaft was changed very soon to the production set-up with the water pump driven off the front and twin ML G4 magnetos. The whole of the overhead valve gear was changed for production, as illustrated, p.67. Note the rear camchest cap, with twin bosses for the rocker rods, which were fitted from the back of the engine. The carburettor is a Claudel-Hobson.**

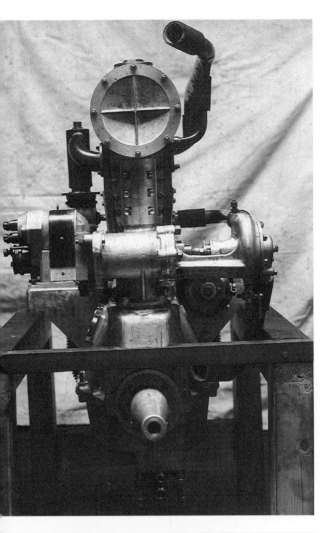

5½″ less than the production cars. It is likely that some of this difference was caused by the positioning of the back axle. On the production cars the axle was located offset to the back of the rear spring, an arrangement said by Nobby Clarke to have been arrived at empirically to reduce clutch judder on take-off. Quite why this offset reduced clutch judder is not explained – WO mentions that the Mercedes used a torque tube while the Peugeot used an open cardan shaft with the torque being taken through the rear springs, (or more specifically, the front anchorage of the rear spring), so presumably this re-positioning of the axle reduced the nature of the physical sensations. This usage of an open propellor shaft with the torque taken by the rear springs was commonly referred to as "Hotchkiss drive", after the French manufacturer (the author has always found it a fascinating sensation to sit in the back of a Bentley and feel the whole of the back of the car dip as the clutch is let in and the torque reaction winds up the back axle on the rear springs!).

On the subject of chassis frame design, Nobby Clarke made the following comments in 1949 in reply to being asked why Bentleys did not use outrigged rear springs until the 8 Litre chassis was designed: "This was partly on account of the influence of the late F.T. Burgess, of 1914 TT Humber fame, who was responsible, with Clive Gallop for chassis design. The latter's additional experience upon the pre-World War 1 Peugeot had an appreciable bearing on 3 Litre Bentley chassis design, together with the fact that –

1. Outrigged road springs introduce torsionals into a channel section frame unless adequately braced by additional cross members.
2. An outrigged frame is thus heavier than the 3 Litre frame.
3. It is more costly to produce.
4. It was furthermore designed 31 years ago.

"The 4½ and 6½ Litres were designed during the lifetime of F.T. Burgess while the 8 and 4 Litre Bentleys were designed after his death. In any case the 4½ Litre was a development of the 3 Litre. The 6½ Litre started life as a 4½ Litre, 6 cylinder chassis, which employed many 3 Litre features." In the circumstances it is hardly suprising that the 3 Litre chassis resembled the GP Peugeot cars!

At the time Bentley Motors owned a TT Humber, reputed to be the car driven by Burgess himself, but although Pennal, among others, recalls the car it has not proven possible to identify it. It is unlikely that it was broken up to provide parts for the experimental cars, as apart from being of value in it's own right, Bentley and Bentley advertised a 1914 TT Humber in *The Autocar* of 9th August, 1919. The whole question of the TT Humbers, and particularly which of the three (or maybe four) Team cars was owned by Bentley, has vexed historians to this day.

What is of interest is that when Wallbank went to Humbers in 1925, they still had a pile of spares in their service department which he was able to acquire. Wallbank also revealed that the whole exercise had cost Humber some £15,000. It is quite conceivable that Burgess might well have used TT

Humber parts in the Experimental cars, and Varley is on record as saying, "If Burgess could contribute components from Coventry where he had been Chief Engineer at Humber, it would have been acceptable to WO. I am also certain that the dies for the front axle beam stampings were not immediately purchased, and makeshifts for many parts were pressed into service till the authentic products and tooling came along." The lack of a front axle is confirmed in papers left by Burgess. In a letter to WO, dated 10th August 1920, relating to the state of work in progress at firms in the Midlands (Burgess was busily obtaining quotes from interested manufacturing concerns), Burgess wrote: "With reference to the front axle dies ordered by BM Ltd and now in hand, an arrangement will have to be come to, enabling the makers of the front axle to control the dies and stampings in question." It would seem that the tooling for the stub axles and steering arms had not been made either. EXP 1 was the only Bentley ever to be fitted with a front axle with that distinctive dropped (and very Edwardian) front axle beam, probably of Humber origin. It seems likely that the front axle beam on EXP 2 was hand-forged.

The Humber apparently left a great deal to be desired, being harsh, noisy and possessed of doubtful handling characteristics. Hillstead expressed relief that the first Experimental Bentley did not inherit any of the Humber's shortcomings. However, one well-known VSCC expert was heard to question how WO could have designed such a lump as the 3 Litre Bentley when he had a TT Humber to look at! Just to add to the controversy, it has also been suggested that Straker-Squire bits might have got into EXP 1 as well.

At last the chassis of EXP 1 was complete, and early in the morning Clive Gallop crept into New Street Mews to try out the chassis before WO's drive. As they edged out Gallop started to disappear downwards as the nest of coats he was sitting on gradually wound round the prop shaft. Walter Easter recalled "Gallop disappearing up the street sitting on a box on the new No. 1 chassis – with all the other workers standing equally spaced on the kerb each with a Pyrene fire extinguisher at the ready." WO drove EXP 1 later that day, describing it thus: "I have an all too clear memory of my first run in the prototype 3 Litre in 1919. I was quite appalled by the noise; that was my first and most lasting impression", later saying "The noise and roughness of this machine were almost indescribable." EXP 1 was then supposedly delivered to Harrisons of Drummond Street who fitted

The engine of EXP 1, soon after it was put on the road. Note that the nearside plugs are blanked-off, and the very obvious oil pipes to the dry-sump tank fitted under the scuttle.

The dashboard of EXP 1, with CAV switchplate and ebonite knob in the centre of the steering wheel for the ignition control. This arrangement was used on EXP 1, 2, and the 'Fire Engine', and probably EXP 3 as well.

light four-seater coachwork with a detachable scuttle. Shortly after EXP 1 had taken to the road, WO wrote to Burgess at Coventry:

"Sunday.
Dear Burgess,

On Saturday after running about a bit, we, Gallop, HM, and myself, went to St. Albans and the round about way back, in the evening I took it about 20 miles and today we went 80 miles.

It ran very well indeed and quite like an old car. The front springs are too weak, but the springing is excellent.

The brakes are wonderful and can be put on hard on grease without any side skid. Foot brake very

good possibly could do with more leverage. The engine keeps much too cold and never got to within $\frac{1}{4}$" of the dial of the thermometer.

The carburation is not good at present, the induction pipe being too cold. When stopped after running you can hold lightly onto the insulation of the plugs.

She is wonderfully smooth (the engine) and you can pull slowly up hills on top without any signs of labouring – 10 mph on top easily. I got up to about 60 but the carburation is not good at full throttle and generally on three cylinders and dropping back.

Holds the road very well – steering is nice but a bit low geared for me. Top gear cams rockers etc

look fine and drain all right. Change speed plunger springs are too strong and make it hard to get out of gear.

Pump or bottom bevels still make rather too much noise. Everything else seems OK.

Finally she is going to be the goods. Easy to handle, very fast and as easy in traffic as a six – easier than a DFP.

More when I see you. I am very pleased.''

EXP 1 was fitted with a four-seat open body, reputedly the work of Harrisons, but the body is somewhat mis-matched, particularly around the scuttle. It is noticeable that neither of the doors has hinges, and indeed that neither of them can be opened because there are tool and battery boxes in the way. It is exceedingly improbable that WO would have let Harrisons have the only Bentley in existence for several weeks to build a body, and then for it to emerge with so many peculiarities. Further, it is strange that Walter Easter recalled the wings and bonnet as being the work of Ewarts. The body on EXP 1 bears more than a passing similarity to that on the post-war DFP demonstrator shown in plates 27 and 28 of *Fifty Years with Motor Cars*, and it seems a fair guess that the body on EXP 1 was a DFP body fitted to the back part of the chassis and then with a crudely made-up detachable scuttle to fill the gap.

While EXP 1 was being tested EXP 2 was gradually assembled for the 1919 Olympia Show, in it's first, largely mocked-up form. As manufacturers and members of the SMMT, Bentley Motors were able to take a stand in a cramped corner with just one exhibit – the mocked-up version of EXP 2. The stand was at least in the Main Hall at Olympia, rather than in the annexe at White City. The Bentley stand, no.126, was conveniently positioned opposite the rather more spacious DFP stand, no.107, which had the room to exhibit a 12/40 chassis and two complete 12/40 cars – a two seater and a four seater. Potential Bentley customers presumably did not have their attention drawn to the carefully finished wooden crankcase – the real one was still being machined. The starting handle was located by a dowel only, to the discomfiture of some bright spark who swung the handle vigorously to test the compression and threw himself over one of the wheels. The atmosphere of that first post-war show was typical of the "boom" mentality of the era and the embryonic Bentley Motors could have sold many chassis – had they been in production! As it was, HM and Hillstead had to be content with taking names of potential customers to create a list of future possibles. Any number of manufacturers were setting up in that era, to produce motor cars, good, bad and indifferent to cash in on demand – but many were very shortlived.

Hillstead recalled that show vividly in *Those Bentley Days*; the leisurely early morning start with the flash of magnesium lights of commercial photographers, then the gradual filling up as the public crowded in. The whole atmosphere became hotter and noisier as the salesmen put out more and more

LEFT **EXP 1 complete, with her rather crude coachwork with detachable scuttle. According to Walter Easter, the wings and the bonnet were made by Ewarts. The body looks far too crude to be the work of a firm of coachbuilders such as Harrisons, and bears a remarkable resemblance to the body on the first DFP demonstrator . . .**

. . . seen here (right) with HM.

catalogues for small boys – with huge queues for the restaurants and bars which had invariably sold out if one was patient enough to out-live the queue. But there was no shortage of demand, Bentleys having to turn away customers with money because there could be no guarantee of delivery. The catalogue itself proved to be a bit of a liability, in that no photographs of the complete car were included, only a colour frontispiece of Gordon Crosby's painting of a "Bentley at Speed", (Crosby was also responsible for the design of the famous winged *B* badge, which on the original has a different number of feathers on one side than the other, reputedly intended to combat fraudulent dealings), and a number of pencil sketches of components of the chassis. This led to vicious rumours circulated by competitors that the Bentley did not actually exist! Lesser rumours circulated that the Bentley was an "assembled" car, put together from other manufacturers' parts. Nothing could be further from the truth, but it was, nevertheless, a rumour that cut deep. This is not as far-fetched as it would seem today, as one or two eccentric manufacturers showed way-out chassis at Olympia and the Paris Salon, while initially never producing any for the public – Bucciali were perhaps the prime example of this with their extravagant Voisin V12 powered creations.

After the overwhelming response of the public to the Olympia Show, it was clearly necessary for cars to be produced as quickly as possible, and for this, premises were urgently needed. Members of the Board suggested the Midlands – in *Those Bentley Days* Hillstead refers to a shareholder in the very early days who "owned a works (in no way connected with engineering) some two hundred miles from London. He undertook – for a consideration, of course – to clear out some sheds which, he said, would provide ample space for the Bentley to be built. He met with a very firm refusal." For his part, WO was determined to establish manufacturing in or near London. The prospect must have been discussed of finishing the two experimental cars and then approaching one of the large manufacturing concerns with adequate capacity to take over and produce the design (this was suggested by some of the shareholders), but by this stage WO and his dedicated team had the bit firmly between the teeth and were determined to set up and run the company themselves.

However, the amount of capital originally subscribed (presumably the £20,000 of the first Company set up in January 1919) was intended only to get the first two Experimental cars on the road, at which point it was intended to raise further funds. It was confidently expected that such a course of action would present few problems. However, as yet EXP 1 was only just on the road and was very far from being a production prospect without substantial development work; and there was no more money. Hillstead: "The thought of having to buy, lease or build a works, literally made our hair stand on end."

Suitable premises proved to be unavailable, until

47

The first building to be erected at Cricklewood, towards the end of 1919/early 1920. This constituted the sole manufacturing facility for several months until the office block and the other shops were put up. EXP 2 and 3 were built in here. Parked outside is WO's DFP. Unfortunately, no photographs of the inside of this shop can be found from this period, so the interior must remain a tantalising mystery.

Hillstead and HM put in an offer for Tangmere aerodrome, in Sussex. This large complex had hangars, electric generating plant, on a vast ground area. Within days of putting in the offer, reaction set in and it was realised that Tangmere represented a massive liability that Bentley could in no way sustain. Negotiations then ensued to release the company from their obligations, which fortunately only incurred minimal legal costs. It is conceivable that they could have taken on Tangmere as a business proposition in its own right, selling off some of it or renting and leasing parts to other firms, but even if that option had been considered, there was no infrastructure to support the manufacturing organisation.

WO then announced he had found a plot of land at Cricklewood by the Welsh Harp (now less romantically known as the Brent Reservoir), bordered by Oxgate Lane and the Edgware Road. In order to pay for it, a mortgage was taken out with the London Joint City and Midland Bank Ltd.

covering "amounts owing on current account" secured against "freehold hereditaments and premises situate at Willesden in the County of Middlesex." On this land in early 1920 the builders erected a brick building into which the entire concern fitted – assembly, test, stores and Nobby Clarke's office as Works Manager. Soon afterwards, a wooden Army hut was put up for office accommodation. WO and Burgess occupied the north end, various offices were located down the centre, with Varley and the Drawing Office being

**The brand new EXP 2 circulating Dorset Square
c. March 1920.**

situated at the south end nearest the original brick
shop. Staff were taken on – Hassan, the Saunders
father and son, and others who became well-known
players in the Bentley story. Peter Purves came in
as Buyer from the Cosmos Engineering firm in
Bristol, which did much of the early work on the
components of EXP 1 and 2, and other firms were
found to manufacture parts. It was not until 1929
when Bentleys first acquired machine shop facilities that any substantial in-house manufacture was
to be undertaken.

The policy of buying-in components manufactured by outside contractors to Bentley drawings
was expensive, and led to an increase in chassis
price from £750 in the 1919 publicity material, to
£1150 at the White City Show in London in
November 1920. Machining facilities in the early
days consisted of one lathe, which was essential for

building the cross-shaft turret assembly. The end
float in the gears has to be set up by machining the
faces of the bushes, as there is no provision for adjustment. It seems rather suprising that WO did
not use the eccentric slot and pinch bolt scheme on
the cross-shaft gear along the same lines as the top
bevel gears, back axle and steering column.

WO was unable to concentrate on the manufacturing as much as he would have liked because of
the boardroom rows over the financial position of
the company which went from bad to worse. In
December 1919 the nominal share capital was
increased from £100,000 to £200,000, but again
this was only an optimistic move to give them the
capacity to use that much money were anyone prepared to put it up. With no imminent sales there
was no income, and costs were rising rapidly. WO
wanted a Heenan and Froude dynamometer to
check engine horsepower, but was blocked by the
board; as WO put it, a ridiculous position to be in,
to be building a high-performance sports car and

not be able to measure the power output of the engine! He got his way, but the pettiness of the board in its sundry forms during this period seems difficult to believe. However, apart from that fact we only have WO's account, as an engineer who never expressed any interest in money, and would probably have been unable to understand or appreciate the mentality of those whose primary motivation was making more money out of the Bentley concern. Any investor putting significant sums of money into the Bentley Motors of 1919/1921 must have been seriously worried about the prospect of ever seeing a return given the mounting costs, increasing chassis prices and the gradual waning of the post-war boom. Indeed, the shareholders and the Board must have been aware of the possibility of losing everything if the firm went under, as there were virtually no assets, so they had little option but to continue somehow in the hope of it all coming right. It is to WO's enormous credit not just that he managed to produce a remarkable sports car, but that he achieved it against monumental odds.

In January 1920 Sammy Davis drove EXP 1 and wrote a glowing account in *The Autocar* of the experience. Davis later recalled that the article and the photographs were presented to imply that the car had been tested in France, because one or two published accounts of illegal speeds had attracted attention from the police. This implication comes over quite strongly, one sentence even reading "In England, as a short run proved, the machine can travel without protest, chatter, or difficulty in top gear at 10 mph or under." The part of the article where "As the speed increased to over 70 mph, the landscape leaped at us, wind shrieked past the screen...as the roar of the exhaust rose to its full song" is definitely constructed to suggest that the test was carried out on Continental roads! Road tests in those days were far less objective than now, and Hillstead rued the lack of production chassis

ABOVE LEFT **EXP 2 on test at Brooklands, circa March/July 1921. EXP 2 won her first race, the Whitsun Junior Short Handicap, on 16th May 1921 in this form.**

LEFT **HM with EXP 2 (soon after that car was first put on the road) with acetylene lights and experimental engine with the prototype valve gear and Claudel carburettor. The body is just the back end of something else grafted onto the chassis, with a gap where the back half doesn't match the panelling over the scuttle.**

that they could have sold many times over on the strength of that first road test. Davis commented only on a "peculiar grating noise from somewhere in the engine", probably caused by the scavenger pump of the dry sump system, and the quantity of oil sprayed over the outside of the engine – problems inseparable from the production of a new design that would be solved before series production started.

WO favoured the dry sump layout with the scuttle-mounted oil tank and scavenger pump below the main oil pump (see GA drawing, patent drawing and photos of EXP 1 engine). However, when starved of oil, as happened frequently, the scavenger pump cavitated noisily, and the sump acted as a sounding box, magnifying the noise. There was an attendant disadvantage with the tank for the dry sump being mounted under the scuttle, as it had a tendency to emit oil froth onto the passenger's feet! WO considered silence in a motor car as being of paramount importance and very quickly the wet sump system, with an almost separate oil tank below the main sump, was designed. This remained basically unchanged until the introduction of the one-piece sump used on the 3 Litre in late 1926 and then on the 4½ Litre. Several arrangements of the two-piece sump were tried with cast-in cooling channels – but these became rapidly blocked with road dirt and were consequently abandoned. This was not the end of problems with sumps. On the very first cars the bottom half of the sump was fabricated from Manchester plate (tin steel as used for petrol tanks) and this was found to magnify engine noise to quite unacceptable levels. New bases were made from aluminium castings which solved the problem and fitters were sent round the coachbuilders to ensure that all of these were changed before any cars were delivered.

Many years later Varley recounted the reason for gear noise with the early two-piece sumps. "The second incident concerns the now legendary scream of the oil pump driving gear and this is the true story. In the original design the oil pump driving shaft, carrying the gear which mated with the one on the crankshaft, had its axis parallel to, but sloping back towards, the centre of the engine. This brought the oil pump near to the middle of the engine lengthwise, but well to one side due to the centres of the two spiral gears. This meant that the pumps could be starved of oil if the engine was standing on a severe camber. So I suggested to WO that we swing the pump round the crankshaft gear to bring the pump nearer to the centre line of the engine, where it would stand a better chance of always being served with plenty of oil. WO agreed and we proceeded to alter the sump to take the new position of the oil pump. But the helix angles of the gears were not altered and so the gears, correctly calculated for the first angle, were put into production for the early engines. Gear experts were called in, more gears were made by another firm, but no-one was able to solve the problem or quieten the noise. Eventually I realised I had made a mistake in assuming that the gear on the oil pump was just being rolled round the crankshaft gear, whereas it should have been treated as an exercise in solid geometry with a correction made for the second angle, which was about seven degrees. New gears were made to this corrected figure and all was peace."

On the 19th January, five days before Sammy Davis' road test appeared in *The Autocar's* issue of the 24th, WO wrote to Burgess from Conduit Street:

"Dear Burgess,

Very glad to hear about the sump, and I am looking forward to seeing it.

I shall not be going to Coventry this week – Gallop is in Paris, and Enstone [Company Secretary] is in bed with a bad cold. I also have a cold, and my case comes off today week [The Royal Commission on Awards to Inventors – see p.32].

The gear ratios I want are:-

Wheels	875
Top	3.75
3rd	5.0
2nd	6.0
1st	10.75

I hope we can get these.

Went up Aston Hill yesterday – we won't do so badly."

They managed to obtain the ratios requested, the gear ratios corresponding almost exactly to an A box and a 3.78 (14/53) back axle, the ratios of which are 3.785, 5.03, 6.17 and 9.99. The smaller wheels used (820 instead of 875) would be more or less compensated for by using a 3.53 (15/53) back axle, which the very light, early cars, would have pulled fairly easily. Nobby Clarke recalled Burgess as saying "I designed this gearbox [the A box] on my dining room table at Coventry before I left to come to London."

EXP 3 was completed at Cricklewood in 1920 and fitted with all-weather coachwork by Harrisons. EXP 3 was much nearer to the production specification than EXP 1 and 2, with the now familiar production frame and a 9'9½" wheelbase. WO

took over "The Cab" as it was called, for his own personal transport. At the time "The Cab" was a source of some concern at the Works because WO refused to let anyone wash it until it had done something like 20,000 miles! EXP 3 was Bentleys' sole exhibit at the 1920 White City Show, described as a particularly well-finished saloon of the vee-fronted type, but Bentleys announced that they would only supply chassis and inspect the finished car before issuing the guarantee.

For the first time mention is made of the well-known Five Year Guarantee, covering any part that might fail due to defective materials or workmanship. The price of the complete EXP 3 was quoted as £1700, the specification including ML magnetos and a Claudel-Hobson carburettor. A speed of 80 mph was guaranteed with a light body and deliveries were expected to start in June 1921. The chassis price was reduced by £100 to £1050 in April 1921, with a guarantee of no further cut in prices before 31st August, despite the fact that no cars had been delivered!

By the middle of 1920, with the Works established at Cricklewood, the Company's registered

office was moved from Conduit Street to 3, Hanover Court. WO observed "Hanover Street had brought us fortune in the past and it seemed a good omen when we took a lease on excellent premises there." The showroom had been a dress-shop, so HM and Hillstead had to improvise with their own furniture and devise a way of getting the polished chassis into the window. The sole contents of the showroom were two tables, two chairs, a small pile of catalogues and reprints of Sammy Davis' road test, augmented by a BR cylinder mounted on a wooden plinth. This austerity must have caused passersby to wonder what went on in there. The cylinder was subsequently replaced by a complete BR2 engine loaned to WO by the Air Ministry. With the exhibit of the polished chassis at last, Bentley Motors' sales team could acquire a modicum of dignity as purveyors of high-class sports cars.

Gallop, remembered for the "pou-fou" breather pipes on the crankcase of EXP 1, had by this time left to work for Zborowski at his racing stable at Higham, and was (in a manner of speaking) succeeded by Frank Clement – a professional racing

EXP 3 at Brooklands. The side view shows the first pattern of wet sump, with cooling ducts cast in longitudinally.

driver in his own right and a competent development engineer. Clement had worked for Vauxhall and Napier before racing for Straker-Squire and after serving in the Royal Engineers was working on aero engines, which delayed his joining Bentleys. WO himself, of course, was as good a racing driver as the rest of them, but due to insurance policies on his life and matters related to the Company, was unable to race again after the 1922 TT (he did, though, drive in various hill-climbs and speed trials). WO was also a formidable expert with carburettors.

With the main buildings put up, the Works was organised accordingly. The Experimental Department moved out of the brick shop across to the new shop, with the Engine Erecting Shop going into the adjacent shop. Production, which had consisted of just R.S. Witchell, the Works Manager, and Fred Conway, in charge of the Stores, was expanded rapidly. Conway set up the Stores along the north end of the shop fronting the Edgware Road, with the Erecting Shop at the Oxgate Lane end. The Running Shop moved into the last of the new shops, at the Oxgate Lane end of the Works opening onto the Works yard. The drawing office staff were already established in the wooden shed in the yard. Nobby Clarke was appointed Works Superintendant, reporting to Witchell. The arrange-

ABOVE LEFT **EXP 3, 'The Cab', was the first 3 Litre to be built on a 9′9½″ production pattern frame. The chassis was then fitted with a Harrisons Allweather body, for WO's personal use. 'The Cab' is seen here outside the showroom at Hanover Street, with HM just visible behind. The BR aero engine can be seen in the window.**

LEFT **'The Fire Engine' outside the Hanover Court showroom, with HM (left) and Hillstead in the doorway.**

ABOVE **Building the first production shops, *c.* Autumn 1920. The buildings were relatively cheap, steel-framed with breeze-block walls and corrugated roofs. The old brick shop on the left became the engine test shop.**

BELOW **HM with 'The Fire Engine', outside the new Works at Cricklewood. To the left is the wooden office block, then the experimental shop, the original brick shop, and the running shop to the right. The boiler house has yet to be built.**

ment of the Works in the early days is illustrated (ibid). The final strand of the management was that of General Manager, to deal with day-to-day running of the office management and the like. Guy Peck joined Bentleys from Airco in the period 1920–21 to fill this role, described by Hillstead as that of *"universal shock absorber. He had to tie the strings together when there were no ends. He had to combat the inherent differences in outlook between Works, Sales and Service"* (*50 Years with Motor Cars*).

Just before the end of 1920, on 10th December to be exact, the Company issued a prospectus for potential investors:-

"1. BENTLEY MOTORS LIMITED was registered on the 10th January 1919, with a nominal share capital of £20,000, of which 15,000 shares for £1 each were issued, 7,000 for goodwill and 8,000 for cash.

2. The present Company was registered on the 10th July 1919, with a capital of £100,000 (one hundred thousand pounds), and divided into 100,000 shares of £1 each. The capital has since been increased to £200,000.

3. Under the Vendor's Agreement, the whole undertaking of the original Company was purchased for 30,000 shares of £1 each, fully paid. No cash was paid to the Vendors.

4. A site of four acres, with double road frontage, has been purchased freehold, with the frontage on the Edgware Road about 200 yards on the London side of the Welsh Harp, and a factory is nearing completion on this site for the erection of the chassis. The cost when completed, including the cost of the freehold, will be about £22,000.

5. A 21 years lease of showrooms at No. 3 Hanover Court, Hanover Street, W., has been secured at a very low rental.

6. Deliveries will commence in June 1921, and provision has been made to produce five chassis per week from that date onwards.

7. After allowing for all costs of production and all overhead charges, including the costs of selling and advertising, there is an average net profit of £200 per chassis.

8. The profit for one year, on the above basis of production, is thus estimated at £50,000, which will enable the Company to pay a substantial dividend and to make an adequate reserve.

9. SALES. The Sales Department took over a car for demonstration purposes during November 1920, and trial runs have only, as yet, been given to a very few members of the Motor Trade. The Company had, nevertheless, on 9th December, orders for 73 chassis (though confirmation in respect of twenty two export orders has not yet been received, but is expected shortly).

10. DESIGN. The car has been designed by Captain W.O. Bentley, the designer of the BR1 (Bentley Rotary 1) and BR2 (Bentley Rotary 2) aeroplane engines, the latter of which, fitted in the Sopwith Snipe, made the finest fighting aeroplane at the time of the Armistice. Many thousands of these engines have been made, and with the exception of the Rolls-Royce firm, Captain Bentley was the only English designer who produced a satisfactory service aeroplane engine.

At the beginning of the War he was responsible for the introduction of aluminium pistons for use in aeroplane engines, and for having these tried on Rolls-Royce, Sunbeam, and Clerget engines.

Before the War, Captain Bentley was associated with the DFP car, and as a result of alterations made by him to the 12–15 HP model, evolved the 12–40 HP model, with which he still holds the Class B records at Brooklands.

His performance in the Tourist Trophy Race of 1914 is well known, and in addition he has won a very large number of Hill Climbing and Speed events."

With reference to No.4, the following figures are from a notebook of WO's, dated August 1920:

> "Main buildings £11200
> Experimental Shop £1000
> Design £4850
> Front road £750
> Heating £950"

One can only say that the foregoing shows a great deal of optimism, and in the event their analysis of £200 profit per chassis was unrealistic – the heavy costs of sub-contract manufacturing assured that. In order to assure a profit of £200 per chassis, the price had risen from an estimated £750 in January 1919 to £1150. Burgess had finally put together enough quotes by September of 1920 to set out a realistic costing for the first production chassis, as follows:

"Sep. 14 1920
Coventry

Costs based on C.O.W. [Coventry Ordnance Works] and Hotchkiss figures to date and other costs obtained by BM Ltd.

	£	s	d
Frame and dumb-irons	34	0	0
Springs – shackles and pins	14	0	0
Radiator and brackets	18	0	0
C.O.W. engine parts	200	0	0
Crankshaft camshaft and gears	50	0	0
Sump and oil pipes	10	0	0
Water pipes and clips and hose	5	0	0
Starting and lighting	35	0	0
Clutch gear, pedals and flywheel complete	44	10	0
Gearbox complete	75	0	0
Back axle less brake drums	150	0	0
Front axle	43	0	0
Steering gear – say	30	0	0
Handbrake gear	25	0	0
Wheels, hubs and brake drums	33	0	0
Tyres	50	0	0
Dash	10	0	0
Tank, bonnet, Autovac and silencer	20	0	0
All work and sundries at Cricklewood	75	0	0
Total:-	£921	10	0

Say chassis tested, painted leaves BM Works at £950.''

Most of the front and back axles, the engine parts and the steering gear along with the handbrake gear were quoted for by the Coventry Ordnance Works. It seems, though, that the it was not C.O.W. but the Automotive Engineering Co. of Twickenham that got the job. WO in *The Cars in My Life*: ''In the end we solved our production problems mainly through a firm at Twickenham which had been on war work and was suddenly, and rather mysteriously, at a complete loss for orders. They did some eighty per cent of the car.'' The axles were to be delivered to Cricklewood complete and ready to bolt to the springs, the hubs being supplied by the Dunlop Rim and Wheel Company. The engine parts were delivered to Cricklewood, and the engine was assembled there. Stirling Metals cast the blocks, cam casings and covers, and the gearbox casing.

Hotchkiss quoted for manufacturing the gearbox internals, but it seems likely that all the gears were made by ENV. WO *The Cars in My Life*: ''Another firm took on the back axle, the gearbox gears and the bevel gears for the engine. By this time – in 1922 – we had become not manufacturers of motor cars as we had intended, but assemblers and testers, and that remained our function almost until the end. In fact the only car for which we did any machining ourselves was that curious and little lamented 4 Litre car. We were also, as we were to learn to

AFC Hillstead, salesman for Bentley and Bentley and later Bentley Motors, at the wheel of 'The Fire Engine'. Although virtually to production pattern, this car still has the ebonite knob on the steering column for advance/retard only.

our cost later, entirely in the hands of these two firms. We had no alternative means of supply and they appreciated this dilemma only too keenly, charging us stiff prices, which became inflated again if we ordered even the most minute modification!''

The Experimental 3 Litres

The whole question of the Experimental cars is somewhat blurred and it is interesting to look at the cars on the road at the end of 1920.

– EXP 1 (registered BM 8287 on 11th December, 1919) had reputedly been relegated to the Showrooms as a polished chassis.
– EXP 2 (registered BM 8752 on 25th March, 1920) was on the road, still fitted with its first 2 seat body.
– EXP 3, the first production-pattern car, was put together for the 1920 White City Show and registered BM 9771 in the name of Bentley Motors on 16th September 1920.

This, then, is the accepted wisdom; but, as with so many aspects of Bentley history, it does not stand close scrutiny. It seems likely that there were more cars around than these three. This hypothesis is supported by Nobby Clarke's recollection that three sets of experimental bits were made up and

that the 1919 Show chassis was made from "manufacturer's scrap" (ie, the basis for four cars). The 1919 Show engine was partly a wooden dummy and there are discrepancies in the details of EXP 1's engine and that of the polished chassis.

It is conceivable that the situation was as follows:

EXP 1 reg'd BM 8287 9' 4" frame EXP engine no. ?

The first prototype, described earlier. Registered BM 8287 with Bedforshire County Council on 11th December 1919 by Bentley Motors Ltd. Road tested by Sammy Davis in January 1920, and used for experimental work. EXP 1 is shown on p. 32, 33 and 44 of *EXP 2* by T.D. Houlding; it was originally thought that these photos were of EXP 2 wearing EXP 1's coachwork, because of the dating to Winter 1920/21. However, it was realised soon after publication, from the front axle, that the photos are, in fact, of EXP 1. (See photograph above.)

It seems likely that EXP 1 never became a display chassis at all, that it was kept as a running car until it was rebuilt late in 1920 as "The Fire Engine". "The Fire Engine" was registered BM 8287 in January 1921, fitted with a dark grey four-seat body. The chassis number was listed as 1, the engine number as "unknown"; the year of the

Frank Clement in EXP 1 at Brooklands, thought to be late 1920. Note the dip in the centre of the front axle beam – compare with the photos of the front of the TT Humber, pp.22–23.

engine was given as 1920. Its claret paint scheme was registered in April 1921, and in May a change of engine to no.17, listed as a 1921 engine. "The Fire Engine" was then sold to W.J. Dowding in 1923, as chassis 1, engine 17. However the Service Records list Dowding as owning chassis 20, engine 17, a 1922 short chassis registered BM 8287; van Raalte, of course, supposedly owned chassis no. 1.

The previously accepted position of three Experimental cars and three experimental engines is shown to be misleading from the following extract from WO's notebook:

"Log of No.1 car.

	Miles
Cyls taken off to ease valves	2,000
New camshaft fitted, lubrication failure	13,500
New engine fitted old one OK	15,000
New tight fit pistons fitted	15,540
New mileage indicator fitted	16,103
Real reading for indic, mlge reg. +500	
Bentley type shock absorbers fitted	17,245
New engine fitted (old one OK transferred to Show car)	19,000"

Of EXP 1, Nobby Clarke wrote in 1970: "Experimental Chassis No.4 ["The Fire Engine"]...it was Experimental No.1...fitted with a new open touring body, and had a production type engine installed...As the chassis was used as a demonstrator it would have been gradually modified to production standards as the bits and pieces became available.

"Re index number BM 8287 – there is a photograph in Hillstead's book which clearly shows, under the caption of the "Fire Engine", the car bearing this registration number.In regard to chassis no. 20?...engine no. 17...I have no recollection whatever of this switch...or number fiddle...and what is more, I do not think it would have been tolerated by the Company. I believe that we had three sets of parts, enough for three cars, and that the Show chassis engine was made up from manufacturer's scrap...Summing up – I have no knowledge of EXP 4 unless it was EXP 1 rebuilt. I have no knowledge of EXP 1, or for that matter, EXP 4 being renumbered production chasssis no. 20."

EXP 2   reg'd BM 8752   9'4" frame   EXP engine no. ?

Built using parts of the chassis exhibited at 1919 Show. Registered BM 8752 by WO on the 25th March 1920, used by Clement for experimental work (referred to as "John Willie" by Clement). It is significant that the Show chassis was very much a dummy, and Pennal comments in *The Other Bentley Boys* that "we built the second and third experimental cars (EXP 2 and 3) in that first shop of the Works." The Works was not built until well after the 1919 Show, implying that at the very least the Show chassis and EXP 2 embodied considerable differences. Hillstead commented that "The Show chassis was not available, as many of the parts were required to complete EXP 2 then under construction." (See comments below on the polished chassis.) It would seem that the working, post-Show engine came from EXP 1.

EXP 2 was later sold to Foden as a complete running car, on the original 9'4" chassis frame – see *EXP 2* by T.D. Houlding.

EXP 3   reg'd BM 9771   9'9½" frame
Production engine

Registered BM 9771 to Bentley Motors on the 16th September 1920. Used by WO as his own car, known as "The Cab". This car was tested at Brooklands:

"No.3 complete with two spare wheels, etc. Weight 3305lbs = 29½ cwt

Tests on Brooklands. January 27th 1921. Weather fine. No wind. Track dry.

1 Mile standing start   ½ secs = av speed 50
½ Mile standing start   41½ secs = av speed 43.9

BELOW **'The Fire Engine', the company's sales demonstrator. The chassis features are virtually to the final production pattern, with well-equipped touring coachwork. The claret/polished aluminium colour scheme seems to have been used on virtually all the pre-production 3 Litres except EXP 1, which was grey.**

TOP **HM** at the wheel of 'The Fire Engine' outside Emscote Grange. This shows the first pattern of wet sump, a deep square casting with cast-in cooling ducts. This sump was rejected because the ducts filled up with dirt, negating their cooling effect.

ABOVE **EXP 2** at the works, *c.* March/July 1921. Nobby Clarke stands to the left, in the doorway to the engine erecting shop. The door to the right leads into the detail shops.

TOP **EXP 3 on the test hill at Brooklands in January 1921 with WO driving and Sammy Davis in the passenger seat. This shows the production pattern frame and front axle, but the starting handle is still an experimental part.**

ABOVE **Wally Hassan with EXP 2 inside the experimental shop, in stripped form, with a special twin carburettor installation.**

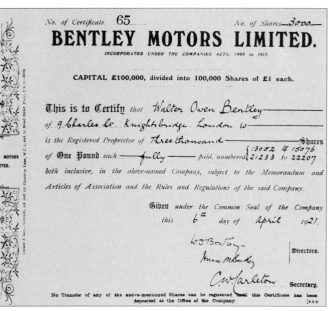

½ Mile flying start   27¼ secs = av speed 66
1 lap flying start   2 min 37 secs =
                              av speed 63.44

20 mph to 60 mph on Top Gear   43½ secs
0 to 60 mph standing start   40 secs

Test Hill   14½ secs = av speed 16.6

Engine had low compression. 160 induction pipe. No.1 camshaft and had done 6,000 miles since cyls. were off.

From a standing start to 60 mph takes about ½ mile.
Also from 0 to 60 mph on top takes about ½ mile."

EXP 3 was sold to E. Taylor in 1924, as a 1924 Model with production engine no. 1158; a five year guarantee was issued on the engine only. EXP 3 was last heard of in 1932, owned by R.S. Sikes in Langley, Bucks.

EXP ?   None   9′4″ frame   EXP engine no. ?
Polished Showroom chassis, probably built from bits of the 1919 Show chassis and various scrap bits and pieces. WO suggests in *My Life and My Cars* that this chassis was exhibited in the old DFP showrooms in Hanover Court Yard before 3 Hanover Court had been set up; so it seems likely that the polished chassis existed concurrently with both EXP 1 and EXP 2.

(EXP 4, used by Burgess for experimental work, particularly in relation to four wheel brakes, probably superseded EXP 2. This car is later than the above cars, with what is a mid 1922 registration, ME 2431.)

ABOVE LEFT **Hillstead topping up EXP 3's radiator at Nonancourt. This photo was taken in 1923, during WO, Clement and Hillstead's trip to the French Grand Prix at Tours (see p. 127). EXP 3 was used by WO as his personal transport, and for experimental purposes, until 1924.**

FAR LEFT **WO's share certificate for the new company.**

LEFT **The polished showroom chassis, referred to above, on display in the showroom. Hillstead is on the left, HM on the right. It seems likely that this chassis was shown in the old DFP showroom at the back of Hanover Court, and it is not clear whether this photo shows the old DFP showroom or the new Bentley Motors showroom in Hanover Court.**

Technical and engineering aspects of the 3 Litre.

While the motor cars themselves do not form the central part of this story, they are nevertheless of fundamental importance and the development work on the cars reflects much about the Company itself. The 3 Litre design formed the basis of all the subsequent models, some components remaining virtually unchanged right through to the 8 and 4 Litres. The Bentley design philosophy can be expressed in two distinct model ranges, the four cylinder and the six cylinder cars. The 4½ Litre and 4½ Litre Supercharged cars were basically developments of the original 3 Litre, and in the same way the 8 Litre can be considered as a development of the 6½ Litre. The 4 Litre was not such an aberation as might at first be thought, as it attempted to amalgamate the chassis sophistication of the 8 Litre with a more modern engine design intended to supersede the four cylinder chassis.

Unfortunately, the only remaining drawing of the first experimental 3 Litre engine is a sectioned general arrangement, but it does show some interesting details. The oil pump drive is very similar to that of the Grand Prix Mercedes engine, as is the camshaft drive, both of which facts were acknowledged by WO. The oil pump with its scavenger below was the subject of patent number 148656, and a very similar principle of using a pump in two parts with a lower scavenger was revived on the supercharged cars in 1930. Lubrication was by dry sump with the oil tank fitted under the scuttle. The dry sump layout offers advantages in that the oil can be kept cooler more easily. The prototype camshaft installation had to be inserted from the front of the engine with the crownwheel in front as well, there being no means of adjustment for end float apart from the dimensions of the closing plate – good theory, but hopeless for manufacturing! Every assembly would have had to be individually fitted by machining, or by shimming between the thrust race and the camshaft crownwheel. The rocker rods had to be inserted through holes drilled in the bulkhead of the car, which again would have been hopeless for routine maintenance – apart from the inconvenience of having to withdraw them from inside the car, they would probably have fouled the dashboard. The camshaft housing had to be bored by the same process as is used for gun barrels and the bearings assembled onto the shaft before offering up the whole assembly through the front.

As far as is known only the engines of EXP 1 and

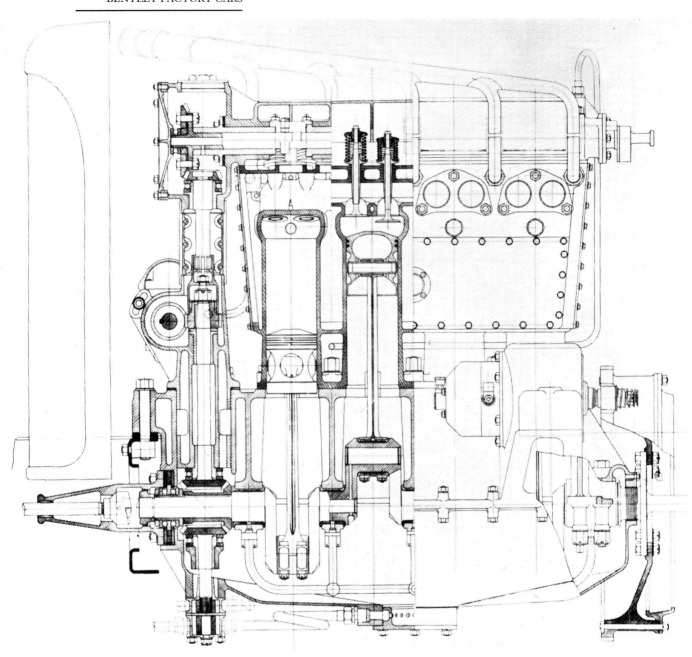

ABOVE **The General Arrangement drawing of the Experimental 3 Litre engine – this is probably (C) of those listed in the agreement between Bentley and Bentley and the first Bentley Motors. Note the lack of adjustment for meshing the top bevels other than by machining or shimming and the principle of using bevel gears as seen on the Mercedes engine. Note also the double oil pump with scavenger for the dry sump layout. The crankcase, crankshaft, location by double-row thrust race, rods, pistons and cylinder block were all virtualy right first time – similarly the clutch and outer ring. The studs holding down the block to the main bearing caps are similar to the Mercedes and greatly increase the rigidity of the bottom end of the engine.**

RIGHT **Vertical section through the 3 Litre engine, c. 1921. Note the production pattern two-piece sump, hourglass pistons and the long main bearing studs through to the foot of the block.**

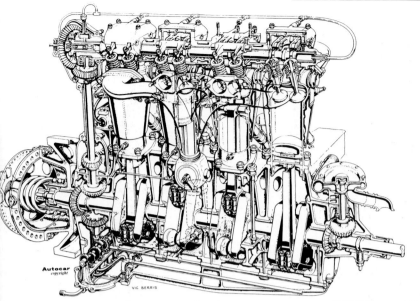

LEFT **The 1914 Mercedes GP engine. Note the use of bevel rear drives, and forked rockers operated direct from the camshaft. Note also the long studs through the crankcase to the main bearing caps.**

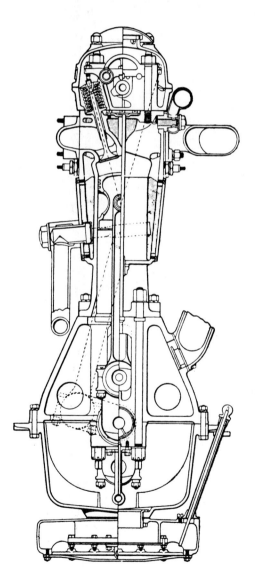

EXP 2 were made in this manner. It is clear that EXP 1 and 2 were very much intended as test-beds and after the soundness of the original design concept had been proven, the engine was extensively redesigned and cleaned up externally for production. The basic engine design proved satisfactory in terms of the bottom end configuration, cylinder block and valve gear configuration, and provision for accessories. An extensive redesign was necessary to render the valve gear practicable, to provide a facility for dual ignition and a water pump drive, to provide adequate oil cooling, and to simplify and tidy up the external appearance of the engine.

Redesigning the valve gear led to the production camshaft drive, in which the crownwheel is behind the top bevel gear. This is mounted on a vernier flange to aid timing the engine with end float adjustment by means of two large 26 TPI slotted nuts. This latter arrangement was patented by WO, Burgess and Bentley Motors on 10th August 1920. The top bevel gear runs in a long bronze bearing, backed up by a thrust race. This bronze bearing has two inclined slots with pinch bolts for setting the gear meshing, the same principle used in the steering column and back axle. In order to aid meshing of the bevel gears, two inspection plugs were incorporated into the camcasing early on into the production.

The camshaft runs in five split bronze bearings, the lower halves sitting in the camchest, the upper halves nipped by the bridgepiece. The bridgepiece is a long aluminium casting, holding down the rocker shafts and provided with recesses to locate the camshaft bearings axially by means of dowels in the bearings. The lower halves are similarly located by dowels into the camchest. The camcover was a

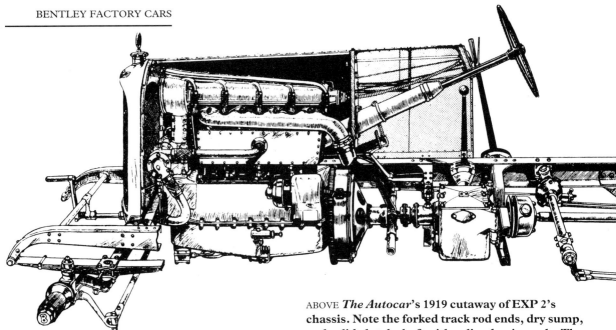

ABOVE *The Autocar*'s 1919 cutaway of EXP 2's chassis. Note the forked track rod ends, dry sump, and solid clutch shaft with splined universals. The gearbox is the standard A type, but the petrol tank and rear universal on the prop shaft differ from the production chassis.

BELOW **The first sketch of the 3 Litre chassis, published in *The Autocar* early in 1919.**

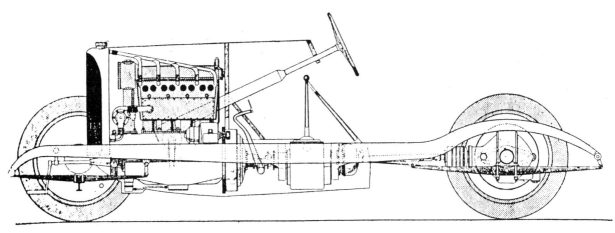

simple aluminium casting, held down by four nuts and provided with a central breather. While being a far superior arrangement to the original, this design incorporates one poor design feature in that the camshaft is located axially by a thrust race and pad sandwiching the forward end of the camchest and the camcover, the 0.004″ end float being set up by means of slotted nuts. In theory, the forward edge of the camcover needs to be exactly the same thickness as the forward edge of the camchest itself, and the two align exactly when the cover is fitted. In practice, the engines seem to run perfectly well despite discrepancies between the two components.

The vertical drive assembly from the magneto turret up to the camshaft crownwheel was also patented by WO, Burgess and Bentley Motors, again on the same date. This patent specifies slotted holes for the cylinder block foot to allow axial movement of the block, which shows an interesting chain of thought. The datum for the position of the block should be the vertical drive shaft from the bottom bevel gears, as the gear train up to the top bevel gear is the only part of the top end of the engine which can not really cope with misalignment. The camshaft (and hence the top bevel wheel) can be moved axially by means of the slotted nuts, and provided the rockers still ran on the full width of the camshaft lobes such movement would be acceptable. Sideways float on the small ends of ⅛″ plus is specified in "Technical Facts" so movement of the block of the order of up to maybe

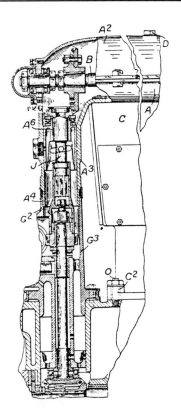

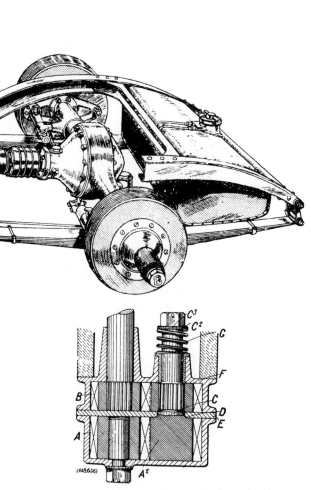

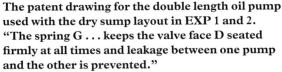

ABOVE **The patent drawing for the whole of the bevel gear train, as fitted in EXP 3 and the production engines. There was a provision for slotting the cylinder block holes (C^2) that was not adopted on the production cars.**

The patent drawing for the double length oil pump used with the dry sump layout in EXP 1 and 2. "The spring G . . . keeps the valve face D seated firmly at all times and leakage between one pump and the other is prevented."

$\frac{1}{16}''$ to line up the bevel shaft seems a sensible provision. In practice this was abandoned and the cylinder block holes are only a standard clearance fit. However, it is standard practice with the non-located top bevels, if the gear does not drop into the splined coupling, to slacken the cylinder block holding-down nuts and turn the engine over until the gear drops in.

The configuration of four valves per cylinder with a pent roof combustion chamber and a cross flow head is remarkably modern in concept, and keeping the exhaust system and carburettors on opposite sides of the engine eliminates a possible fire hazard. Ideally the spark plug should be placed in the top of the combustion chamber, but with the overhead cam this was not possible and the plugs

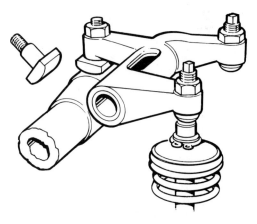

ABOVE **The early pattern of identical rockers used for both inlet and exhaust on EXP 1 and 2. The production engines used two single exhaust rockers and a symetrical forked inlet rocker. The production camshafts had three lobes, as opposed to two on EXP 1 and 2. The collet design with circlips was not used, the collets being split cones fitting a waisted portion of the valve stem.**

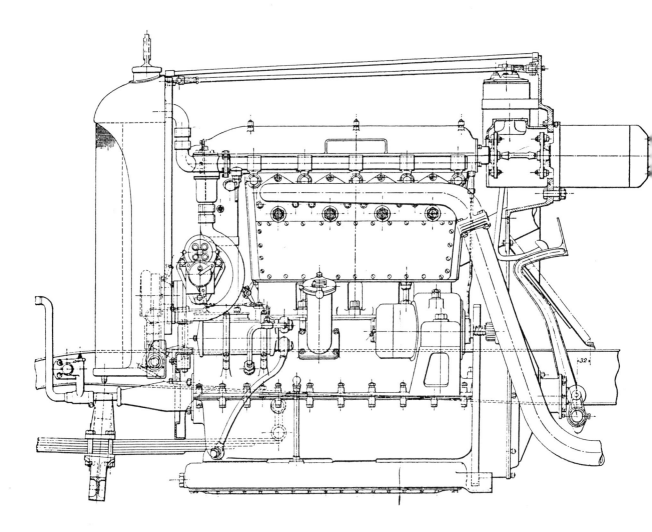

are positioned one on each side. The very first photos of EXP 1 engine show a single ML G4 magneto mounted on the offside, with the water pump mounted on the nearside, driven off the other end of the cross-shaft. This magneto was initially used with the nearside plug holes blanked off, until WO had sufficient data to evaluate a dual ignition system. This was effected by using a double-spark Bosch magneto, believed to have been taken from one of WO's Brooklands DFPs. This experiment showed that there were considerable benefits to be gained from a dual ignition system, so provision for this was incorporated in the re-design of the engine for production. As the Bosch double-spark magneto was unobtainable, twin ML G4 magnetos were specified, which necessitated relocation of the water pump. The solution to this was to drive the magnetos off each end of the existing cross-shaft gear and drive another fore-and-aft gear off it to drive the water pump. This cross-shaft set-up has

long been regarded as the Achilles heel of the 4 cylinder engine design, but strangely the problems of the 1950s and early 1960s did not occur when the cars were new or even before the Second World War. The original gears were made from BND, a Firth-Derihon air hardening 100 ton steel, (described for some reason as "impact hardening" and reputedly somewhat traumatic to the gear cutters), a material that was used for almost all the gears on Bentleys.

Later on BND was superseded by 5% CHMS, but on the Team cars Nobby Clarke always used BND. Tests by Shell years later showed that the original gears were fine when used with straight 50 oil and few problems occurred providing the magnetos and particularly the water pump could be turned fairly easily. Most cross-shaft gear failures are probably attributable to excessively tight water pumps. The water pump gland consists of a square-section string packing between two coned

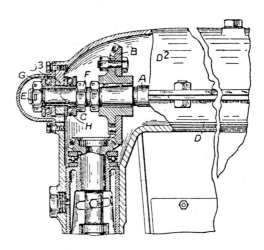

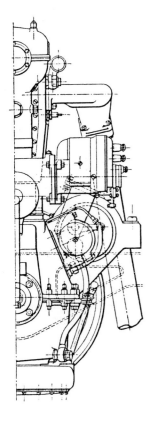

LEFT *Engineering's* **drawing of the 3 Litre engine** *c.* **1923. This is the full production version, as used on the two wheel braked cars – the later cars had the sump drain plug moved from the front to the side, as it fouled the track rod when taken out. Note the fitting of a thermostat and the direct driven dynamo, superseding the earlier dynamo driven by a 3:1 gearbox. ML G4 magnetos and the small water pump can be seen.**

ABOVE **The patent specification for the revised top bevel arrangement. Note the use of slots and pinch bolts to raise or lower the top bevel gear and nuts to move the camshaft and crownwheel axially. The rear of the two thrust races (H) was later deleted. The specification calls for exact alignment of the front faces of the camchest and cover, which would be difficult to achieve in practice.**

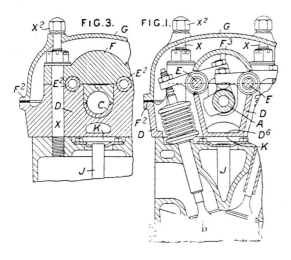

LEFT **The patent specification for the production valve gear. Splitting the camchest allows the rockers to be threaded onto rods that can be dropped in from above, rather than inserted from the back. Note also the drain tubes to drain oil from the camchest back to the crankcase – on the Experimental engines, the drain had had to be plumbed in externally. Apart to changes to duralumin rockers with rollers and replacing the long bridgepiece casting with separate rocker boxes, the valve gear design remained basically unchanged through to the 8 Litre.**

faces, the gland being tightened by compressing the string. It has been observed that if the water pump does not leak, then it's too damned tight! It was found at Le Mans in 1929 with one of the 4½ Litres that even though the pump leaked while the car was stationary at the pits, it did not leak at speed. It is always worth recalling that the original Bentley components and design lasted rather well at Le Mans, Brooklands, Ulster, Montlhéry and various other places. Problems did start to occur in the 1950s when the original gears were worn out, but the reason for many failures was the usage of the wrong materials – sets made in higher grade steels required far thicker oils than those normally encountered in car engines and use of 20/50 in place of straight 50 merely aggravated the problem. The solution was to make the centre gear of the trio out of bronze, which operates perfectly on modern multi-grades and has in recent years made cross-shaft gear problems once more a thing of the past. Interestingly, when ENV were approached after the war to make new cross-shaft gears, they declined. One of the reasons given was that the lapping jig had been destroyed. The early water pumps lacked axial location for the impellor, which tended to chew its way backwards into the casing. Under racing conditions the very high wear rates of the early pumps proved to be a problem, so the impellor spindle was fitted with a 26 TPI nut bearing onto the end of the driving gear spindle, which is adjusted and split-pinned so that the gear floats in the pump and does not contact the casing. The pump itself was also enlarged in later years.

The cylinder block design itself was the subject of a patent, number 140265, again taken out by WO and Burgess. WO had had the lessons of good cooling in engine design ingrained in him by the experiences of the obturator ring in the Clerget engine, and the lessons learned are clearly shown in the design of the cylinder block casting. The main advantage of four small valves in place of two large ones is the ease of cooling the seats. The 3 Litre cylinder casting is very open around the inlet and exhaust ports promoting free water circulation and good cooling. In order to ensure that all the sand was cleared away and to ease the positioning of the cores, the sides of the block were left open and later aluminium closing plates were fitted. This design worked well on the 3 Litre, but the extra 20 mm on the bores of the 4½ Litre engine in the same physi-cal space as the smaller engine results in slightly inferior cooling and in recent years competition-minded owners have modified the cooling system, particularly around the exhaust ports, to avoid the hairline cracks between valve seats that can occur. Four small valves also need lighter springs than two large ones, which reduces the hammering effect and builds less inertia into the valve gear permitting higher speeds.

On the racing engines, further attempts to reduce the inertia of the valve train were made by drilling the valve spring caps, and from 1926 onwards by the use of duralumin for the rockers themselves on all the engines. Nevertheless, as later experiences were to show (particularly the 1926 Montlhéry record attempts) the valve gear was the limiting factor in engine speeds, with excessive revs resulting in dropped valves. In more recent years, modern technology has resulted in 4½ Litre and 8 Litre engines revving to 5000 RPM without loss of reliability, using needle roller bearing rockers – the 5/8″ Hoffman rollers (fitted to the duralumin rockers – the steel rockers of the earlier 3 Litres have hardened steel pads) used by Bentleys have a tendency to skid and wear the cam and the rocker itself rather than roll. In more recent years, the cylinder block design has proved to be a problem, due to seat recession after long use. It has proven very difficult and expensive to fit seat inserts to the Bentley block, a problem that will be greatly aggravated if leaded fuels become unavailable. However, WO can hardly be blamed for not considering the problems of owners 60 years on!

The general impression given by the design of the 3 Litre is that of a car designed without any great consideration of manufacturing, or certainly in the case of all the Bentley chassis with much of an eye to economy in production at all. Frostick's remarks concerning WO's time at Gwynnes are interesting in this context: "Gwynnes, it would seem, had been unable to make head or tail of 'that Officer Bentley', largely because he took little thought about production. His ideas, as ideas, were excellent, but he expected someone else to sort out the problems that might arise in putting them into practice". Several examples of this can be found on these pages, principally the camshaft design of EXP 1 and 2, the lack of adjustment on the cross-shaft gear assembly and the cast aluminium deckboards.

5 *Production Starts*

ASSEMBLY OF THE first batch of production chassis proceeded at such a rate that in June 1921 Bentley Motors proudly announced that deliveries would commence the following month. Noel van Raalte, as an engineer and a gentleman of high social standing, was selected to receive the first production car. In his autobiography WO recalls taking him over to a low, grey coupé (although for the upright 2-door saloon/coupé coachwork to modern eyes the term "low" seems something of a misnomer!), and reminding him of the Five-Year Guarantee. Van Raalte was a major shareholder in KLG Plugs and owned Brownsea Island. His motor car experiences included some of the more notorious student races at Cambridge. Van Raalte won a race from the Market Square to the station on a Sunday morning in a 140HP Minerva, beating Rhodes-Moorhouse in a 90HP chain driven GP Fiat – the loser to pay the fines! Van Raalte had also raced Sunbeams; all in all an effective, but perhaps a little risky, first owner.

That Van Raalte was a suitable customer was shown by his defence of the Bentley in correspondence soon after taking delivery, offering the opinion that as far as steering, suspension, road-holding, brakes, gear change and engine efficiency were concerned, the Bentley was in advance of any of the other cars he had owned, including two Rolls-Royces. So pressed were they to get the 3 Litre into production that some of the support activities were neglected – it was not until some customers asked for a handbook that it was realised that such a thing did not exist. WO wrote the first one over several evenings, which was duly typed up and presented with the chassis. Another oversight pointed out by a customer was that the bottle jack

in the toolkit was too tall to go under the front axle.

In the Drawing Office the drawings were produced as General Arrangement drawings (prefixed "A"), which were then broken down into detail drawings of the individual components. The detail drawings were often prefixed "E" (for experimental) before full production, at which point the drawings acquired a "BM" prefix to the part number and were issued for manufacturing. Drawings with a "B" prefix were for coachbuilding, either drawings of the chassis with dimensions marked on for the coachbuilders giving such details as clearances for rise and fall of axles and movement of brake levers, etc, or presumably body drawings for Bentley Standard coachwork. Certainly Vanden Plas built bodies to drawings from the Cricklewood Drawing Office. Drawings with an "R" prefix referred specifically to components for the Racing Shop.

A schedule was assembled for each model, listing in detail the breakdown of all the components. Thus the schedule for each model would take the major assemblies (axles, engine, gearbox, etc) and break these down to their component parts (right down to nuts and washers). Obviously these schedules were revised frequently to take account of design changes and upgrades in model specifications, as illustrated. Feedback from the shops to Design seems to have been very much an oral process, helped by the fact that Burgess was to be seen on the shop floor virtually every day. WO was also a frequent visitor.

The drawings and schedules reproduced here (pp.74–85) are selected to show the systems used by Bentley Motors.

Early production – this must be one of the very first production cars, photographed in Hanover Street with WO at the wheel. The 'body' appears to be the back end from EXP 2, pressed into service again. The pedal shaft brackets are distinctly odd, but otherwise the chassis is to the early production specification with DN shock absorbers and plain bonnet with Boyce Motometer in the radiator cap.

1. Chassis schedules.

a. Front cover for 0/700, a specification for a two wheel braked 3 Litre 10′10″ Long Standard, issued to Fred Dewhurst (Chief Draughtsman) by Frank Ayto.

b. Cover sheet, for the Standard Six (Smith-Bentley 55BVS carburettor).

c. Sub-section divider, this time 4½ Litre.

d. Sub-section, 4½ Litre radiator and bonnet. Note particularly the "remarks" column. "BOF" stands for Bought Out Finished. As virtually everything was sub-contracted, most parts fell into this category.

e. As d, but this time "Instrumentation" for the Supercharged 4½ Litre chassis.

f. This is for the 3 Litre Light Standard, or Light Tourer. This shows moves from the four bolt con rod BM1768 to the two bolt shell rod BM3105 thence to the two bolt direct metal rod BM4017, in each case "used until present stock is exhausted." All the production modifications seem to have been incorporated in this manner, as old stocks were used up.

g. Standard Six again, showing the use of the order no. 0/1000 and 0/1200 as a reference to a chassis specification. 0/700 and 0/900 refer to 3 Litres. The significance of the "0/..." numbering system remains unclear, but it is thought to refer to the Buying Office.

h. 4 Litre, showing the use of the project code "SV" for this model.

2. *Drawings.*

a. BM111, a steel washer used between the differential side plate and ball race. Drawn by H.F. Varley, traced by Lillian Atkinson. Passed by WO and Burgess, and then for production on 24th February 1921. As with all the early drawings, there is no project code, as they were all used on the 3 Litre in its earliest chassis form.

b. BM4005, reproduced to show the use of the code "SS" for the 9' wheelbase 100 MPH "Supersports" chassis. In this case, a specially short pot-joint propellor shaft.

c. BM5037, a component for a 6½ Litre. Drawn by Eric Easter and checked by Burgess. Note the date – February 1925. 6½ Litre drawings start at BM5000 and have the project code "S".

d. BM3402, a very early 4½ Litre drawing referring to the model as "4 cylr 100 bore". At this date, July 1926, the new 4½ Litre was very much in the experimental stage, and if there was a chassis specification it would have been little more than that for the Long Standard 3 Litre except for the engine and gearbox. Drawn by C.W. Sewell and checked by Burgess. Note that although the drawing is numerically before (c), the date is much later.

e. BM4698; by this stage, the 4½ Litre had aquired it's own project reference. This part was for use with the heavy pattern bolted strut gear, and in fact was only fitted to the Team cars and the Supercharged chassis.

f. BM4836, 4½ Litre Supercharged piston rings. Reproduced to illustrate the connections between Bentley Motors and the Automotive Engineering Co., who employed Weslake and Hewitt.

g. E3266/BM7181. This is an experimental (E) drawing passed for production and given a BM number. Note that although this part is for a hypoid back axle used only on the 8 Litre, the project code is 1931 Sports, which would be a 6½ Litre (see p.294). The chassis specification/ model usage would, of course, have been a moving target depending on all sorts of considerations. Note also the date – February 1928; this is too early for the hypoid axle to be intended specifically and only for the 8 Litre.

h. BM 7211. A part for an F box – but note the project code is 6½ Litre or SV (SV is sidevalve, referring to the 4 Litre). The F box was never fitted to the 6½ Litre, but see p.294, as (g) above.

i. BM7503, the con rod for the SV model, the 4 Litre. Note the date; April 1929, and the fact that the drawing was passed by none other than Burgess. Notice also the remarkable number of alterations made, indicative of the delays and problems that dogged the gestation of the model.

j. B36, a very early body drawing for the coachbuilder, to fit the exhaust cut-out used on the early chassis to the floor-boards. Note the clearance instructions to allow the pedal and floorboards to be lifted, and to facilitate removal of the gearbox with its flanges; in this case, an A box. The lever on the B and C type gearboxes is positioned further forward (p.82 below).

3. *Gear Ratio and Speed Charts*

a. Speed Charts Index. Note the "R" (racing) bevels, for use with the 6½ Litre racing differential. Also note the reference to chart No. 30 for B box and 3.92 axle, reproduced at (e) below.

b. C box layout. These were produced for all the boxes.

c. Associated chart to (b) for speedo drive gears, for all combinations of wheel sizes and back axle ratios (the C box was the "universal" box, used on 3, 4½, Standard and Speed Sixes).

d. A typical speed chart, No. 53, for a C box with a 3.9 back axle and 5.25x21" tyres.

e. Speed chart No. 30 for 0/1000 – 0/1000 refers to a Standard Six build (see (f.) of Chassis Schedules above) and this speed chart would seem to be for an early Standard Six with balloon tyres, B box and 3.9 back axle.

f. More charts, but whereas (d) and (e) are Works charts, these are typical of the charts published in the handbooks.

4. *Standards*

Drawing to show dimensions of standard spark plugs. There are similar drawings for unions, bolts, studs, etc, as well as standard charts for things like bearing fits, conversion charts, standard material specifications, etc.

1a

1b

SPECIFICATION INDEX

PAGE	COMPONENT	SUB COMPONENT	PAGE	COMPONENT	SUB COMPONENT	PAGE	COMPONENT	SUB COMPONENT
1	BACK AXLE	CASING & SPRING SADDLE	41			81	GEARBOX	MAINSHAFT
2		BEVEL GEAR HOUSING	42			82		REVERSE GEAR
3		BEVEL GEAR HOUSING	43	ENGINE	CAMSHAFT & BEARINGS	83		SELECTOR MECHANISM
4		BEVEL GEAR HOUSING	44		CAMSHAFT CASING	84		STRIKING GEAR
5		BACK UNIVERSAL JOINT	45		CAMSHAFT DRIVE & DAMPER	85		CHANGE SPEED LEVER & GATE
6		CARDAN SHAFT WHEELBASE	46		CRANKCASE	86		SPEEDOMETER DRIVE
7		AXLE SHAFTS	47		CRANKSHAFT & FLYWHEEL	87		UNIVERSAL JOINT
8		BRAKE GEAR	48		CRANKSHAFT DAMPER	88		
9		BRAKE GEAR	49		CYLINDERS	89		
10		BRAKE SHOES	50		CYLINDERS	90	RADIATOR & BONNET	RADIATOR
11		HUBS & WHEELS	51		DYNAMO DRIVE	91		DASHBOARD
12			52		EXHAUST MANIFOLD	92		DASHBOARD
13			53		FAN	93		OIL & PETROL PIPES
14			54		HOLDING DOWN BOLTS	94		OIL & PETROL PIPES
15	FRONT AXLE	FRONT AXLE BED	55		INDUCTION PIPE (SS CARB)	95		AUTOVAC
16		STUB AXLE	56		DISTRIBUTOR CROSS-SHAFT	96		
17		FRONT BRAKE GEAR	57		DISTRIBUTORS & HT HM CARRIER			BONNET & FASTENER
18		FRONT BRAKE GEAR	58		MAIN BEARINGS & SERVICE PIPE	98		DECKBOARDS
19		HUBS & WHEELS	59		OIL FILLER & LEVEL INDICATOR	99		RADIATOR (CROWNED SIDE)
20		TIE ROD	60		OIL FILTER - PRESSURE SIDE	100		BONNET & FASTENER
21		FRONT BRAKE OPERATING SHAFT	61		OIL (RELEASE GEAR)	101		INSTRUMENTS
22			62		OIL PUMP	102	EXHAUST PIPE & SILENCER	CONNECTING PIPE
23			63		OIL PUMP	103		EXHAUST PIPES
24	BRAKES	COMPENSATING GEAR	64		OIL PIPES	104		EXPANSION CHAMBER
25		COMPENSATING GEAR	65		PISTON & CONNECTING ROD	105		SILENCER
26		COMPENSATING GEAR	66		ROCKER MECHANISM	106		SUPPORT
27		ROCKING LEVER - FRONT	67		STARTER	107		
28		THRUST RODS	68		SUMP	108		
29			69		THERMOSTAT	109	STEERING	STEERING
30		EXPANDER CYLINDERS & BRAKE ADJUSTER - FRONT	70		TUNNEL BASE	110		STEERING
31		BRAKE ADJUSTER (HAND BRAKE)	71		VALVE GEAR	111		STEERING
32		HAND BRAKE LEVER	72		WATER PUMP	112		STEERING
33		HAND BRAKE LEVER	73		WATER PUMP	113		STEERING
		REAR BRAKE RODS - ROCKING LEVER	74		WATER PIPES	114		CONTROLS
			75			115		
	PLATE CLUTCH	CLUTCH & SHAFT 76	76	GEARBOX	GEARBOX CASING	116		
		WITHDRAWAL GEAR 77	77		GEARBOX FRONT SUPPORT	117	CONTROLS	
		PEDAL SH & FITTINGS 78	78		NOSE PIECE & COVER	118		

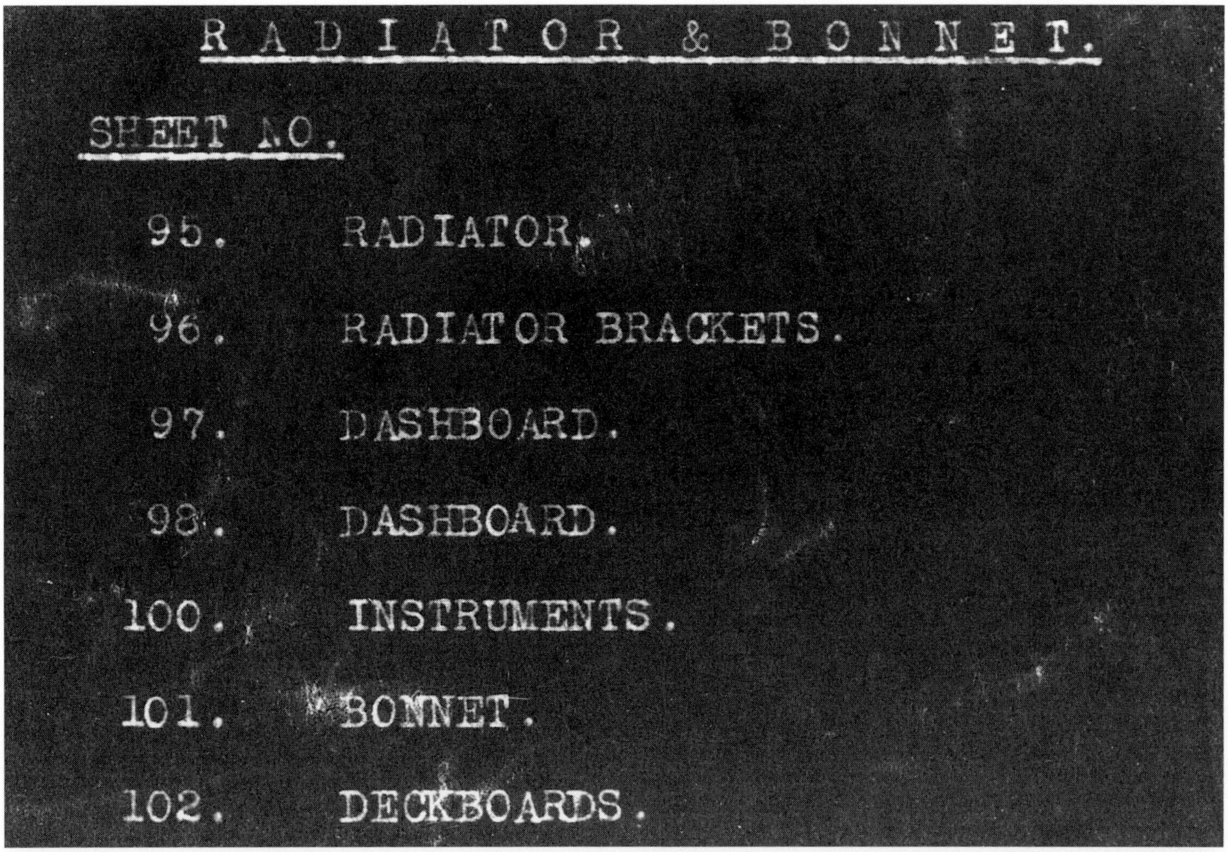

RADIATOR & BONNET.

SHEET NO.

95. RADIATOR.
96. RADIATOR BRACKETS.
97. DASHBOARD.
98. DASHBOARD.
100. INSTRUMENTS.
101. BONNET.
102. DECKBOARDS.

1c

1d

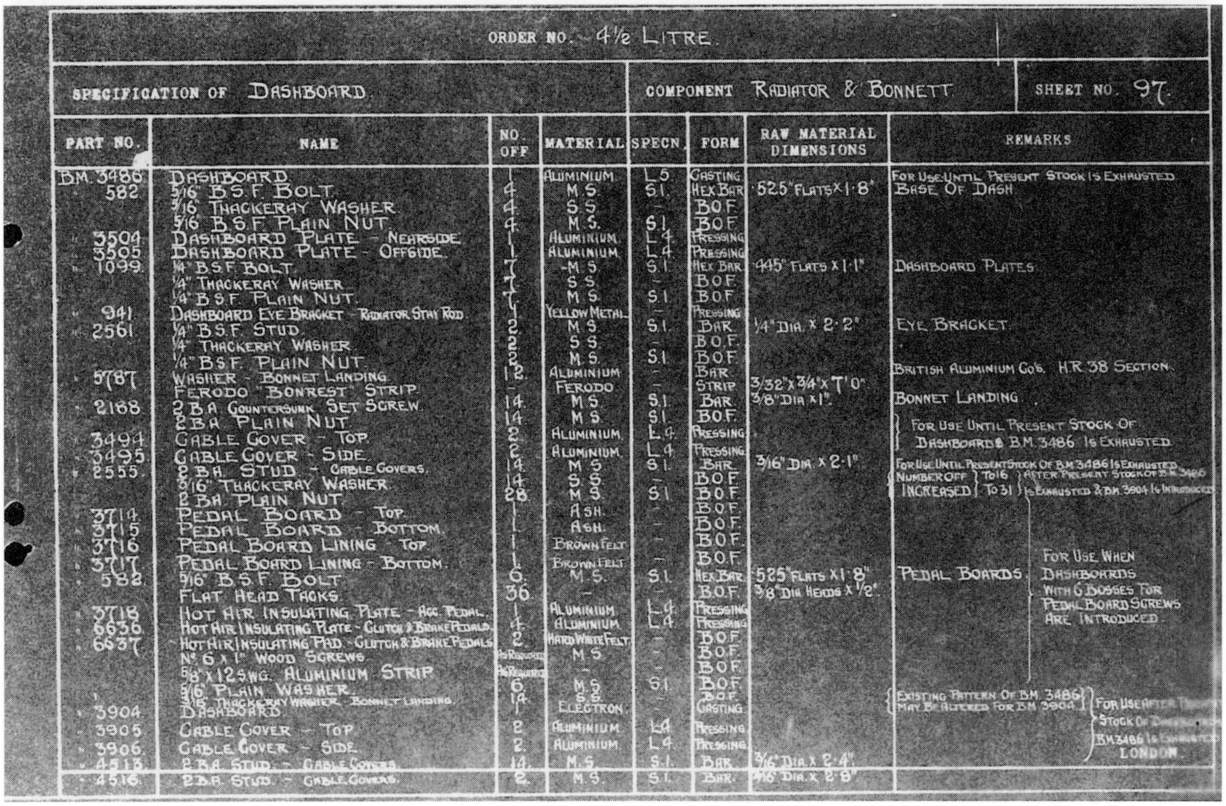

ORDER NO. 4½ LITRE.

SPECIFICATION OF DASHBOARD. COMPONENT RADIATOR & BONNETT. SHEET NO. 97.

PART NO.	NAME	NO. OFF	MATERIAL	SPECN.	FORM	RAW MATERIAL DIMENSIONS	REMARKS
B.M. 3486	DASHBOARD	1	ALUMINIUM	L5	CASTING		FOR USE UNTIL PRESENT STOCK IS EXHAUSTED
582	5/16" B.S.F. BOLT	4	M.S.	S.1.	HEX BAR	.525" FLATS × 1·8"	BASE OF DASH
	5/16" THACKERAY WASHER	4	S.S.	–	B.O.F.		
	5/16 B.S.F. PLAIN NUT	4	M.S.	S.1	B.O.F.		
3504	DASHBOARD PLATE – NEARSIDE	1	ALUMINIUM	L.4	PRESSING		
3505	DASHBOARD PLATE – OFFSIDE	1	ALUMINIUM	L.4	PRESSING		
1099	1/4" B.S.F. BOLT	7	M.S.	S.1	HEX BAR	.445" FLATS × 1·1"	DASHBOARD PLATES
	1/4" THACKERAY WASHER	7	S.S.	–	B.O.F.		
	1/4" B.S.F. PLAIN NUT	7	M.S.	S.1	B.O.F.		
941	DASHBOARD EYE BRACKET – RADIATOR STAY ROD	1	YELLOW METAL	–	PRESSING		
2561	1/4" B.S.F. STUD	2	M.S.	S.1.	BAR	1/4" DIA. × 2·2"	EYE BRACKET
	1/4" THACKERAY WASHER	2	S.S.	–	B.O.F.		
	1/4" B.S.F. PLAIN NUT	2	M.S.	S.1	B.O.F.		
5787	WASHER – BONNET LANDING	12	ALUMINIUM	–	BAR		BRITISH ALUMINIUM CO'S. H.R. 38 SECTION.
	FERODO "BONREST" STRIP	–	FERODO	–	STRIP	3/32" × 3/4" × 7'0"	
2188	2 B.A. COUNTERSUNK SET SCREW	14	M.S.	S.1	BAR	3/8" DIA × 1"	BONNET LANDING
	2 B.A. PLAIN NUT	14	M.S.	S.1	B.O.F.		
3494	CABLE COVER – TOP	2	ALUMINIUM	L.4	PRESSING		FOR USE UNTIL PRESENT STOCK OF DASHBOARDS B.M. 3486 IS EXHAUSTED
3495	CABLE COVER – SIDE	2	ALUMINIUM	L.4	PRESSING		
2555	2 B.A. STUD – CABLE COVERS	14	M.S.	S.1.	BAR	3/16" DIA × 2·1"	FOR USE UNTIL PRESENT STOCK OF B.M. 3486 IS EXHAUSTED. NUMBER OFF INCREASED TO 16 TO 31 AFTER PRESENT STOCK OF B.M. 3486 IS EXHAUSTED & B.M. 3904 IS INTRODUCED
	5/16" THACKERAY WASHER	14	S.S.	–	B.O.F.		
	2 B.A. PLAIN NUT	28	M.S.	S.1	B.O.F.		
3714	PEDAL BOARD – TOP	1	ASH	–	B.O.F.		
3715	PEDAL BOARD – BOTTOM	1	ASH	–	B.O.F.		
3716	PEDAL BOARD LINING – TOP	1	BROWN FELT	–	B.O.F.		
3717	PEDAL BOARD LINING – BOTTOM	1	BROWN FELT	–	B.O.F.		
582	5/16" B.S.F. BOLT	6	M.S.	S.1	HEX BAR	.525" FLATS × 1·8"	PEDAL BOARDS. FOR USE WHEN DASHBOARDS WITH 6 BOSSES FOR PEDAL BOARD SCREWS ARE INTRODUCED
	FLAT HEAD TACKS	36	–	–	B.O.F.	3/8" DIA HEADS × 1/2"	
3718	HOT AIR INSULATING PLATE – ACC. PEDAL	1	ALUMINIUM	L.4	PRESSING		
6636	HOT AIR INSULATING PLATE – CLUTCH & BRAKE PEDALS	4	ALUMINIUM	L.4	PRESSING		
6637	HOT AIR INSULATING PAD – CLUTCH & BRAKE PEDALS	2	HARD WHITE FELT	–	B.O.F.		
	Nº 6 × 1" WOOD SCREWS	AS REQUIRED	M.S.	–	B.O.F.		
	5/8" × 12 S.W.G. ALUMINIUM STRIP	AS REQUIRED	–	–	B.O.F.		
	5/16" PLAIN WASHER	6	M.S.	S.1	B.O.F.		
	3/16" THACKERAY WASHER – BONNET LANDING	14	S.S.	–	B.O.F.		
3904	DASHBOARD	1	ELECTRON	–	CASTING		EXISTING PATTERN OF B.M. 3486 MAY BE ALTERED FOR B.M. 3904. FOR USE AFTER [illegible] STOCK OF [illegible] B.M. 3486 IS EXHAUSTED
3905	CABLE COVER – TOP	2	ALUMINIUM	L.4	PRESSING		
3906	CABLE COVER – SIDE	2	ALUMINIUM	L.4	PRESSING		
4513	2 B.A. STUD – CABLE COVERS	14	M.S.	S.1.	BAR	3/16" DIA × 2·4"	LONDON
4516	2 B.A. STUD – CABLE COVERS	2	M.S.	S.1.	BAR	3/16" DIA × 2·8"	

ORDER No. 4½ LITRE SUPERCHARGED.

SPECIFICATION OF INSTRUMENTS | COMPONENT RADIATOR AND BONNET. | SHEET NO. 100.

PART NO. B		NAME	NO. OFF	MATERIAL	SPECN.	FORM	RAW MATERIAL DIMENSIONS	REMARKS 8
3042	1	NAME PLATE - For use with H.V. and G.5. Type S.U. Carburettor	1	Brass	-	Pressing.		Made from B.M.3044 Plate.
247	1	4 B.A. BOLT	4	M.S.	S.I.	Bar.	5/16"dia. x 1.4"	Black Finish.
		5/32" PLAIN WASHER	4	M.S.	S.I.	B.O.F.		
		4 B.A. PLAIN NUT	4	M.S.	S.I.	B.O.F.		
		DYNAMO SWITCH	1	-	-	B.O.F.		Complete with Screws.
		IGNITION SWITCH	2	-	-	B.O.F.		Complete with Screws.
		60lb (300% Overload) Oil Pressure Gauge	1	-	-	B.O.F.		Black Finish - Face Black - Letters White.
		"JAEGER" SPEEDOMETER	1	-	-	B.O.F.		Complete with 4'6"flex and Driving Cable.
		SMITH'S BEZEL SWITCH & AMMETER	1	-	-	B.O.F.		
		SMITH'S STARTER SWITCH - Hand.	1	-	-	B.O.F.		
3381	1	CASING - Rev. Counter Drive	1	G.M.	B.2.	Casting.		
3230	1	COVER	1	G.M.	B 2.	Casting		
3231	1	DRIVING GEAR	1	M.S.	S.I.	Bar.	1 5/8"dia. x 1.4"	
3382	1	DRIVEN GEAR	1	C.H.M.S.	S.14	Bar.	1 1/8"dia. x 2.2"	
3233	1	THRUST SCREW	1	C.H.M.S.	S.14.	Bar.	5/8"dia. x 1.5"	
		1/4" B.S.F. LOCK NUT	1	M.S.	S.I.	B.O.F.		
3234	1	LOCKING RING	1	Spring Steel-		Wire	16 Gauge x 4 5/8"	
3235	1	THRUST WASHER .10"	1	C.H.M.S.	S.14	Bar.	1"dia. x .3"	
3236	1	THRUST WASHER .075"	1	C.H.M.S.	S.14.	Bar.	1"dia. x .3"	
3237	1	THRUST WASHER .05"	1	C.H.M.S.	S.14.	Bar.	1"dia. x .3"	
2510	1	2 B.A. STUD - Cam Casing.	6	M.S.	S.I.	Bar.	3/16"dia. x 1.2"	
		3/16" SPRING WASHER	6	S.S.	-	B.O.F.		
		2 B.A. PLAIN NUT	6	M.S.	S.I.	B.O.F.		
"Jaeger"		REV. COUNTER COMPLETE WITH CABLE AND FLEX	1	-	-	B.O.F.		
942	1	1/4" B.S.F. BOLT - Casing.	1	M.S.	S.I.	Hex.Bar.	.445"flats x 1.6"	
		1/4" THACKERAY WASHER	1	S.S.	-	B.O.F.		
		1/4" B.S.F. PLAIN NUT	1	M.S.	S.I.	B.O.F.		

B.M. LTD
LONDON

ORDER NO. LIGHT STANDARD.

SPECIFICATION OF PISTON & CONNECTING ROD. | COMPONENT ENGINE. | SHEET NO. 71.

PART NO.	NAME	NO. OFF	MATERIAL	SPECN.	FORM	RAW MATERIAL DIMENSIONS	REMARKS
B.M 3259.	PISTON. (80 DIA.)	4.	ALUM.	L.8.	CAST'G		4.75 COMPRESSION RATIO } FOR USE AFTER PRESENT STOCK OF B.M. 306 IS EXHAUSTED.
" 3258.	PISTON RING	12	C.I.	K.6.	CAST'G.	(STANDARD BRICO RING B.S.62)	
" 3279	PISTON RING - SCRAPER TYPE	4.	C.I.	K.6.	CAST'G		
" 2159.	GUDGEON PIN.	4.	5% C.H.N.S.	S.17.	BAR.	7/8" DIA x 3.1"	
" 2452.	GUDGEON PIN BUTTON.	8.	ALUM.	L.8.	DIE CAST'G		
" 1768.	CONNECTING ROD.	4.	F.N.C.	-	STAMP'G		FOR USE UNTIL PRESENT STOCK OF
" 1769.	CONNECTING ROD CAP.	4.	F.N.C.	-	STAMP'G		STAMPINGS B.M.1768/9. IS EXHAUSTED.
" 1770.	CONNECTING ROD BRASS.	8.	WHITE METAL P.B.	B.8.	CAST'G.		
" 913.	CONNECTING ROD BOLT.	16.	3% N.S.	S.8.	BAR.	5/8" DIA x 2.4"	
" 558.	NUT - CONN: ROD BOLT.	16.	3% N.S.	S.8.	HEX BAR	.525" FLATS x .6"	
	5/64 x 1" SPLIT PIN.	16.	M.S.	-	B.O.F.		
" 560.	DOWEL PIN - CONNECTING ROD BRASS.	4.	M.S.	S.1.	BAR.	1/4" DIA x .4"	
" 1646.	GUDGEON PIN BUSH.	4.	ALUM.	L.8.	BAR.	1" DIA. x 1.6"	
" 3105.	CONNECTING ROD.	4.	B.N.D.	-	STAMP'G		FOR USE AFTER PRESENT STOCK OF
" 3106.	CONNECTING ROD CAP.	4.	B.N.D.	-	STAMP'G.		STAMPINGS B.M.1768/9. IS EXHAUSTED.
" 3107.	BIG END BRASS.	8.	PH. BZE. WE. HOYT.	B.8. No.11.	CAST'G.		FOR USE UNTIL PRESENT STOCK OF
" 3108.	CONNECTING ROD BOLT.	8.	ALLOY STEEL	S.2.	BAR.	5/8" DIA x 2.7"	STAMPINGS B.M.3105/6. IS EXHAUSTED.
" 3109.	NUT - CONN: ROD BOLT.	8.	3% N.S.	S.8.	HEX BAR	.6" FLATS x .8"	
	3/32" x 1 1/4" SPLIT PIN.	8.	M.S.	-	B.O.F.		
" 4017.	CONNECTING ROD.	4.	B.N.D.	-	STAMP'G.	{WHITE METAL LINING.}	FOR USE AFTER PRESENT STOCK OF
" 4018.	CONNECTING ROD CAP.	4.	B.N.D.	-	STAMP'G.		STAMPINGS B.M.3105/6 IS EXHAUSTED.
" 3272.	CONNECTING ROD BOLT.	8	ALLOY STEEL	S.2.	BAR.	5/8" DIA x 2.6"	
" 5210.	NUT - CONN: ROD BOLT.	8.	3% N.S.	S.8.	BAR.	5/8" DIA x .8"	
	3/32" x 1" SPLIT PIN.	8.	M.S.	-	B.O.F.		
" 3311	PISTON (SUPER-LIGHT) 4.75 COMP. RATIO	4	ALUM	L.8	CAST'G		FOR USE AFTER PRESENT STOCK OF B.M.3259 IS EXHAUSTED.

B.M. LTD.
LONDON.

ORDER NO. °/1000 & °/1200.

SPECIFICATION OF CYLINDERS.					COMPONENT ENGINE.		SHEET NO. S 49

PART NO.	NAME	NO. OFF	MATERIAL	SPECN.	FORM	RAW MATERIAL DIMENSIONS	REMARKS
B.M. 5270	CYLINDERS.	1	G.I.	K.II.	CASTING.		
„ 5002	CAM CASING HOLDING DOWN STUD - LONG.	8	3% N.S.	S.8.	BAR.	5/16" DIA. X 5·1"	
„ 2513	5/16" B.S.F. STUD.	16	3% N.S.	S.8.	BAR.	5/16" DIA. X 4·6"	CAM CASING.
„ 174	1/4" B.S.F. STUD.	4	M.S.	S.1.	BAR.	1/4" DIA. X 1·5"	
„ 2519	5/8" B.S.F. STUD.	18	M.S.	S.1.	BAR.	3/8" DIA X 1·8"	EXHAUST MANIFOLD.
„ 584	5/16" B.S.F. STUD.	24	M.S.	S.1.	BAR.	5/16" DIA. X 2·2"	INDUCTION PIPE.
„ 5003	TOP WATER COVER.	2	M.S.	S.3.	PRESSING		
„ 5004	JOINT - TOP WATER COVER.	2	HALLITE.	-	B.O.F.		
„ 675	2 B.A. CHEESE HEAD SET SCREW.	44	M.S.	S.1.	BAR.	5/16" DIA. X ·8"	
„ 5737	CYLINDER END WALL - FRONT.	1	R.H. ALUM⅔ ALLOY	Nº 58.	CASTING.		
„ 5010	CYLINDER END WALL - REAR.	1	R.H. ALUM⅔ ALLOY	Nº 58.	CASTING.		
„ 5023	JOINT - CYLINDER END WALL.	2	HALLITE.	-	B.O.F.		
„ 5012	CYLINDER SIDE WALL - TOP.	2	ALUM⅔ SHEET.	L.4.	PRESSING		FOR USE UNTIL PRESENT STOCK IS EXHAUSTED.
„ 5014	CYLINDER SIDE WALL - BOTTOM.	2	ALUM⅔ SHEET.	L.4.	PRESSING		
„ 6638	CYLINDER SIDE WALL - TOP.	2	STAYBRITE STEEL	-	SHEET.	14 S.W.G.	FOR USE AFTER PRESENT STOCK
„ 6639	CYLINDER SIDE WALL - BOTTOM.	2	STAYBRITE STEEL	-	SHEET.	14 S.W.G.	OF B.M. 5012 & B.M. 5014.) IS EXHAUSTED.
„ 6160	JOINT - CYLINDER SIDE WALLS - TOP.	2	HALLITE.	-	B.O.F.		
„ 6161	JOINT - CYLINDER SIDE WALLS - BOTTOM.	2	HALLITE.	-	B.O.F.		
„ 5022	JOINT - CYLINDERS & CAM CASING.	1	BROWN PAPER	-	B.O.F.		
„ 3713	WASHER - CYLINDER WALL SCREWS.	158	COPPER	-	PRESSING.		TO BE SOFTENED AFTER PRESSING.
„ 2511	2 B.A. SET SCREW - CYLINDER WALLS.	158	STAINLESS STEEL	-	BAR.	5/16" X ·9"	
„ 5721	WATER DISTRIBUTOR BOX.		SHEET BRASS.	-	PRESSING		
„ 5005	CAM OIL DRAIN PIPE - TOP.	4	W.S.T.	T.6.	TUBE.	7/16" O.D⁷ 14 S.W.G. X 10·5"	
„ 5151	CAM OIL DRAIN PIPE - BOTTOM.	4	W.S.T.	T.6.	TUBE.	3/4" O.D. X 3/8" I.D. X 3"	
„ 5152	WASHER - CAM OIL DRAIN PIPE.	4	RUBBER INSERTION.	-	B.O.F.		
„ 5153	WASHER - OIL DRAIN PIPE.	4	M.S.	S.1.	BAR.	3/4" DIA. X ·25"	
	1/8" X 1 1/4" SPLIT PIN - OIL DRAIN PIPE.	4	M.S.	-	B.O.F.		

B.M. LTD. LONDON.

ORDER NO. "S.V." MODEL

SPECIFICATION OF BEVEL GEAR HOUSING.					COMPONENT BACK AXLE.		SHEET NO. GV e

PART NO. B M		NAME	NO. OFF	MATERIAL	SPECN.	FORM	RAW MATERIAL DIMENSIONS	REMARKS
		DOUBLE THRUST WASHERS 4 1/8" X 2 7/8" X 2 7/16"	1	-	-	B.O.F.		Hoffman Type W.S.17.
5423	1	DIFF. CASE CENTRE.	1	Steel Alloy	-	Cast'g.		To be made of Steel Alloy after present stock of Elektron castings is exhausted.
5424	1	DIFF. CASING SIDE PLATE - Near-side	1	40 Ton. C.Steel	S.26.	Stamp'g.		
5425	1	DIFF. CASING SIDE PLATE - Off-side.	1	40 Ton C. Steel	S.26.	Stamp'g.		
5426	1	CROWN WHEEL 50/12	1	Firth's 5% C.H.N.S	-	Stamp'g		
5427	1	BEVEL PINION 12/50 4.166 Ratio	1	B.N.C.	-	Stamp'g.		
8732	3	CROWN WHEEL 55/12	1	Firth's 5% C.H.N.S	-	Stamp'g	May be made from stamping B.M. 5426
8733	2	BEVEL PINION 12/55 4.58 Ratio.	1	B.N.C.	-	Stamp'g.	B.M. 5427.
	1	CROWN WHEEL	-	Firth's 5% C.H.N.S.	-	Stamp'g.		
	1	BEVEL PINION		B.N.C.	-	Stamp'g.		
5432	1	DIFF. GEAR WHEEL 32/12	2	Firth's 5% C.H.N.S.	-	Stamp'g		To be made of Firth's 5% C.H.N.S. After present stock of B.N.D. Stampings is exhausted
5433	1	DIFF. PINION. 12/32	4	Firth's 5% C.H.N.S	-	Stamp'g		
5434	1	BEARING - Diff. Gear.	2	Phos.Bze.	-	Press'g.		
5435	1	PIN - Diff. Pinion.	1	Mesmeric	-	Bar.	1"Dia. x 3"	
5436	2	CROWN WHEEL BOLT.	12	F.N.C.T	-	Hex.Bar. 6"Flats x 5.4"		
5437	1	NUT - Crown Wheel Bolt.	12	M.S.	S.I.	Hex.Bar. 6"Flats x .9"		
		3/32" x 1 1/4" SPLIT PIN.	12	M.S.	-	B.O.F.		
5438	1	LOCKING PLATE - Diff. Housing R.H.	1	M.S.	S.3.	Press'g.		= As Required.
5439	1	LOCKING PLATE - Diff. Housing L.H.	1	M.S.	S.3.	Press'g.		
5440	1	LOCKING PLATE - Adjusting Nut.	1	M.S.	S.3.	Press'g.		
174	1	1/4" B.S.F. STUD.	1	M.S.	S.I.	Bar.	1/4"Dia. x 1.5"	Locking Plates
		1/4" SPRING WASHER.	2	G.S.	-	B.O.F.		

B.M. LTD. LONDON.

Sheet one - Three Sheets.

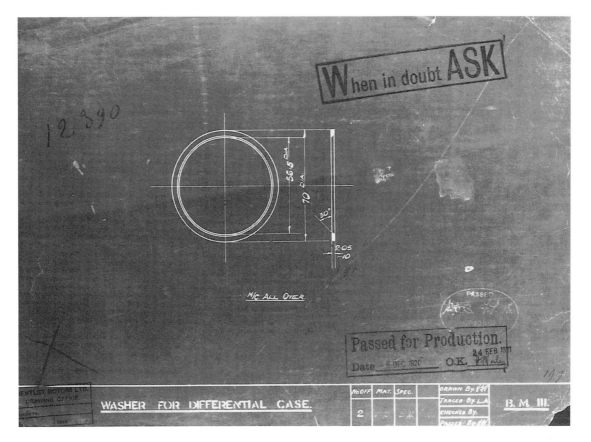

2a

2b

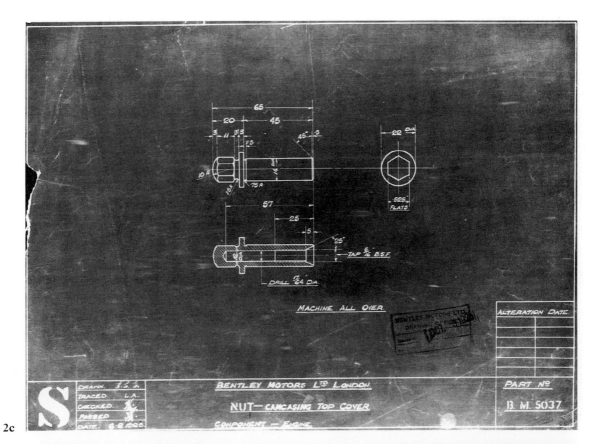

2c

2d

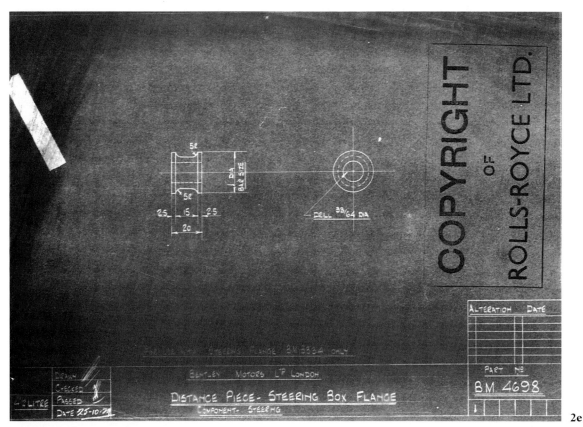

2e

2f

2g

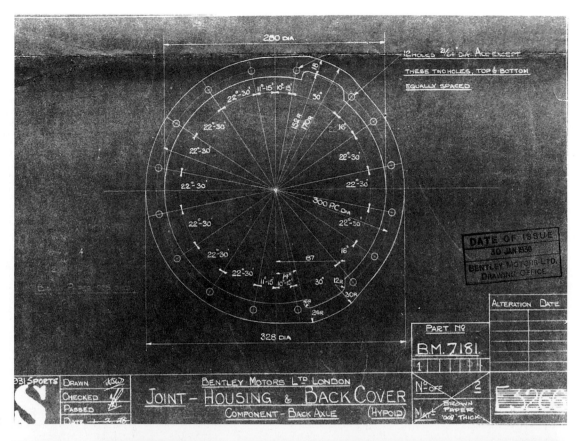

2h

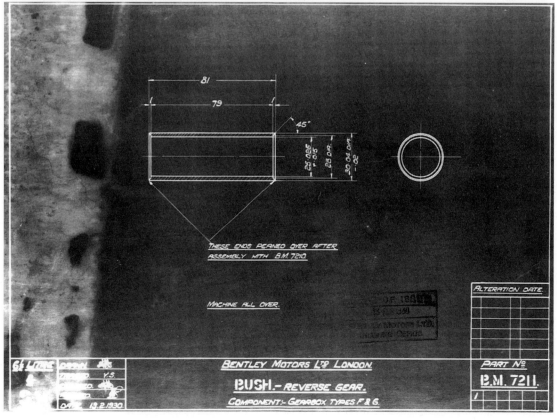

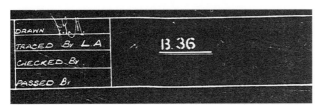

DETAILS OF FOULING POINTS & CLEARANCES 9—9½ WHEELBASE

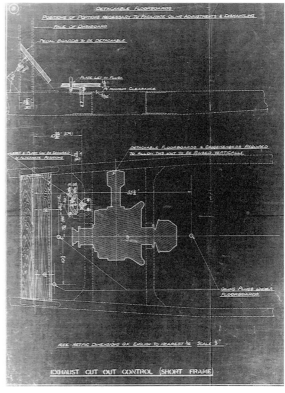

2i

2j

3a 3b

3c

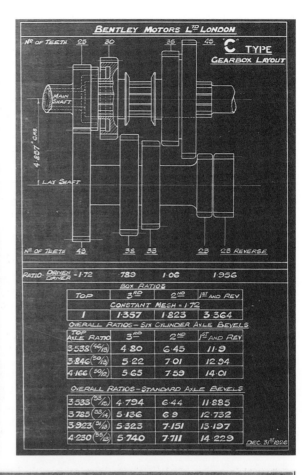

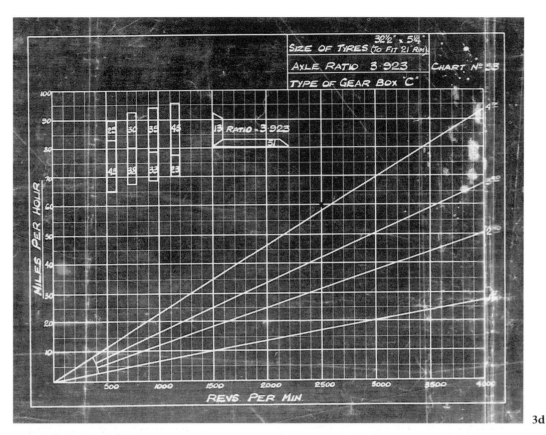

3d

3e

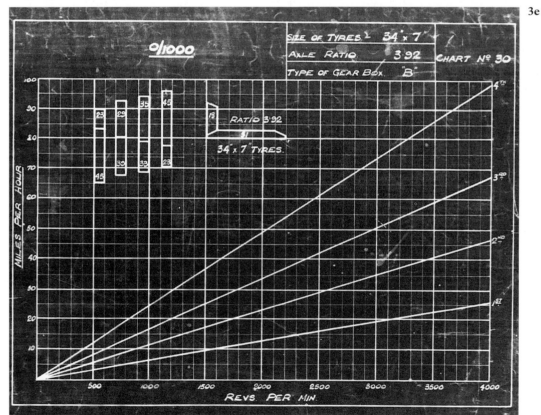

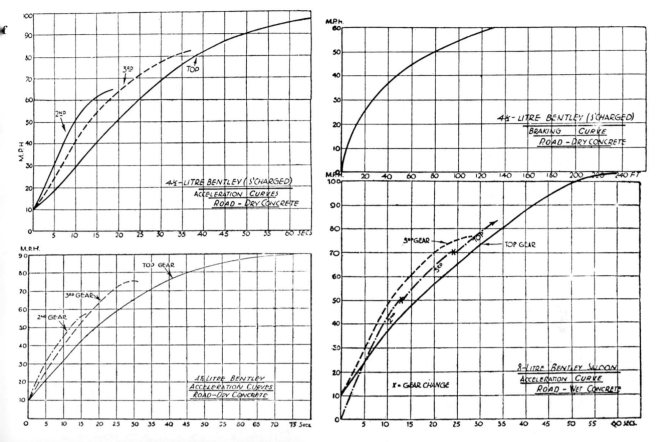

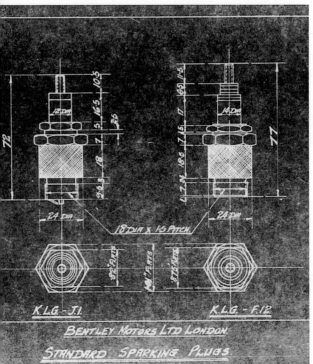

The schedules were then used by Peter Purves, the Chief Buyer, for ordering components. As Bentleys were very much an assembly firm, the production control facility must have been fairly well developed to co-ordinate the manufacturing and delivery to Cricklewood of the right bits when needed, but very little is known about it.

Parts made outside then came back to the Inspection Department. These were inspected individually or on a percentage basis, depending on the part concerned, under the direction of Jack Packer, the Chief Inspector. In 1930/31 the Inspection Department was re-organised to meet the needs of the new Machine Shop. From Inspection parts would go into the Stores. Conway, the Chief Storeman, formed an integral and vital part of the production process. The parts were then issued from Stores to the shop floor as necessary. The division of labour between Design and Production was summed up by Nobby Clarke in *The Other Bentley Boys*: "As far as we were concerned in the shops, the plans and prints for a new model would be produced. We knew it was a better motor car than its predecessor, or else it fulfilled a different purpose."

As far as is known, in the very early days gear-boxes, axles and steering columns were assembled away from the Works and supplied as built-up assemblies. This assembly work was soon brought in-house, and these items were built up in the Detail Shops. These sub-assemblies then went to the Erecting Shop either direct or via the Stores, to be fitted up to their respective chassis. Production in the early days was planned at two chassis a week, with chassis being built in batches of six (chassis were later built in batches of 25). The frames were received from Mechans of Glasgow fully rivetted and assembled (except for the 8 and 4 Litres) and were first of all laid out in the Erecting Shop for drilling. The frames were marked out by stretching a piano wire between the distance tubes between the front and rear dumb-irons and all lateral dimensions taken from this datum. Dimensions along the chassis were taken from the front face of the front cross-member. In many cases, drilling jigs were made up, for such things as the bulkhead, deck boards, pedal shaft brackets and head light brackets.

Because of the method of manufacturing all the major parts were toleranced to be interchangeable,

BELOW **Layout of the Works at start of production in 1922, and key to the photographs, pp.87–89.**

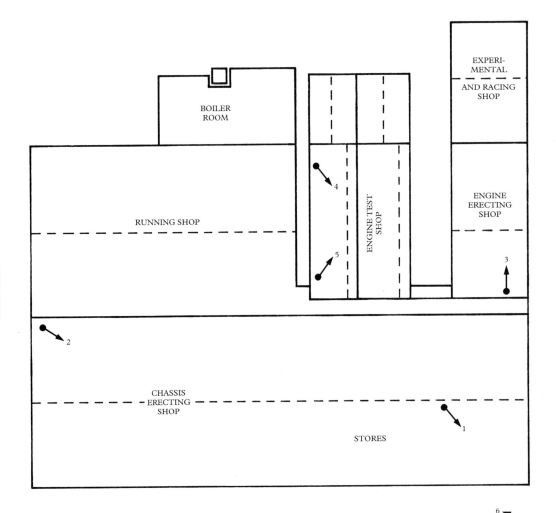

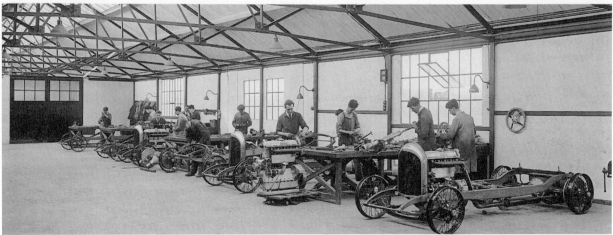

TOP 1 **The interior of the stores, Conway's domain, with him in attendance. With close study, many of the components can be identified; petrol tanks, exhaust pipes, silencers, undertrays, wheels, locking rings, accumulators, gaskets, bulkheads, steering columns, magneto turrets, clutch/brake pedals, radiator trunnions, clutch components, road springs, brake rods, radiators. The photograph is interesting in that in the case of small parts particularly (the quantity of clutch springs is very evident) they were ordering large quantities well in advance.**

ABOVE 2 **The interior of the chassis erecting shop. At this stage in the company's development, as far as is known, all the major assemblies, consisting of the axles, steering column, gearbox and differential, were assembled outside and delivered complete to Cricklewood. The engines were put together and tested at the Works, the whole lot then being delivered to this shop for the components to be put together. Note the practice of assembling axles with the frame inverted. The chassis are being assembled on new wheels with new locking rings.**

COMPANY STRUCTURE AT START OF PRODUCTION, 1921/22

Managing Director – W. O. Bentley

General Manager – G. A. Peck

| Company Secretary C. W. Carleton | Chief Buyer P. R. Purves | Design Chief designer F. T. Burgess Chief Draughtsman H. F. Varley | Sales H. M. Bentley A. F. C. Hillstead | Production Works Manager R. S. Witchell Works Superintendent R. A. Clarke | Experimental–Racing F. C. Clement | Services Service Director H. Pike |

| Engine Erecting Shop | Engine Test Shop F. Holloway | Stores F. Conway | Road Test R. Tomlins |

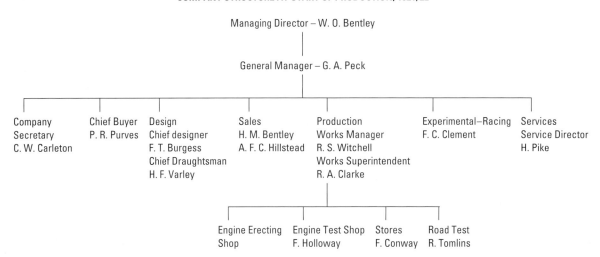

ABOVE 3 **The interior of the engine erecting shop, showing many of the components of the early 3 Litre engine. From left to right are Ben Padbury, Len Plumley, working on the top bevels of an engine, Arthur Hotchkiss and Fred Hollister by the coat rack, Villetorte smoking a cigarette working on the magnetos of another engine, Reg Wall, Fred Potts, and Nobby Clarke. At the bench nearest the camera are, from the front: Arthur Trott, Stan Ivermee, Billy Rockell and Wally Hawgood. On the other side of the bench are not known, Fred Lusted and finally Jack Cridland on the far right. At this time, engines were the only component known to be assembled at Cricklewood.**

RIGHT, 4 & 5 **The works in late 1921/early 1922, after the start of production. These photos were taken to be used in catalogues, possibly to dispel the rumours that the Bentley was an 'assembled' car. This photo shows the interior of the engine test shop, with three 3 Litre units on test. On the right is the Heenan and Froude dynamometer. To its left is a 3 Litre petrol tank on the wall, above the flow meter for measuring petrol consumption. Note the G4 magnetos, four bolt rocker covers with breathers, dynamo drive gearbox and extended oil filler neck. This building was the first to be put up at Cricklewood (see p.48) and initially housed the entire manufacturing concern during the completion of EXP 2 and 3 in early 1920.**

BENTLEY MOTORS, Ltd.	Chassis No.

Chassis Test Job Card No.

Tester's Name	Date of Test

3142-PL

DESCRIPTION OF WORK	CHECK NO.

Chassis test card. This was started by the mileage man, and then filled in by the tester until the chassis was passed off, when it would be filed. It would seem that all the completed cards were destroyed years ago.

and in general major components can be moved from car to car without too much difficulty. However, as the cars were hand-built, there was some degree of hand-fitting. This also led to competitions between the fitters to produce the best turned-out chassis, or the most powerful engine.

In later years, the bulkheads were assembled in the Erecting Shop, separately from the chassis. The castings were fettled and then, on all models except the 3 Litre, the covers for the wiring conduit would be dressed up and fitted. The studs for these, bonnet tape and fastenings, and other detail fittings were assembled up as well. Steering drag links and other parts were brazed up in the Erecting Shop, on benches near the forge by the back of the Engine Test Shop. Bonnets were individually fitted by a sub-contract fitter from Ewarts, who also fitted the dumb-iron cowlings on the late Speed Six and 8 Litre chassis.

After the chassis were drilled, the axles, steering gear, suspension, gearbox etc etc would be carried in and assembled, this process taking about five

days on a 3 Litre. Drilling was manual only with the well-known "gutbuster" drill, one of Nobby Clarke's favourite tricks being to lean on the fitter with his arms around him and say, "Harder lad, harder!" Axles were normally fitted with the chassis inverted and slave wheels fitted for moving the chassis around. The chassis fitter would fit the dynamo, cut-out, fuse box, etc. The wiring was then fitted by the electricians in the Erecting Shop. Foreman in the Erecting Shop was Norman Fawkes – George Horsley was the Foreman Electrician.

The engine was assembled by a fitter and his mate in the Engine Erecting Shop, the mate drawing all the parts from stores and normally being responsible for fettling and cleaning up all the castings, while the fitter built up the engine itself (this sequence was later broken down – see p.285). In the early days the only part of the engine to be built as a sub-assembly was the vertical drive turret, for which all the bushes had to be machined on the lathe that was virtually Bentley's entire machine shop for some time. Once the engine was complete, still bolted to its steel trolley, it would be wheeled down the slope to the Engine Test Shop (in the original brick building that had been the Experimental Shop). In the Test Shop, as far as is known, the engines were run for 12 hours then stripped and checked, with particular attention paid to the bearings. The engine was then reassembled and run for another 12 hours, before final tuning on the dynamometer. After the engine had been passed off as satisfactory it would be mated to its completed chassis in the Erecting Shop and then finally delivered to the Running Shop, supervised by Tomlins. By the time the chassis left the Erecting Shop, it was fully wired up and ready to run. Engines found to be significantly more powerful were generally earmarked to be taken to the Racing Department. Many years later, one fitter recalled having an engine failed because the split-pins in the main bearing bolts were not all facing forward – according to the old RFC dictum of "heads to the wind".

In the Running Shop the chassis would be fitted with seats and a scuttle and run in for 100 miles by the mileage man. It would then be driven by the tester to check all aspects of the chassis. The road testing seems to have consisted of thrashing the chassis around the lanes while devising wilder and wilder antics. 360 degree spins on wood-block road surfaces and games of "follow my leader" were the order of the day, on one occasion leading to a phone call back to Cricklewood for help after one 3 Litre ended up stuck in a ploughed field! It was said that the testers did not have to be mad, it was adequate qualification if there was insanity in the family.

George Hawkins, road tester from 1928 to 1931, described the test procedure as follows: "There were two tests that had to be carried out, whatever the weather. First, up Brockley Hill full throttle in third gear, when the chassis was expected to reach 60 mph at the top. At the top the road curved to the right and went on to Elstree and on the verge outside the Royal Orthopaedic Hospital was a stone water trough where draymen watered their horses after the haul up Brockley. Many times I nearly scraped that trough getting round that bend. To give an idea of Brockley Hill, on a clear sunny day the glass of the Crystal Palace across London could be seen glinting in the sun. The next test was a full throttle test down one of the by-passes, the Watford or the Barnet, where the chassis had to reach 90 mph. If it exceeded it, that was a bonus. Then go where you pleased carrying out braking and steering tests, and a thorough inspection of the chassis. I got to know the countryside very well indeed. When satisfied the tester submitted the chassis to

The exterior of the new shops, with a test-bodied 3 Litre chassis about to go out. Driving is Bob Tomlins, the head tester, accompanied by Jackson. Nobby Clarke is standing on the left in the doorway, with C.W. Carleton, the company secretary.

the Chief Tester for final approval, accompanying him on the Final Test."

One of the testers, a man called Middleton, left his notebook to the Bentley Drivers' Club, a book which gives an interesting insight into the consistency of the numbering of the chassis components. Components found to be unsatisfactory on test would be removed and sent back and other parts fitted, which is one reason why the numbers on vintage Bentleys frequently do not tally with the chassis number.

After being passed off by the Running Shop the finished chassis would be delivered to the coachbuilder, who would spend an average of three to six months building the body of the customer's choice. Much of this time, of course, was taken up by the coachpainting, which could itself run into weeks. In the early years some real monstrosities were built, particularly by some of the less experienced coachbuilders. Most coachbuilders had their roots firmly in the carriage trade and even in the Edwardian era had still been able to get away with building high, heavy bodies of rigid construction. These methods of coachbuilding were totally unsuited to the typical vintage sports car chassis with its fairly rigid suspension and flexible chassis, and it was not unknown for splits and cracks to appear very rapidly. The coachbuilders blamed the chassis manufacturers and vice-versa. Bentleys were soon forced to issue drawings to the coachbuilders giving clearance dimensions around the braking, steering,

axles and gear/brake levers. On return from the coachbuilders to the Works all these clearances would be checked before the issue of the guarantee. At the same time, the complete car was weighed and rear springs compatible with the weight of the body would be fitted. Finally the car would be passed off and the Five-Year Guarantee issued.

A combination of these problems with the coachbuilders and the appearance of some of the finished cars (fitting full four-seat open coachwork to the 9'9½" 3 Litre chassis without the rear passengers sitting uncomfortably high was not easy), led to the introduction of Bentley Standard Coachwork. HM and Hillstead at the Sales end agonised endlessly over ways of producing attractive full four-seat touring coachwork on the 3 Litre. Vauxhall solved this problem on the 30/98 with the Velox. However, in February 1921 the Velox bodied 30/98 was catalogued at £1675, while the 3 Litre sports tourer was catalogued at £1295 in late 1922. WO commented that "it was interesting to hear the reports of [30/98] owners who had shifted allegiance to the 3 Litre Bentley. After the Vauxhall's bigger engine, of course, they noticed the rather inferior acceleration and top speed of the Bentley but they always commented that they could travel faster from point to point over English roads in our car with its superior braking and road-holding." Nowadays any 3 litre car, or indeed, a 3 Litre Bentley, is seen as being quite a large car, but in those days 3 litres was seen as the smallest engine from which reasonable performance could be extracted.

HM finally solved this problem, sketching out the body design which was then made by Harrisons. Hillstead notes that the prototype was fitted to a 3 Litre sold to Stuart de la Rue. That car was chassis 154, registered PM 1585 delivered to de la Rue in December 1922, and indeed fitted with that classic body style. *The Motor* road tested PM 1585 in their issue of 6th March, 1923, commenting that "From the point of view of appearance the Bentley is good looking and the four seater body is wide and roomy". Park Ward produced this self-same body style on Hillstead's demonstration chassis 183, a design which then seems to have been taken over lock, stock and barrel by Vanden Plas. The first Vanden Plas body built on a Bentley chassis was one of a batch of Allweathers fitted to chassis 58, to the order of Guy Peck. Peck was the General Manager at Cricklewood at the time and Hillstead intimates that Peck took the body design to Vanden Plas because Harrisons were very expensive (as were Park Ward) and too busy.

Bentleys gradually evolved more or less standard body designs covering the full range of styles, and in later years employed their own body man. It is perhaps significant that Vanden Plas did not have a design department of their own, and that many of the standard body designs emanated from the drawing office at Cricklewood.

Electrical equipment on EXP 1 was by CAV, but very soon Smiths electrical equipment and instrumentation were used. In 1921 Smiths (Motor Accessories) Ltd moved from Speedometer House in Great Portland Street to new premises in Cricklewood, barely a mile from Bentleys and supplied lighting sets (5 lamps), dynamos, starters, fuse boxes and cut-outs, clocks, oil pressure gauges and bezel switch/ammeters for most of the production cars. The ammeter with a bezel to operate the lights was patented by two of Smiths' engineers, Dover and Rogers, in 1917. AT (Auto Tempo) speedometers and rev counters were usually fitted to 3 Litres and early 4½ Litres, before the fitting of Jaeger chronometric instruments. Establishments Ed Jaeger (UK) was set up in premises at Willesden Junction in Cricklewood in 1924, and by 1927 Smiths had acquired a controlling interest in the company. It has been suggested that Bentleys moved over to Jaeger because Smiths would not extend them any more credit.

The decision to race the 3 Litre must have been made practically at the outset. WO's own advocacy of racing as a means of improving the breed and gaining publicity went back to the DFP days of 1912-14, and EXP 2 was the obvious choice. EXP 2 was very much Clement's baby and was used extensively through 1920–21 for experimental work. Suitably prepared, EXP 2 came to the line at Brooklands on Easter Monday, 28th March 1921, for the Essex Short Handicap for cars over 1700 cc over 5¾ miles, with Clement driving and A.F.C. Hillstead as passenger. The Bentley oiled a plug soon after leaving the line, finishing on 3 cylinders. EXP 2 was entered for two events at the Brooklands Whitsun meeting on 7th May. The Bentley again oiled a plug on the line in the first race, the Twenty-Fourth 100 MPH Short Handicap over 5¾ miles, with Hillstead as passenger. However, in the last race of the day, the Whitsun Junior Sprint Handicap, Clement on his own in the car took first place – "the Bentley getting high up on the banking got past the field, and going to the front in the straight, it won by four lengths." (*The Auto-Motor Journal*, 19th May 1921). However, as was pointed out to Hillstead in the showroom, the Bentley's average of 72.5 mph was not very impressive compared to Barnato's average of 74.5 mph in his side

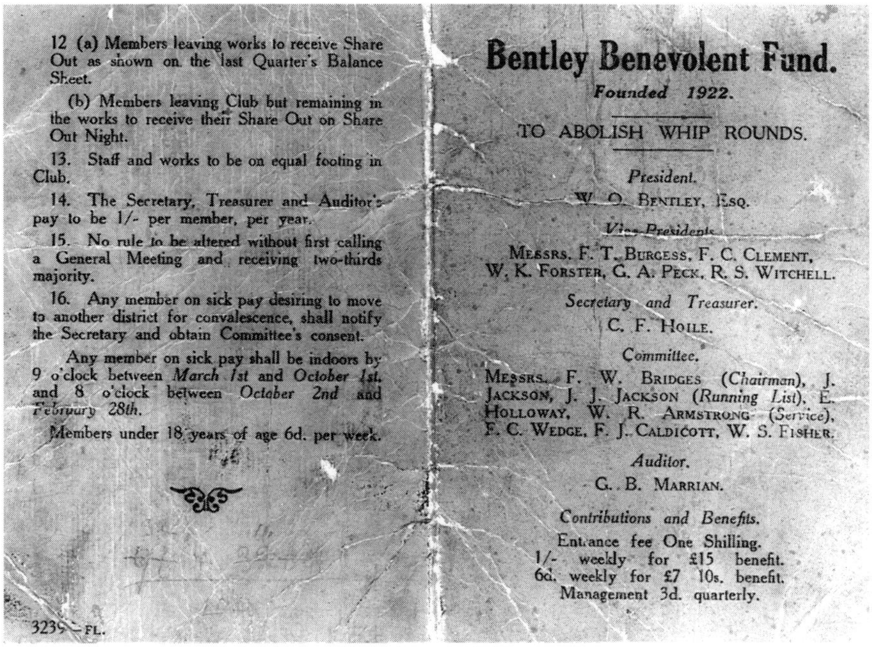

12 (a) Members leaving works to receive Share Out as shown on the last Quarter's Balance Sheet.

(b) Members leaving Club but remaining in the works to receive their Share Out on Share Out Night.

13. Staff and works to be on equal footing in Club.

14. The Secretary, Treasurer and Auditor's pay to be 1/- per member, per year.

15. No rule to be altered without first calling a General Meeting and receiving two-thirds majority.

16. Any member on sick pay desiring to move to another district for convalescence, shall notify the Secretary and obtain Committee's consent.

Any member on sick pay shall be indoors by 9 o'clock between *March 1st* and *October 1st*, and 8 o'clock between *October 2nd* and *February 28th*.

Members under 18 years of age 6d. per week.

3239—FL.

Bentley Benevolent Fund.

Founded 1922.

TO ABOLISH WHIP ROUNDS.

President.

W. O. Bentley, Esq.

Vice-Presidents.

Messrs. F. T. Burgess, F. C. Clement, W. K. Forster, G. A. Peck, R. S. Witchell.

Secretary and Treasurer.

C. F. Hoile.

Committee.

Messrs. F. W. Bridges (*Chairman*), J. Jackson, J. J. Jackson (*Running List*), E. Holloway, W. R. Armstrong (*Service*), F. C. Wedge, F. J. Caldicott, W. S. Fisher.

Auditor.

G. B. Marrian.

Contributions and Benefits.

Entrance fee One Shilling.
1/- weekly for £15 benefit.
6d. weekly for £7 10s. benefit.
Management 3d. quarterly.

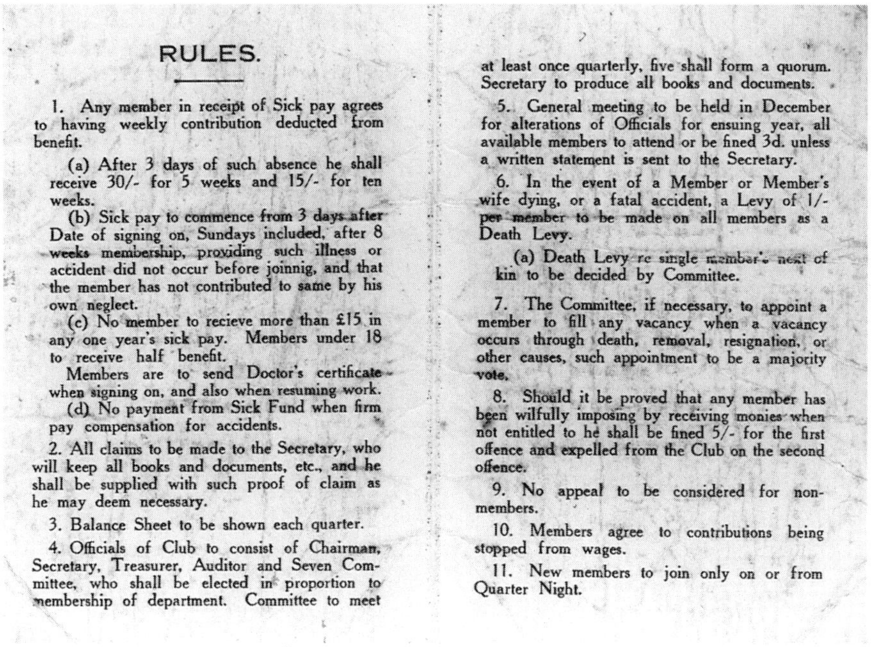

RULES.

1. Any member in receipt of Sick pay agrees to having weekly contribution deducted from benefit.

(a) After 3 days of such absence he shall receive 30/- for 5 weeks and 15/- for ten weeks.

(b) Sick pay to commence from 3 days after Date of signing on, Sundays included, after 8 weeks membership, providing such illness or accident did not occur before joinnig, and that the member has not contributed to same by his own neglect.

(c) No member to recieve more than £15 in any one year's sick pay. Members under 18 to receive half benefit.

Members are to send Doctor's certificate when signing on, and also when resuming work.

(d) No payment from Sick Fund when firm pay compensation for accidents.

2. All claims to be made to the Secretary, who will keep all books and documents, etc., and he shall be supplied with such proof of claim as he may deem necessary.

3. Balance Sheet to be shown each quarter.

4. Officials of Club to consist of Chairman, Secretary, Treasurer, Auditor and Seven Committee, who shall be elected in proportion to membership of department. Committee to meet at least once quarterly, five shall form a quorum. Secretary to produce all books and documents.

5. General meeting to be held in December for alterations of Officials for ensuing year, all available members to attend or be fined 3d. unless a written statement is sent to the Secretary.

6. In the event of a Member or Member's wife dying, or a fatal accident, a Levy of 1/- per member to be made on all members as a Death Levy.

(a) Death Levy re single member's next of kin to be decided by Committee.

7. The Committee, if necessary, to appoint a member to fill any vacancy when a vacancy occurs through death, removal, resignation, or other causes, such appointment to be a majority vote.

8. Should it be proved that any member has been wilfully imposing by receiving monies when not entitled to he shall be fined 5/- for the first offence and expelled from the Club on the second offence.

9. No appeal to be considered for non-members.

10. Members agree to contributions being stopped from wages.

11. New members to join only on or from Quarter Night.

Rule card for the Bentley Benevolent Fund – "to abolish whip rounds".

valve Calthorpe of less than 1500cc in winning the Thirteenth 75 MPH Long Handicap over 8¼ miles, and indeed Cocker's 72.23 mph in winning the Sixth Light Car handicap over 5¾ miles in the primitive Crouch! EXP 2 was raced at various hill-climbs and sprints during 1921 and 1922, mostly driven by Clement, with EXP 3 sometimes competing as well, driven by WO. Hill climbs and sprints were legal on public roads until 1924, and many private owners entered their Bentleys. Principal among the sprints were the Blackpool meetings and the Madresfield Speed Trials.

Contrary to production and racing successes, the financial aspects of the Company continued to fester and periodically erupt at the Board meetings that WO was so loath to attend. WO came from a professional background with the particular sort of upbringing that leads almost to a distaste of finance and the manipulators of it. It is obvious that Royce's opening remark at their only meeting "I gather you are a commercial man, Mr Bentley," rankled. WO did not regard himself as a business man and was much happier at the Works surrounded by engineers and his workforce, in all of whom he inspired a remarkable loyalty.

The first mortgage for the land at Cricklewood was paid off on the 18th May 1920, presumably out of money raised by the sale of shares. The first return on shares for 1919/1920 listed 16 shareholders holding 40,400 shares, which, allowing for the 10/- call, represented £20,200 (the list does not include shares paid up by means other than cash). The Company's first full meeting was held on the 3rd June 1920, at which point 40,900 shares had been issued for £18,575 (£1,875 remained unpaid) and a further 30,000 shares issued for considerations other than cash. These, of course, represented the assets of the first company which related to nothing more than the perceived value of the design work and WO's services. This was a somewhat desperate position – a year away from receiving any income, and the selling price of only 18 chassis in the bank. At least they didn't owe any money! Walter Sharrat Keigwin, a London merchant who had known HM at Clifton, joined the Board, the office at 16 Conduit Street was disposed of in August, and the registered offices moved to 3 Hanover Court.

Bentley and Bentley moved to 36 North Audley Street, WO resigning as a director on 31st October 1920. WO's shares (in Bentley and Bentley) went to A.H.M.J. Ward and Hugh Kevill-Davies, who was later to become a salesman for Bentley Motors. HM, also still a director of the company, sold his shares to the same duo, and resigned on the 22nd November 1922. Bentley and Bentley finally went into liquidation on 11th December 1922. The DFP concession remained in Ward's hands, in the guise of Ward & Driskell (J.K. Driskell drove DFPs in trials and speed events), but the parent concern came to an end in 1926 when they were taken over by Lafitte.

As an indication of the desperate position in which Bentley Motors found themselves, Keigwin received 100 shares in September 1920 for procuring shareholders and more money – Keigwin's commission being "not exceeding 10 per cent of the nominal amount of the shares in respect of which it is payable." In October Charles Frederick Stead and Charles Frederick Boston joined the Board in place of A.H.M.J. Ward. Boston was chairman until the minor depression of 1923 when

he was succeeded by Stuart de la Rue. Although the company was under-financed, it is difficult to find any evidence of economy, partly because those in charge expected a certain lifestyle and were used to it. However, in pointing out the need to maintain appearances, Hillstead also mentions a lack of expenses, so perhaps much of the social whirl was financed out of the participants' own pockets. Certainly companies were run and managed in a very different way from the present and the approach at Bentley Motors seems almost feudal. John Moore, in charge of back axle and gearbox assembly, expressed this well in 1969: "The atmosphere in the works was unusual and very different from that in most factories. The top brass kept itself to itself and gave one the impression of being amateur industrialists, not quite knowing the correct attitude to adopt towards the serfs. A sort of benevolent Lords-of-the-Manor and to hell with industrial relations that had not been invented. It was a non-union shop, although some individuals held A.S.E. union cards to protect their pension rights." In truth, the Company seems to have been shot through with a pioneering spirit, that they were breaking new and exciting ground.

In April of the new year (1921), they were back for more money. In this case £7,500 was forthcoming from three gentlemen by the name of Savill, one of whom already held 500 shares. This loan was secured against the land at Oxgate Lane and all the buildings thereon.

The first three Experimental cars, EXP 1, 2 and 3 (BM 8287, BM 8752 and BM 9771 respectively) continued their chequered careers. EXP 1 had become "The Fire Engine", the old chassis going to Brocklebank's Peugeot and the old experimental engine lying around at Kingsbury until the early 1930s. Presumably it was scrapped at some time subsequently, as it has never been seen since. EXP 2 was rebuilt on a 9'9½" frame and fitted with a production engine, and sold to Foden in 1923. Many years later she passed to Tony Townshend who dismantled her for a major rebuild and she is now being restored by Gordon Russell, still incorporating many original features. EXP 3 was retained until 1924, when she was sold to E. Taylor. It is a fairly safe assumption that EXP 3 was retained by WO until the prototype 6½ Litre car was available for his own transport.

Burgess also had his own experimental car, registered ME 2431 fitted with an open 4-seat body. In 1927, this car was rebuilt at Bentleys with a 4½ Litre bulkhead, radiator, bonnet, etc. It was then fitted with a fabric saloon body, still on a 9'9½"

wheelbase frame. The engine fitted had a big sump and must have been a prototype 4½ Litre engine. It is possible that this was either engine EXP 6 or EXP 7. The identity of this car has long been a mystery as it does not appear in the Service Records, apart from an entry against a very late 4½ Litre chassis FS 3625, when the latter car was burnt out in November 1931. It was rebuilt by Bowler, using ME 2431, the resulting car being an amalgamation of the two. The Service Records note that ME 2431 was Burgess' 1922 Experimental car. Subsequently, ME 2431, fitted with a Vanden Plas style body was raced by Margaret Allan at Brooklands. Recently the present owner, Keith Schellenberg, found the chassis number EXP 4 stamped on the front cross-member. As recounted earlier, Nobby Clarke always maintained that there was no such car as EXP 4. The "BM" prefixes on the experimental cars were not a coincidence – they were deliberately registered in Bedfordshire. It is said that WO applied to Beds County Council for a batch of "BM" numbers that they could then issue to the production cars at leisure, but Beds CC were not prepared to play ball.

WO's policy of using only combinations of tried and tested design philosophies in the design of the 3 Litre, being painfully well aware that they had neither the time nor the money to cope with development work, seems to have paid off. Apart

ABOVE AND LEFT **Although 'The Fire Engine' has been referred to as EXP 4 in the past, this car is the real EXP 4. Registered ME2431, this car does not appear anywhere in the Works Service Records, apart from an entry against a 4½ Litre in 1931. This 3 Litre was built for experimental purposes, and was used extensively by FT Burgess between 1922 and 1926/27, when it was rebuilt as a 4½ Litre. EXP 4 is seen here outside Burgess' house at 16 Ravensdale Avenue, North Finchley. The body is that first built on Chassis No.5.**

LEFT **This photo of EXP 4 is included because of the petrol tank arrangement. The rear view of EXP 4 earlier shows a standard petrol tank with number plate attached to the rear tie bar, with filler neck on the offside. As seen here, EXP 4 has a special tank extended beyond the rear tie-bar, with the number plate painted on the back, and the tie-bar going through the tank itself. This system was adopted for big tanks on the team cars in later years, where the tanks were extended backwards with the tie-bar going through the tank. The reason for the extended filler neck is not known.**

from the oil pump problems, which were sorted out before the first cars were delivered, there seem to have been remarkably few teething problems. The only known major problem was in the gearbox, with the selectors. The three selectors (top/third, second/first, and reverse) are high tensile bronze forgings sliding on steel rods inserted right through the gearbox (in the case of the A and D boxes – two of the three selector rods on the B, BS and C boxes only go part way), the gear positions being defined by the gate and the selectors held in position by spring-loaded balls locating into notches milled into the selector shafts.

The problem centred on the top/third selector. In this case, the radii were insufficient, resulting in the selector bending and only pushing the gear more or less two-thirds into mesh despite the plunger engaging, which of course resulted in jumping out of gear under load. Burgess' initial design had been made too light to save weight, the new high tensile bronze that the selectors were made from not being as strong as had perhaps been thought. Burgess re-designed the selector, but rectifying the cars already delivered made it an expensive business. WO and Burgess spent some time working on a syncromesh box later on, but as usual, there was no time to develop it.

At the 1921 Motor Show, Bentleys were again at White City, not the main concourse at Olympia, exhibiting two cars and the polished chassis. One of the cars had a 4-seat body finished in polished aluminium by the Curtis Automobile Company and the other an Allweather body by Harrisons – probably EXP 3 again. As the articles in the press showed chassis number 5, Hillstead's demonstrator, fitted with its first 4-seat body before the TT replica body, it is likely that this was the Curtis tourer. The general impression given by the press was that the Bentley was in a class of its own, "a specialised super car, and I doubt if anything finer in its class is produced in the world" as the *Daily Express* put it in their issue of 3rd November 1921. The polished chassis was used for several years, being progressively updated, and fitted with a glass lid to the A type gearbox. Arthur Saunders, later Racing Shop foreman, was brought into the Company initially to build the polished chassis. However, the debts mounted as a further £7500 was borrowed from Richard Martin Joseph Burke and William Thomas Treglohan. The stress had evidently proved too much for the first company secretary, ex Camel pilot Jimmy Enstone, as he had been replaced by C.W. Carleton. Hubert Pike and Guy Peck (General Manager) joined the Board.

3 Litre Chassis Development.

It is very difficult now to view the 3 Litre objectively and compare it with its contempories. The design is firmly rooted in pre-Great War racing practice. In fact, for its time, the 3 Litre was close to a road-going racing car, needing more manipulation of the gearbox to achieve performance than had been common in earlier days. Engine development during the Great War led to efficient, well-designed engines, and in those days 3500 rpm, the safe maximum of the Bentley engine, was pretty high.

It is more significant to consider the effect of that 149 mm stroke – piston speeds at 3500 rpm then become comparable with modern engine practice. Indeed with the advent of the light alloy pistons that WO had pioneered in the DFP, piston speeds ceased to be a limiting factor in engine design and bearing capacities and valve gear became the problems. Piston speeds alone, of course, were not the only factor. There is also the matter of instantaneous load reversal on the connecting rod at each end of the stroke. The original rods tapered towards the little end, but were changed to a parallel design for manufacturing reasons. It is certainly advisable with Bentley engines to keep the piston weights down to reduce the loads on the con rods. 80 x 149 mm was quite a common Coupe de L'Auto dimension to get under 3 litres (capacity was 2996 cc) and the long stroke engines so typical of the vintage era were, in part at least, engendered by the RAC horsepower rating that the taxation system was based on. This system related to bore only and not to stroke, so the 3 Litre registered at 15.9 hp with an annual tax of £16. By comparison, the far cheaper Model T Ford with an engine of 3½ litres attracted a tax rating of 25 hp or £25. The dimensions of 149 x 80 mm, or 1.9:1 stroke/bore ratio approximates to the pre-Great War formula for Grand Prix cars of a stroke/bore ratio of roughly 2:1.

The chassis frame, the beam axles and suspension were typical of the day, but the high geared steering – with one turn from lock to lock – with beaded-edged tyres was direct, sensitive and self-centred nicely, a feature not all that common at the time. Such features as the central accelerator between the clutch and brake, right-hand gear lever operating in a gate and right-hand handbrake (outside on many of the open cars) were normal on high-quality chassis. The appearance of the handbrake itself is no accident – WO told Harry Varley to model it on a locomotive's reversing lever!

The gearbox itself, the A type in the early cars,

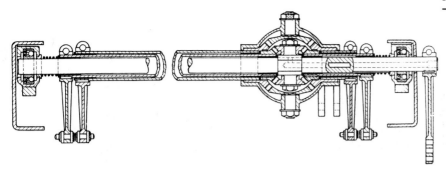

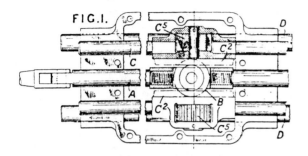

ABOVE **The bevel gearbox design of compensator used on the early 3 Litres with rear wheel brakes only. Note that the handbrake had its own geared compensator. The chassis bearings incorporate self-aligning races, superseded by aluminium castings with plain steel balls. The main take-up for the footbrake is behind the compensator, not in front as on the later four wheel braked cars, and is similar to the adjuster fitted to the handbrake.**

ABOVE **It is known that WO and Burgess worked on an hydraulic four wheel brake set up and this rack and spur gear compensator obviously represents a step in their thinking towards a mechanical four wheel brake set up. The compensator compensates side-to-side by the gear attached to the main brake rod acting on racks attached to the pairs of rods for each side, the rack-and-gear** allowing relative movement. Front/back compensation is then achieved by a similar spur gear acting on racks attached to the front/back rod respectively, allowing relative movement between the front/rear rods for each side of the car. The idea is simple but has obvious disadvantages in the side loads exerted on the spur gear pivots and backlash in the gear train.

had high, close ratios and had to be used to extract the best from the car. The 4½ Litre has enough low speed torque that it can really be driven more or less anywhere in top, but not the 3. The A box, with its 0.020″ eccentric adjustment on the layshaft for meshing the constant mesh gears, was very much Burgess' idea. The box was designed without a speedo drive, as the speedo was belt driven from a pulley on the propshaft. When a speedo drive was needed, it was incorporated straight into the front of the box near the top edge on the nearside to avoid altering the casting patterns and causing the minimum amount of extra machining. The drive suffered from the aluminium gear chewing up, the drive used on the B box fitted into the nearside of the box in a machined recess suffering in a similar manner. This problem was finally eliminated with the fibre gear used on the C and D boxes.

One of Hillstead's favourite tricks for demonstration purposes was to drop into second at sixty, accelerate and then change up again – at a time when it was commonly thought more or less impossible to change gear at over 30 mph. Burgess was confident about the box, and was heard to say "Use it and don't be afraid to crunch it – you won't break it." But then Burgess commented on his drive in the 1914 TT – "That's nothing. You should have heard me in the TT. When I changed gear on the mountain the people in Ramsey had to stuff their fingers in their ears!" A maximum speed of the order of 70–85 mph depending on the model was fast for the early 20s if a long enough stretch of open road could be found to attain it, with cruising speeds in the 50s and 60s on a road system bearing no relation at all to present motorways.

The clutch was of the cone variety and went though several variations before a successful scheme was adopted. Initially a very light cone with an angle of 9 degrees was used, with the lining (of leather) carried on the outer ring with no clutch stop. This was changed to 11 degrees with the lining carried on the cone itself and finally the outer ring was split and formed into first engagement fingers to smooth the application of the clutch. The increase in weight of the moving parts showed up the need for a clutch stop and the first was made up by Arthur Saunders in the Experimental Shop. This consisted of a horseshoe laminated up from steel hoops lined with Ferodo, acting on a steel disc mounted on the clutch shaft. While proving satisfactory, the later design used on the 4½ Litre with a thumbwheel adjuster was a considerable improvement.

The pot-joint propellor shaft proved satisfactory, as long as it was kept well lubricated – preferably every week, but was superseded by the proprietary Spicer shaft as soon as it became available in 1927. The problem with the pot-joint was that it tended to throw the oil out, unless the joints were very carefully made with lapped metal-to-metal faces. Experiments were made on the EXP cars with various patterns of gaiters, but none of these proved satisfactory, so a cast aluminium stone deflector was made up and fitted just in front of the rear joint. The Spicer shaft was a great improvement on the pot-joint in service and has since been fitted to virtually every chassis not so delivered.

Brakes on the early cars on the rear wheels left something to be desired and any oil on the cast-iron liners reduced the available retardation almost to nothing. Experiments on four-wheel brake systems using hydraulics were carried out on EXP 2 in 1920/21 but the fluids of the day were inadequate and there was no time to develop such a system. WO's intention was to get even pressures to all four wheels, but they did not get as far as the operation of the brake shoes themselves. According to Nobby Clarke, an hydraulic system was developed in late 1922/early 1923 that showed promise, but occasionally refused to release after heavy application due to pressure build-up. "A.P.B." in *The Times Weekly Edition* of 24th March 1922 commented that: "The brakes are good. ...their effect is certain, and their proportions allow of efficient working, but with a car that can do 80 mph I personally would prefer braking on all four wheels."

The two-wheel brakes used a compensator with a bevel gearbox to compensate the braking pull on each wheel. This compensator was the subject of patent number 150423 taken out by WO and Bentley Motors Ltd, registered in May, 1919. The main drawback of a geared compensator is that, should there be any seizure of one side, all the braking force will be applied to the other side only and tend to lock that wheel. As WO and Bentley Motors applied for a further patent on 11th January 1921 for another design of compensator using spur gears and racks that never saw production, WO was obviously not happy with the principle of the bevel gear compensator.

On designing the four-wheel system, WO adopted Perrot principles at the front with twin universal joints and a sliding shaft assembly with

keys to take the brake torque. With this went a novel balance beam compensator with friction built into the system, so that in the event of a sticking brake, some force would be exerted to apply it rather than complete load transfer to the other wheels as happens with non-friction types of compensators. However, unlike most manufacturers in the UK offering four-wheel brakes in the early post-war era, the Bentley system was powerful – many firms fitted rudimentary front brakes, often with substantial failings. Common among these was a tendency to stick on at full lock and not release until after the accident. The Perrot type of front brakes largely avoided these problems, but would have been better had the set-up possessed the constant velocity properties of modern front-wheel driveshafts. The lack of this leads to a compromise in setting the Bentley brakes between pedal travel and binding on full lock. Brooklands, as a proving ground, did not encourage the development of braking systems, but the Le Mans race certainly did. The first system was to be fitted to

The patent specification for the front brake backplate.

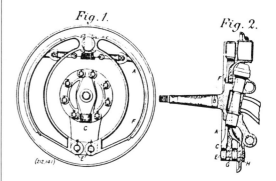

MOTOR ROAD VEHICLES.

212,141. Bentley Motors, Limited, London, and F. T. Burgess, London. Brakes. (2 *Figs.*) April 14, 1923.— The invention relates to brakes for use on motor vehicles, and it has for its object to provide a very substantial pivotal attachment of the brake shoes upon the anchor plate. According to the invention, the plate is formed with a radial web which is spaced slightly from the wall of the plate, and the anchor pin for the shoes is mounted in this web, projecting at both sides thereof. The anchor plate A is dished and is formed with a large central hole, around which it is faced for attachment to

Fig. 1.   *Fig. 2.*

a flange on the axle B. The plate A is dished so that the edges lie considerably to one side of the plane of attachment, and at the bottom the plate has formed integral with it a depending web C, which lies in the plane of the centre of the plate. This web is spaced slightly from the rim of the plate, and it is formed with bosses G to receive the anchor pins E for the shoes F. The pins may also enter bosses H in the dished edge of the plate A and be fixed there by a split pin. The shoes F are forked at their pivotal ends, and these forked parts pass one each side of the web C and are held in place by the anchor pins E. (*Sealed.*)

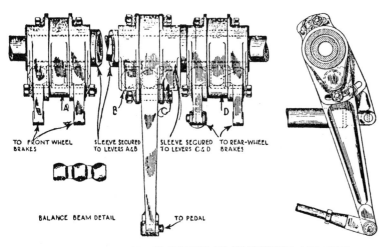

TO FRONT WHEEL BRAKES

SLEEVE SECURED TO LEVERS A&B

SLEEVE SECURED TO LEVERS C&D

TO REAR-WHEEL BRAKES

BALANCE BEAM DETAIL

TO PEDAL

THE PATENTED BENTLEY SYSTEM OF FRICTIONAL COUPLING FOR BRAKE COMPENSATORS.

LEFT **This is the final design of mechanical compensator invented by WO and Burgess and used on all the chassis including the 8 and 4 Litres. The bolts nip the compensators, with the central compensator balancing the brakes side-to-side and the individual compensators front-to-back.**

Burgess' car, ME 2431 chassis EXP 4 and tested extensively before being passed for production.

The first design of Perrot shaft was known as the Stage 1. This shaft has a pin and block type universal joint at the inboard end and a claw and ball type of universal joint at the outboard or backplate end. The brake torque is taken through a pair of keys, but unfortunately these keys were also designed to slide to take plunge in the shaft due to vertical movement of the axle. In practice the torque tends to lock the keys and can pull the shaft out of the frame bearing, in which it is only retained by a washer and circlip. The design was later improved slightly by lengthening the frame bearing to obviate the high wear rates, and locking it with a collar and pin with a spring washer to give some degree of flexibility. This design is often referred to as the "Stage 1½". The shaft was next modified to work in the way it was trying to, by pinning the keys and allowing it to slide in the frame bearing – the Stage 2 set-up. Subsequent designs (the Stage 3) abandoned the sliding key centre portion and went over to a solid shaft. The illustrations of the four-wheel system used in Bentley's 3 Litre catalogue for 1924 are of Burgess' experimental car ME 2431 and show a rather different arrangement from that used on the production cars. This seems a surprising oversight by Bentleys. The early cars used two identical forgings for the clutch and brake pedals, with a swinging arm on the front gearbox cross member to align the brake rods. This arrangement was replaced in toto by a new cranked brake pedal with the rod going direct from pedal to master lever on the compensator, with a single adjustment to take-up all the brakes simultaneously.

The compensator itself was a completely new design by Burgess and WO, largely overcoming the problems of the earlier bevel gearbox compensator of the rear wheel braked cars. The basic idea was covered by patent number 215178, applied for on 14th April 1923, and in practice proved to be highly effective. An 8 Litre on a rolling road recently achieved braking efficiency figures comparable to a Ford Granada fitted with ABS. The front axle beam was redesigned and made much heavier, and the front springs stiffened up. The axle movement had to be substantially limited because of the Perrot shafts. The front brake shoes were mounted on pivot pins on a cast aluminium backplate, patent (no. 212141) being granted to F.T. Burgess and Bentley Motors in respect of an invention related to brakes for motor vehicles, also dated 14th April 1923.

In theory, of course, the compensator should not do anything at all. If the brakes are correctly set up so that all the brake shoes touch the drums simultaneously, then the compensator does not move. The distribution of the braking load front/rear is dealt with by the leverage, being a function of the lengths of the operating levers. Under very light braking, the pull-off springs will affect the front/rear distribution, higher pedal pressures negating this effect. The compensator beam itself is massively constructed to reduce whip: because the master lever is on the off-side, any wind-up in the shaft will cause the off-side brakes to be applied first and

ROYAL AUTOMOBILE CLUB

Demonstration No. 12.

(UNDER THE OPEN COMPETITION RULES OF THE R.A.C.)

BENTLEY FOUR-WHEEL BRAKES

25th October, 1923.

REPORT.

Mr. W. O. BENTLEY of 3, Hanover Court, Hanover St., London, W.1. entered a three litre Bentley car for an Official Demonstration of the efficiency of the four-wheel brakes. The car having four cylinders 80 mm. x 149 mm. and weighing (unladen) 3164 lb. (28 cwt. approx.) (front axle 1573 lb., back axle 1591 lb.), was fitted with a light four-seater body. The load carried consisted of either four or two passengers (including driver). The braking system was such that the depression of the brake pedal, applied the brake to each wheel, both front and back. Another pair of brakes was also applied to the back wheels only, by movement of the side lever. During the Demonstration these brakes were not used.

The Demonstration was held on Brooklands Track, the concrete surface being quite dry. The tyres fitted to the car (820 mm. x 120 mm.) had all-rubber treads, the surfaces of the treads being broken up by an indented pattern. The Demonstration was intended to show the stopping power of the brakes, and any effect on the steering of the car which they might have. A number of tests was taken at various speeds, in the first instance with four passengers in the car, and in the second instance with two passengers. Particulars of the tests are given below:—

Speed	Stopping Distances	
m.p.h.	With four passengers (laden weight 4036 lb.)	With two passengers (laden weight 3700 lb.)
10	3 ft. 6 in.	4 ft. 3 in.
20	19 „ 9 „	22 „ 0 „
30	46 „ 0 „	42 „ 0 „
40	86 „ 6 „	91 „ 0 „
50	144 „ 9 „	† —

† The distance in this test could not be recorded, owing to the failure of the apparatus used to record the time of the application of the brakes.

The weight of the two extra passengers was almost wholly over the back axle.

A test was also made to show what effect the application of the brakes had upon the steering of the car when the car was being driven in a circle. The car was driven in a circle of 80 ft. diameter at a speed of 19/20 m.p.h., and with four passengers, stopped in 14 ft. 0 in. With two passengers, the car stopped within 12 ft. 9 in. The steering did not appear to be affected. During another similar test the offside back tyre left the rim. In all the tests there was locking of the back wheels, but not of the front wheels, when the brakes were applied.

The action of the brakes was tested upon the Test Hill at Brooklands (gradient one in four), and it was found that the car was brought to rest and held at rest in both directions.

(Signed) F. P. ARMSTRONG, *Secretary.*
Pall Mall, London, S.W.1.
30th October, 1923.

(Signed) ARTHUR STANLEY, *Chairman.*
(Signed) G. H. BAILLIE, *Chairman of Technical Committee.*

[Copyright]

The RAC report on the four wheel brake trials on EXP 4, conducted with the paint-brush rig. In its final production form, the 3 Litre was a safe machine with a relatively neutral ride, and the brakes were part of that success.

with more force. It is interesting to note that the early 3 Litre compensators only had one hole in the levers, except the master lever, which had two, while later compensators had three holes in the levers and five in the master levers. Similarly, the idler levers in the system and operating levers at the axles were also provided with more holes. As this occurred later there was a reason for providing these additional holes. It seems likely that the Works set up each chassis individually by using different combinations of holes. As the only way of changing braking distribution is by altering the ratios of force applied front/rear and side/side, a 12'8½" wheelbase 2½ ton Speed Six saloon would have a different braking set-up from an 11'2" wheelbase tourer weighing maybe 35 cwts with a different front/rear weight distribution.

In terms of handling, the 3 Litre was very neutral and safe, and extremely difficult to roll. On its introduction the 3 Litre was something of an event, its qualities inspiring a small but dedicated following. On the negative side the performance, particularly when fitted with the Smith's carburettor, was inferior to that of the 30/98 Vauxhall. The Claudel-Hobson, offered as an option on the early cars, despite offering better performance, was a temperamental instrument and had a tendency to catch fire when starting from cold. Fuel capacity of 11 gallons was also distinctly ungenerous, although with a fuel consumption of 30 mpg at 30 mph the long-stroke engine was pretty economical. It was also expensive and probably somewhat noisy – between the announcement in mid 1919 and deliveries starting in mid 1921 the chassis price increased from £750 to £1150 before dropping to £1050, largely due to the high cost of sub-contract manufacture of small batches. Noise was one of WO's particular dislikes and throughout production of the 3 Litre, considerable measures were devised to reduce it, even to the extent on some cars of riveting lead sheets to the underside of the rocker cover. While possessing weight advantages, aluminium engines are inherently noisier than those made from cast iron. Development of the chassis largely resulted from lessons learnt in racing – a philosophy that was still valid in the 1920s when racing cars bore more than a superficial resemblance to the road-going product.

6 International Racing

IN THE LATTER PART of 1921, the RAC announced the revival of the pre-war Tourist Trophy race, to be held at Ramsey on the Isle of Man. After long discussions, Bentleys entered a team of three cars in November. In January of the new year Clement was allocated three chassis from stock and in a surprising move a fourth chassis was prepared for the Indianapolis 500 Miles Race. Quite what the embryonic company, with no foothold in the United States, hoped to gain from such an entry is difficult to conceive and it cannot have been a particularly cheap episode. WO commented that "it was our first major error" but that "we were anxious to show the Americans what we were doing."

Browning related the story thus: "One morning towards the end of April 1922, Mr. W.O. Bentley came into the Experimental and Racing Shop (of which I was in charge [presumably Browning was the foreman as Frank Clement was in overall charge of the shop] and said in his cool, considered voice, "I want you to build a car to run at Indianapolis and you and Hawkes are going to go over for the race." And with a slight smile, he walked back to his office.

"I stood there quite unconcerned for a minute and then said to myself "What was that he said? Indianapolis? Where the devil's that?" I think I had read about it somewhere and had forgotten about it. Then the penny dropped! While I was still thinking it over, WO sent for me and outlined the project more fully. Then I began to look around the shop, making the usual mechanic's plans – where to build it, what we should need, where to get things from and so on. There were no suitable trestles available so we started looking for the wood to make them

from. Before long the frame arrived, followed by bits of this and bits of that, and the boys got down to the job of building the car...the deadline date for completion seemed ridiculous – we wouldn't be anywhere near it – we would have to cancel the whole project...But somehow we just scraped through, with the car sufficiently near completion in time to catch the very last boat that could arrive in time. This was the SS *Olympic*, but I had gone over some time before this to prepare for the arrival of the crate with the car and all the spare bits and pieces. It got to our workshop at 10am on the morning of Elimination Day. We were at panic stations! No time to look around, just knock the crate to bits, put on the wheels and fill up with juice, oil and water, fit the battery, which I had been nursing, press the starter – and pray! A couple of puffs of smoke and she was running, push her down to the track, make a few adjustments and run a spanner over the vital bits and it was "OK, come on, Browning" and we were away. After two laps we signalled "OK, for test and elimination" – she went like a bird – and we were through – and then we had three days to get ready for the race itself."

Hawkes, nominated to drive at Indianapolis, was effectively a professional driver. He had driven a Victor before the war, and the 1912 15 litre Lorraine-Dietrich "Vieux Charles" after it at Brooklands. De Palma and one or two of the American drivers took the British team under their wing and made sure that they were ready for the race. Browning recalled later the need for a strong elasticated body belt because of the rough ride on the "Brickyard" and because the times set on the first day of practice determined the grid positions, the Bentley started in quite a favourable position, 19th out of 27. The Americans were astonished by the manner in which the Bentley was uncrated before immediately lapping at fairly high speed. Hawkes drove a steady race, being flagged-in once by the Starter, Rickenbacker, for getting in the way of some of the faster cars. They also made one re-fuelling stop, finishing 13th at 74.95 mph, the last car to complete the full 200 laps. This promising showing of a lone British car far from home with little factory support, made a favourable impression on the American public, but with its four-speed gearbox and high revving, high efficiency engine, the Bentley was not a car the Americans could readily understand. A lack of dealers and spares networks could hardly have helped, and a sales trip some years later in a 6½ Litre confirmed the initial impression. However, some chassis were sold to the States. (Hawkes drove again at Indianapolis in 1926 in an Anzani-engined Eldridge, in which year John Duff drove a Miller-engined Elcar Special after the nominated driver Herb Jones crashed and was fatally injured in practice.)

Immediately after the race, Hawkes and Browning returned to the UK and were back in time to start the TT race on the 22nd June. The drivers for the TT race had already been decided upon, as Hawkes, Clement and WO. However, there was a strong possibility that Hawkes would not be able to make the return journey from Indianapolis in time, so Hillstead was nominated to drive the third car. This was quashed due to company in-fighting. The reasons why remain unclear to this day – presumably there were others who felt more entitled by reason of financial interests than driving ability. Due to the time pressures, Hillstead took a 3 Litre over to the Isle of Man early in April to get his eye in on the circuit.

Preparation of the team cars consisted of fitting a Claudel-Hobson carburettor, hour glass pistons with raised compression, and a small, flat radiator of almost Bugatti-like appearance. The reasons for this radiator are a source of conjecture. It would seem better to have raced with a proper production pattern radiator to emphasise the point that the cars were practically unmodified and it would make them more instantly recognisable as Bentleys. However, the standard radiator presented a larger frontal area and hence more wind resistance, and it is known that WO consulted the National Physical Laboratory on the subject of aerodynamics for the TT cars. Equally, the standard radiator was a great deal more expensive to make, so perhaps it was more of an economy measure.

The press carried announcements that the modi-

First major race – Hawkes and Browning at the wheel of the Indianapolis 3 Litre chassis No. 94 at the track. Note the special steering wheel, fabricated in steel, in place of the standard aluminium wheel. This was because there had been so many steering failures at Indianapolis that aluminium was banned. The Bentley put up a very creditable performance, but was completely outclassed.

fications carried out to the TT chassis were available to the general public for an additional sum of £25 on the basic chassis price. As far as is known, nobody ever asked for the special radiator and it seems that the four made for the team cars were the only ones ever made. This £25 conversion has to be taken with a considerable pinch of salt, as the modifications were more substantial than is implied. The chassis had a lightened flywheel with no ring gear or starter motor, outside exhaust on a special manifold, special bulkhead, bonnet, radiator mountings, shock absorbers (Hartfords, as later used on the Speed Model chassis) with special brackets and a 22 gallon under-floor petrol tank with air-pressure feed. In addition there was a special oil tank mounted between the gear box cross-members, with a switch over tap arrangement to allow the air-pressure pump operated by the riding mechanic to be used to pressurise the oil tank should the oil level drop. The chassis were also fully worked over and meticulously prepared – "Everything split-pinned and nothing too much trouble", as Nobby Clarke used to say.

The chassis were then delivered to Ewarts who built the bodies – a light 2-seater aluminium shell

THIS PAGE **The three TT cars, inside the experimental shop at Cricklewood. No.3 was Clement's baby, and the streamlining to the front axle can be seen. The Claudel-Hobson manifold can just be seen on No.3, with the Factory Acts pinned to the walls above.**

with minimal woodwork and a horizontally-mounted spare wheel in the tail, with a hinged cover. The Indianoplis chassis had a more stream-lined tail with a smaller fuel tank fitted into it and carried no spare. The "Indy" car also had to have other changes to meet the regulations. Principally, these consisted of a special tubular steel steering wheel, as aluminium was not allowed for steering wheels at Indianapolis and special front spring-hangers to stop the axle going back in the event of failure of the master leaf. This was later redesigned and fitted to all the production cars. The front springs had the second leaf rolled round the spring eye to strengthen that part.

It is clear that WO was very keen to emphasize the point that the cars were basically standard pro-duction chassis, and this point was made again in a letter from Bentley Motors in *The Autocar* of 17th June 1922: "During the past week several

THIS PAGE AND FAR RIGHT **Frank Clement in the No.1 TT car, Chassis No.42, which raced as No.3. These photos were taken in the space in the works yard between the original brick shop, by 1922 the engine test shop, the main works building, and the experimental and engine erecting shops. The side view shows the test shop to the right, the door to the main works behind, and to the left rear the double doors into the engine erecting shop. These photos show well the details of the TT cars. Note the special radiator mountings bolted into the frame, and from the side, the production pattern two-piece sump, undershield, oil tank between the gearbox cross-members, and then the 22 gallon saddle tank. The rear view shows well the special double shock-absorber brackets, the lower ones fitted to the carrier plates for the rear brake shoe backplates.**

erroneous descriptions have appeared in the Press with reference to the cars entered by this firm in the Senior Tourist Trophy Race. We would like to point out that, unlike our competitors, we have not built special racing cars for this event, but instead entered three standard chassis drawn from our production stock. The engines have been fitted with high compression pistons, and a special carburettor and smaller radiator are used. In every other respect the cars are absolutely standard, and even the high compression pistons are supplied to customers who desire a sports type car. It is our desire in this race to demonstrate the capabilities of our standard car, and we feel that this is a policy that will commend itself to the motoring public."

Early in March 1922 Clement was photographed in ME 1884, the number 1 TT car chassis number 42. Hawkes was shortly afterwards photographed in ME 4976, the Indianapolis car chassis number 94. Trials at Brooklands had shown the TT and "Indy" cars to be capable of 100 mph and calculations by Cyril Wadsworth in 1972 showed that with the frontal area, available power and weight of the complete car with a 2.87 back axle, 100 mph was indeed possible. The Works chassis had a special 14/48 (3.43:1) back axle and special A type gearboxes with non-standard 5 DP ratios.

For the TT race the three cars were driven from

teams were hurriedly fixing temporary splash guards to the drivers' side front wheels – only Vauxhall had wings already fitted. The formula for the TT race of up to 3 litres capacity was somewhat out of kilter with the formula adopted for continental racing of 2 litres, so the entry was correspondingly thin: 3 Vauxhalls, 3 Sunbeams and the 3 Bentleys. The Vauxhall and Sunbeam teams consisted of out and out racing cars with experienced drivers, of the likes of Chassagne, de Hane Segrave and Kenelm Lee Guinness. The Bentleys by comparison were virtually standard chassis and it was that point that WO hoped to emphasize by a strong team performance. To fill the grid a 2 litre class was added, made up of three Bugattis, three Talbot-Darracqs, an Enfield-Allday, a Hillman and an Aston-Martin, bringing the field up to a total of 18 – not many for a long circuit.

Clement got away to a good start, but Chassagne in the Sunbeam pulled away to an early lead. Guinness' Sunbeam did not come to the line because of mechanical problems. The superior speed of the Sunbeams and Vauxhalls took them into the higher positions early on, but then attrition set in; Park's Vauxhall dropped out, then the Sunbeam of Segrave, and finally Swain's Vauxhall, leaving Chassagne first in the Sunbeam at 55.78 mph, Clement second in the Bentley at 52.21 mph, Payne's Vauxhall third with WO and Hawkes fourth and fifth. The drivers and mechanics were filthy and soaked and Pennal was half-gassed and roasted when the exhaust pipe sheared off at the manifold flange. WO's drive was rendered extremely uncomfortable by the floor board breaking up and having to be thrown overboard, WO being so fed up at having to drive for three hours with no support but the pedals that he ignored the "faster" signal hung out by the pits and missed Payne's Vauxhall by six seconds. Hawkes lost the plug from his radiator and all the water, and was forced to put cold water into a very hot engine amidst clouds of steam – fortunately the car survived and continued at undiminished speed.

Pit organisation was rather less professional than later years – Poggy Orde, the Pit Manager, bought full-length sledge hammers for the locking rings! The team prize went to the Bentley Team, but until after the race there was no such award! In the event, G.R.N. Minchin, who had stayed with the Bentley Team, asked Sir Julian Orde, the secretary of the RAC, if the Club would award a Team Prize if Mr Eric Horniman presented it. After consulting members of the committee Sir Julian agreed, but the cup had to be presented that evening at the

the Works to Liverpool and shipped across. The last minute preparations had been pushed to the point where the white bath enamel painted on the night before failed to dry and had to be washed off in the morning and replaced by quick-drying white undercoat. The team settled in pretty well at Ramsey, the mechanics revelling in the night-life in a style that was to become the norm at Le Mans in later years. Practice had to be conducted in the early mornings when the roads were not closed, leading to perilous avoidances of early-morning farm carts by some of the competitors. WO, with his experience of the race, was already looking for the kind of minor improvements in the cars and procedures that were in no small part the foundation of Bentley's racing successes. Calculations showed that with the extra 11 gallons of a standard tank added to the 22 gallon underfloor saddle tanks, the 3 Litres could run the entire 302 miles, 8 laps of the 37¾ mile circuit, without a refuelling stop. The Works were wired to send the tanks to Liverpool, where Wally Hassan met them at the docks. All the plumbing for the air-pressure feed had to be revised, but by the race day of 22nd June, all these mods had been fully incorporated and the cars fully prepared for all eventualities – except, of course, the British weather. The day dawned wet and cold and both the Bentley and the Sunbeam

awards ceremony by the Governor of the island.

Horniman and Minchin chased around in the latter's Silver Ghost, before rousing the proprietor of a shop in Ramsey and parting him from a large cup intended to be presented to the best beast in the forthcoming agricultural show. This was duly presented and the looks on the faces of the Bentley drivers who knew nothing about it must have been a picture. As had been hoped, the press response was very welcome. As WO later said, he would gladly see his cars circulate in glorious isolation in order to appear on the front page of the *Daily Mail* the following day. The "Bentley Boys" (a name applied to the mechanics well before it was applied to the drivers) were having such a good time that on the last day Saunders had to shovel all the spares, tools and other bits and pieces into a crate which was sealed up and returned to the Works in that state. Conway, the storeman and one of the lynch pins of the Bentley Works, blanched at the prospect of sorting that particular mess out. Minchin himself was more of a Rolls-Royce enthusiast and his firm Peto & Radford was a major supplier of electrical equipment to Rolls-Royce as well as of batteries to Bentley Motors. He did, though, buy a 3 Litre in 1924, a long chassis fitted with a James Young All-weather body which proved to have a rather disappointing performance when he lent it to Rolls-Royce.

The TT race also introduced George Porter, a garage owner in Blackpool, to the Bentley car – "That car of yours is going to be a world beater", as he put it to WO. Porter took on an agency for the North West with a great deal of enthusiasm, buying his own car chassis 221 registered FR 5189 and fitting it with the radiator and body of the number 3 TT car ME 3115 that had been driven by WO and was then owned by Edge. Sadly Porter killed himself in that car in 1925, but it was rebuilt onto a 9' chassis by Bentley Motors and raced in standard and supercharged form with considerable verve by May Cunliffe. WO's acquisition was Bertie Kensington Moir, with whom WO talked all the way back across the Irish Sea on the *Castle Mona* accompanied by substantial quantities of Scotch. Bertie had driven the 1500cc Aston-Martin in the TT before a dropped valve ended his race and had considerable experience racing Straker-Squires. Indeed, Sidney Straker was Moir's uncle. In the end WO offered Moir a job at Bentley Motors, as second in command to Hubert Pike in charge of the Service Station. This was a position Bertie filled with great zest but erratic management for several years before a re-organization to put the Service

ABOVE **The outside of the first service department in Oxgate Lane. These two photos were taken by JR Andrews on 1 May 1924. This was the site of the service department from 1921 to 1925/26, when service moved to Kingsbury. The car outside is George Porter's 3 Litre chassis No. 221, FR5189. This car was looked after by Paul Dutoit and Abey Kates in the service department. On the left is Fred Squibb, a fitter/tester, particularly on ignition and carburettors. On the right is James Gorringe, another fitter.**

ABOVE RIGHT **The staff at the Oxgate Lane service station. Hubert Pike is seated fifth from the left.**

RIGHT **The service crew with FR5189. Captioned by JR Andrews on the back: 'A snap of the most wonderful car in the Universe'! From left to right, F Fox, R Podger, J Gorringe (at the wheel), Fred Squibb (passenger), not known, J Logan, Charlie Hall and B Huggins (far right).**

Station on a rather more professional basis under Nobby Clarke. In the very early days Service was based at the top of Oxgate Lane, with about 40 or 50 staff, before moving to Kingsbury. Pike was one of the central players in the Bentley scheme of things and remained with the firm to the bitter end, putting a lot of his own money into it. Pike held a directorship on the strength of a not inconsiderable shareholding (in September 1921 he had 8000 £1 shares) and in later years was known to buy a 6½ Litre chassis on Friday afternoon so there was enough money to pay the wages.

Pike's first Bentley was chassis number 7, used at the Boulogne Trophy race in 1923, but his usual

TOP **Oxgate Lane service department, circa 1923/24. The car on the far right appears to be chassis no.7 XH9047, in the form in which it was raced at the 1923 Georges Boillot Cup race by Kensington Moir.**

ABOVE **Service department at work – a 1924 model 3 Litre with Vanden Plas coachwork, extensively wrapped.**

car was chassis 238, a Vanden Plas saloon on the long chassis registered PD 40. This car was used for all sorts of experimental work between 1923 and 1929, being fitted with a 4½ Litre engine at some stage, before 1928, because in November of that year it was fitted with the bottom half of engine EXP 6. EXP 6 is very much a mystery, it was presumably a complete car because Middleton of the

Road Test Department's notebook records that the gearbox from EXP 6 was fitted to a 3 Litre chassis in 1927, but no record of EXP 6 to identify it exists, except for an early 4½ Litre frame of unknown origin with "EXP 6" stamped on it. In 1929 the bottom half of engine EXP 6 was transferred to "Old Mother Gun", the prototype 4½ Litre, when the bottom half of "Old Mother Gun's" engine ST 3001 was fitted to 238. The Service Records contain frequent references to Experimental parts being tried out in cars belonging either to members of the Company or the Bentley Boys.

All the TT cars were raced at Brooklands in the late summer of 1922 and the whole team was then offered for sale – but not before a little chicanery was indulged in. Clements' own car chassis 42 was the fastest and most developed of the trio, but as

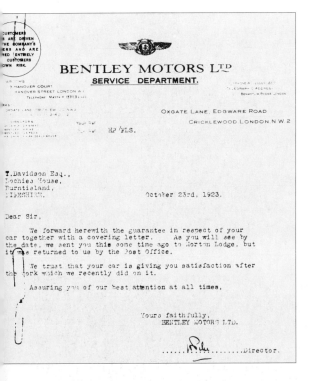

ABOVE **Letter from Hubert Pike to T Davidson, the original owner of Chassis 19, showing the original Service Department address in the old Oxgate Farm buildings in Oxgate Lane.**

RIGHT **Service staff again: this time stores personnel with a 3 Litre, and J Logan and J R Andrews (right) with another 3 Litre tourer.**

usual money was in desperately short supply and selling the team cars was a short term way of (hopefully) easing the cash flow. Understandably, Frank Clement must have been very reluctant to part with "his" car. The solution to this problem was to swap the identities and bodyshells of the Number 1 TT car and the Indianapolis car, so the "Indy" car was effectively sold as the TT car.

Selling the TT cars did not prove to be very easy, possibly because they were too impractical for road use and probably weren't cheap either. The *Auto Motor Journal* of 13th July 1922 (under the heading "How Many Sales per Win?") commented "I have not heard what happened to Bentleys as the outcome of their magnificent team performance, but here again I believe that every Bentley the works can produce is sold long before it is built, as has

ABOVE LEFT **Pike's first Bentley, chassis no.7. This car was raced by Moir in the 1923 Georges Boillot Cup race. Again, the records for this car are somewhat inexplicable. The record for chassis no.7 ends with a major rebuild in February 1926, including 'new nameplate fitted'. As far as is known this car became chassis no. NR520 (see p.152) registered MK2941, the record for which starts in March 1926. Unfortunately, there is no trace of this car after 1937.**

LEFT **Several members of the Bentley staff had cars which were used for experimental and test purposes. PD40 was Pike's car, a 1923 model long chassis fitted with a saloon body by Vanden Plas. Pike kept this car until 1930, when it disappeared, rather mysteriously. This is the car as new, with two wheel brakes.**

ABOVE **PD40 photographed by Joe Weedon, Pike's chauffeur, circa 1928/30. Note that the big sump has been fitted; it is likely that the engine is a 4½ Litre. PD40 was fitted with the bottom half of the engine from EXP 6 in November 1928, and this bottom half was swapped with the engine from 'Old Mother Gun' chassis no. ST3001 in October 1929. The records for 238 end in 1929, with the cryptic note 'See chassis ST3011'. ST3011 was 'damaged in an accident and dismantled', but reappeared in October 1929; 'Chassis converted to 4½ Litre', and it was then registered GF1631 in March 1930. Unfortunately, quite what went on will now probably never be known.**

been the case for a year past, if my information is correct". Would that it had been! Hillstead's and WO's conflicting comments over the fate of the TT cars are very revealing. WO commented: "We then offered the three team cars for sale and they went in a flash, at very satisfactory figures." Hillstead observed later, in *Fifty Years with Motor Cars* that "This, however, is not my recollection of what happened", followed by accounts of Brooklands appearances and sales runs. It is easy to understand the opposite views – WO succeeded in his aim of producing "a good car, a fast car, the best in its class" and probably had trouble understanding why it didn't sell. Hillstead on the other hand daily attempted to persuade customers to part with roughly £1300 (the price of a 3 Litre with Vanden Plas touring coachwork) at a time when the average skilled wage was less than £5 per week. The Bentley was a difficult car for a driver used to the heavy, slow moving, low performance cars that were typical of the time – and that gearbox!

It is no wonder that endless debates raged through the motoring press of the 1920s about one speed or one gear cars, when more skill than the average driver possessed was needed to manipulate the gearbox quietly. To those brought up on synchromesh gearboxes and the (boring) effortless manner needed to drive the typical modern family car, first encounters with the Bentley crash box are almost invariably traumatic. As Stanley Sedgwick recalled on his drive home in his first 3 Litre, "A veil should be drawn over that drive – although it would have taken something much more substantial to cover up the noises from the gearbox." The VdP 3 Litre was also somewhat lacking in creature comforts as a 4-seater – the rear passengers are rather exposed to the elements and in saloon form (particularly with a metal roof), the impulses of the 4 cylinder engine can be quite strong. As a direct competitor the Vauxhall 30/98 with its 4½ litre engine was also somewhat quicker and the 30/98 and its predecessor the Prince Henry were very good cars. Up to the General Motors takeover in 1926, the Luton product was equal to the best sportscars made anywhere in the world.

WO's reaction to Hillstead's comments on the superior performance of the 30/98 was to give him the inlet manifold and Claudel-Hobson carburettor from his own TT car, which gave a useful increase in maximum speed and initial acceleration. However, this also rendered Hillstead's demonstrator, chassis number 5, formerly Burgess' car (before the latter took over ME 2431 as his experimental car), non standard. The Claudel-

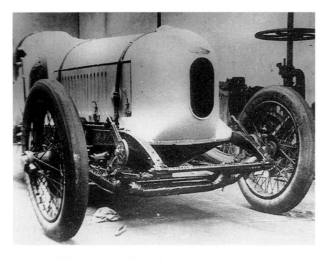

Hobson carburettor was made available as an option to those owners desirous of greater performance, but it was a temperamental instrument needing careful tuning and prone to catching fire when starting from cold. It is not known how many were fitted, but only two seem to have survived – one on Cyril Wadsworth's TT reproduction, reputedly from one of the TT race cars, and a second sup-

plied by Wadsworth to be fitted to chassis 5.

Some customers also chose to fit heavy saloon and Allweather bodies, somewhat to WO's dismay, although as his own Bentleys were almost invariably closed cars, WO's comments on the unsuitability of the four cylinder chassis for such bodies seem somewhat difficult to reconcile. (However, WO did suffer from ear problems, which explains

ABOVE LEFT **The No.1 TT car seen inside, and emerging from, the experimental shop late 1922. Note the extensive taping and streamlining to the chassis fittings. The large wheel seen to the right of the front view is for the drilling machine that Pennal was so scathing about in** *The Other Bentley Boys:* **'there was a huge drilling machine [at Cricklewood] when we got there, a great big thing that was never used. It was far too big, like a coffee grinder with a huge great wheel at the top – quite unsuitable'.**

LEFT **The single seater outside the experimental shop, in later guise with cut-down radiator and wearing the registration plates from the Indianapolis car ME4976. The body has been extensively cut down, and the handbrake reduced to a pull-up lever.**

ABOVE **Frank Clement in the single-seater in final form. No handbrake at all, with the rear drums reduced to the thickness of one pair of shoes. Heavily streamlined and lightened, and still fitted with the special front spring-hanger developed for the Indianapolis car, used on probably all the TT cars and Duff's car chassis no. 141.**

his preference for closed cars.) This state of affairs led to the introduction at the 1922 Olympia Show of the long chassis car with a 10′ 10″ wheelbase, exhaust system with twin silencers, 4.23:1 back axle and B type gearbox. The standard model engine retained the Smiths five-jet carburettor throughout its life and was fitted with lower compression pistons than the TT replica/Speed Model

chassis. The B box with its wide, low ratios was quite different from the A box, with the end caps on the layshaft secured by three studs tapped into the casing, rather than a single long stud through the hollow layshaft as on the A box. The gears were ground, so the eccentric adjustment on the layshaft was reduced from 0.020″ to 0.010″. The B box provided the basis for the BS and C boxes. It seems that the long chassis 3 Litre was only intended to be a stop-gap until a six cylinder car was ready for production and that was certainly WO's view. Plans for the 6½ Litre must have been laid down fairly early, because Varley recalled working on the three-throw camshaft drive well before he left in 1924. The patent for the three-throw drive was first applied for in December 1922, by Bentley Motors, WO, and Burgess. The specification was modified in March 1923 and accepted in March 1924.

A new piston design was developed to reduce the slap of the hour glass pistons from cold. Because of the high-thermal expansion coefficient of aluminium, when cold the pistons were a loose fit in the bores and rattled around. The small contact areas between the hour glass pattern of piston and the cylinder walls exacerbated this problem, which although acceptable on a racing or sports car, was not acceptable on a refined touring or closed car. To overcome this, Varley designed the BHB piston (Bentley-Hewitt-Burgess) which consisted of a crown with an almost separate split skirt supported from the gudgeon pin boss. This overcame the problems of piston slap, and Varley was lent to the Automotive Engineering Company of which H.S. Hewitt was a director, to develop a range of these pistons. Varley, however, always maintained that

he never received a penny for his design. Hewitt ran one of the first long chassis cars, chassis 164 registered MF 1632, which appears to have been used for a certain amount of experimental work. Certainly it was driven by Burgess around 1926, and when it's engine no. 315 came into the author's possession it was attached to an early experimental big sump with "E" part numbers on both the sump casting and the oil pump driving gear.

Hewitt was a talented engineer who had a considerable input into Bentley design work. The duralumin rocker was designed by Hewitt and Burgess, the patent number 209685 being applied for by Bentley Motors, Hewitt and Burgess in August 1923. Along with the duralumin rocker, he was involved in much of the valve gear work. Hewitt's firm, Automotive Engineering, did much of the machining work on the smaller parts (ref. "Inventory of Parts at Bentley Motors, 1926" *Bentley – Cricklewood to Crewe*, p. 291 "Automotive Eng. Co. – Running contract terminable at three months' notice for machining of front axles, pistons, valves and brake gear, etc"). Automotive Engineering did a lot of rig testing work for Bentleys and it is evident that Hewitt and Burgess worked closely together.

WO always emphasised the point that the long 3 Litre was merely a stopgap for a 6-cylinder car, a different direction from that original proposal of getting the 3 Litre into serious production and then producing a bread-and-butter car (see Appendix V). Over this latter point one has an insight into

ABOVE LEFT **Nobby Clarke (right) with a Long Standard chassis on test. This is the 1923 model, with ML CG4 magnetos, thermostat, direct drive dynamo and six bolt rocker cover. Note the double silencer and the Bentley Motors standard tubular front wing irons.**

CENTRE LEFT **More early test chassis, this time the original short chassis model (right) and a long chassis (left). Note the DN shock absorbers on the short chassis, and the fuse box mounted on the back face of the bulkhead.**

LEFT **The Long Standard 10'10" chassis with double silencer and B box was introduced for the 1923 season, the first example being chassis No. 161. This is the third example, chassis No. 164, sold to HS Hewitt of the Automotive Engineering Co. This car was used for experimental work, here with Mrs Burgess outside the Burgess' house in Ravensdale Avenue, North Finchley.**

some of the tensions and strains created inside Bentley Motors in those early years by the constant financial pressures. WO comments in his autobiography: "The policy was a simple one. We were going to make a fast car, a good car, the best in its class; and when we had begun to show a profit and had obtained our own machine shop, then we would make a smaller, cheaper car – a bread-and-butter car in fact – as well." In his *Fifty Years with Motor Cars* published two years later in 1960, Hillstead observes "It was vital for Bentley Motors to have a source of income other than that derived from its quality productions. If not a bread-and-butter motor car, then an aero-engine, or even subcontracted work from some other engineering concern. I mentioned this on several occasions, but invariably received the proverbial raspberry." In WO's *The Cars in My Life*, published in 1961, WO retorted that "There was no question at this time or later, as some people have suggested, of our making a bread-and-butter car as we had no production facilities. It would have been even more ridiculous than suggesting that Rolls-Royce or Hispano-Suiza should go into competition with Morris or Citroen. A bread-and-butter car has to be made in huge quantities and we had neither the capital nor the means to do this." It must be remembered that it was WO himself who first mooted the suggestion.

It is possible that the inclination was lacking as well. As Rolls-Royce found out later, a smaller car can only be made significantly cheaper by lowering manufacturing standards. Otherwise, virtually the only savings are in raw materials (Bentleys themselves were aware of this, as evidenced by the changes to the cylinder head for the 4 Litre to save money). It is perhaps significant that WO never involved himself throughout his career with a small, cheap car. It would have been very difficult for him to alter his perspective, from the enormously expensive cars that the firm was producing, to more mundane utility vehicles.

Although the 3 Litre is best remembered in Speed Model form, more long chassis 3 Litres were sold than any other model made between 1919 and 1931. However, there is some disagreement over the gestation of it. WO implies that it was introduced because of pressure from Sales. Hillstead denies this, stating that he never asked for it (or, for that matter, liked it) and that as far as he knew HM did not ask for it either. Despite this disparity, both in opinions and of the deviation from the original intention to produce a fast, sporting car, the long chassis car outsold the Speed Model and the TT

replica put together. The commercial success of the long chassis when expressed in those terms refutes Hillstead's suggestion that "the long chassis was a mistake that did us quite a lot of harm." The Sales staff was expanded, with Kevill-Davies ending his stint at Bentley and Bentley to take over sales of the long chassis model. The TT replica model was marketed at the same time, with raised compression, Hartford shock absorbers and one or two other minor modifications to capitalise on the success of the original TT cars.

The latter part of 1922 saw the arrival on the scene of the first of the "Bentley Boys"; John Duff. Duff was an adventurer in every sense of the word. At the outbreak of the Great War he made his way overland from China via Siberia to return to the UK and enlist; to be greeted with considerable suspicion by the Recruiting Sergeant, who suspected him of being a Russian spy. He was commissioned in the Royal Berkshire Rifles, as a Lieutenant and then Acting Captain. After the war Duff bought an

ABOVE **Chassis 141 at Brooklands for the Double 12 record, September 1922. WO on the far left, Clement partially obscured behind the bonnet, Duff at the wheel, Arthur Saunders stands at the back of the car, and Jack Besant wields the oil gun. Note the absence of dynamo and starter motor.**

RIGHT **Early 3 Litres in Lancashire, showing the preponderance of sporting coachwork in those days. FR5189 is Chassis No. 221, raced extensively by George Porter and later May Cunliffe. Porter ran the Queens & Brighton Motor Co., and was a major Bentley dealer. His fatal crash in 221 in 1924 was a blow. The flat radiator and TT body on chassis No. 221 came from ME3115 chassis No. 72, WO's TT car No.6 which was rebodied by owner Dick Edge with a Harrison 4 seater.**

ancient aero-engined 59.6 HP Fiat, even going to the lengths of a personal visit to the Turin factory to persuade them to part with spares. Duff raced this Fiat with some verve at Brooklands, until he went over the top of the banking and broke both ankles. The Fiat had previously suffered engine problems caused by the separate cylinder barrels moving, the subsequent spectacular blow-up having a humorous aftermath – Duff and his passenger stepped from the car amidst a shower of bolts and assorted mechanical bits that had settled in every fold of their clothes! Duff had already set up his own company, Sporting Cars Ltd. and it seems that Duff, who had already ordered his first Bentley, chassis 141, had been promised a franchise if he could work with someone who had good premises. Duff spoke to W.J. Adlington at Brooklands, and the latter agreed to set up Duff and Adlington Ltd. in premises already owned by Adlington at 10 Upper St. Martin's Lane in Central London.

After the demise of his Fiat, Duff took to racing Bentleys more seriously, principally chassis 141, which was prepared by Clement in the Racing Shop and fitted with engine 62 from EXP 2. As night driving was not allowed at Brooklands because of complaints from the locals, Duff attacked the 24 hour record in two 12-hour stints, driving single-handed. Number 141 was fitted with a lashed-up 4-seat body with a single bucket seat with no wings or electrical equipment, and was fitted out in much the same manner as the TT cars – the rear seat being removed and replaced by an ad-

ditional petrol tank. The first run was scheduled for 28th August and Class D records were taken at 86.24 mph for the one hour and 100 miles, and 100 kms at 137.91 kph. The run was then abandoned, but on the 29th Duff set Class D records at up to 400 kms at an average of 85mph, before dropping a valve. However Duff was back on the 21st/22nd September with a rebuilt engine and set the British double twelve record at 86.79mph. At the end of the first day on an aluminium bucket seat with virtually no upholstery, his back was so raw he had to be practically lifted from the car. During one refuelling stop Duff disappeared behind a shed for sometime, and with the stopwatches ticking on, WO sent Pennal to investigate. Poor Duff's fingers were so numb he couldn't manage and Pennal was rather shocked at WO's suggestion that they would have to fix him up with a bootlace! The car ran virtually without fault for the whole of the 24 hours, losing only six minutes at an early stage because of choke problems. It was this record attempt and the strong showing of the TT cars that really started to put Bentley Motors on the map as manufacturers of high-quality sporting cars.

It is difficult to retain perspective all these years later and remember that in 1921/22, the Bentley was a new, untried and expensive car built by a newly-formed firm with little known pedigree outside a small body of the cognoscenti who knew of the DFP successes and the achievements of the BR rotaries. Without Duff's record successes and 1923/24 Le Mans entries, which were very much

PREVIOUS PAGES AND LEFT **Bentleys kept a polished 3 Litre chassis for show purposes, the chassis which Arthur Saunders was originally employed to build. This chassis appears in various press reports around 1922–24, and was updated as the specification of the 3 litre was revised. These photos show the chassis being rebuilt to 1923 specification, with direct-drive dynamo, six bolt rocker cover, ML CG4 magnetos and thermostat. The carburettor is the early Smiths 45VS. The gearbox is the A type, with a glass lid. The double silencer system is fitted, with the cut-out between the front and rear boxes. The spare plug carrier/ dynamo cover is the early pattern mounted to the bulkhead by the bottom edge rather than the flange; only one example of this pattern is known to survive. The cone clutch is the 1923 pattern, with hardened steel rollers running on a thrust pad. The petrol tank has no reserve tap arrangement with the pipe going straight up and over the rear cross-member, and the exhaust pipe can be seen going over the back axle. These photos were all taken inside the erecting shop. The view with the pile of frames in the background shows the double doors leading into the running shop, the route taken by the completed chassis.**

private entries, it would have been more difficult to establish the Bentley name and without the subsequent Le Mans victories it would have been virtually impossible to bring the firm to the level of posing a serious threat to such a long-established and eminently successful firm as Rolls-Royce. Despite the cost of the racing, the firm would have been a much less important company without it. WO's policies of using racing for publicity and development of the production cars – a principle frequently emphasized in the sales catalogues – was vindicated by the results. It is also difficult to realize now how much importance was attached to breaking records, be they World's or merely Class records. The attempts in the early years of Bentley Motors received publicity that now seems somewhat disproportionate with respect to what was achieved.

1922 closed with three models established on the market: the long chassis, the TT replica and the original short standard. At the end of the first full year of production, some 150 cars were in customers' hands, with a good record for reliability. On the whole the Five-Year Guarantee seems to have been a fairly well judged compromise between the initial attraction to induce people to buy the car and the costs incurred in maintaining the agreement. The £5 transfer fee charged by the Works for inspecting a car before transferring the balance of the guarantee to a new owner and the income derived from conducting the work generated by such inspections must also have been a fairly lucrative source of income (it is pretty clear from the Jack Barclay sales records that the guarantee was not transferred unless any matters that were felt by Bentleys to be required were performed, and that work had to be paid for). The clause that the guarantee would be cancelled if any work on the chassis was carried out by an outside firm not approved by Bentleys must have sweetened relations with agents somewhat as well. Nevertheless, the Service Station lost on average about £10,000 per year between 1924 and 1931. New buildings erected at Cricklewood gradually filled up, but as yet there were no in-house manufacturing facilities. Exhaust pipes often had to be taken by bus to Willesden for final assembly, as they had no welding plant. Minor changes to the chassis announced at Olympia were also carried through – six bolt rocker covers without a breather and inspection plugs on the camchest. Some of the features of the very early cars, such as the speedo drive belt driven off the prop shaft, had long since been replaced by the more familiar features seen on all the later cars.

The financial position was still as hand-to-

mouth as ever and on 30th May 1922 a special resolution was passed so that the number of directors could be increased to eight. W.K. Forster was now the company secretary, a position he held until 1930. WO, HM, Keigwin, Stead, Pike and Peck were all still on the board, joined by Carl Louis Breeden in place of Boston and Stuart de la Rue of the printing firm, who was also the chairman. On top of the two mortgages for £15,000, a further £8,750 was borrowed from Burke and Treglohan on 23rd January 1923 and anyone would be concerned over the nature of the security: "The undertaking of the Company and all its property whatsoever and wheresoever both present and future including its goodwill and uncalled capital for the time being."

A more extensive racing programme was planned in 1923. Clement had been racing the Number 1 TT car (re-numbered as the Indianapolis car) with a single-seat body before this car was returned to a road-going specification and sold. Its replacement was a 2-seater based on the standard 9'9½" chassis cut to a wheelbase of 9' (the predecessor of the 100 mph 9' wheelbase production cars) and extensively lightened. The chassis was drilled out along both side-members and it was fitted with a highly-tuned engine, number 89. Clement experimented with magnesium alloy pistons, but these tended to disappear down the exhaust pipe after only one short race. This car was registered MF 330 and invariably referred to as the "Number 2 2-seater" with a chassis number of 400. Where the 400 came from is unclear, as production chassis 400 was a long chassis saloon by Chalmer and Hoyer owned by Noel van Raalte – presumably a convenient number was needed for the licensing authorities. This car proved to be fairly successful and was further modified by cutting the centre out of the radiator, which narrowed it considerably. It was this car in its dirty red paint that Dr. Dudley Benjafield, the eminent Harley Street bacteriologist, was to clap eyes on at the Service Station while discussing his latest acquisition with Bertie Kensington Moir. Moir used to deal with cars that were raced by private individuals and there seems to have been a certain duplication of functions between Service and the Experimental Department at Cricklewood which was generally thought of as the Racing Shop.

Benjafield had bought a long chassis saloon and although he was delighted with it, ribbed Moir by saying it couldn't pull the skin off a rice pudding. Bertie's reaction was to ask him if he liked to go fast, leading to an invitation to go around Brook-

ABOVE **The single seater was superseded by the No.2 two seater, seen here with Frank Clement. This car was based on a 9′9½″ frame shortened to 9′, by cutting and plating the frame. The braking arrangements are very non-standard. The radiator was cut down and narrowed, and the engine fitted with three Claudel-Hobson carburettors. This car** was raced by Frank Clement in 1923 and 1924, before being sold to Benjafield.

BELOW **The No.2 two seater at Brooklands with Browning up, painted red and with discs to the rear wheels, and still with outside petrol line and filter.**

lands in the Number 2 2-seater the following day, an offer that under the circumstances Benjy could hardly refuse. The upshot of that wild ride was an agreement to buy the car, which Benjy raced in 1924 and 1925 with some degree of success. This led directly to an invitation to drive at Le Mans in 1925. In the by now well-established Experimental/Racing Department, Clement took over chassis 7 (Hubert Pike's car) and chassis 246 and prepared these for the Boillot Cup at Boulogne, an event in which Bentleys never achieved any real success. But before that came the first of the series of races that really made the Bentley internationally famous – Le Mans.

The No.2 two seater inside the service department at Oxgate Lane, where it was looked after by Bertie Browning and Freddie Settrington under Moir's direction.

7 The Le Mans Era Opens

EARLY ON IN 1923 the Automobile Club de L'Ouest decided to run a 24-hour race for road-going touring cars on the old French Grand Prix circuit near Sarthe. All cars had to carry a hood and full weather and electrical equipment and had to meet certain regulations relating to the overall dimensions of the body. In addition, all tools and spares had to be carried in the car and all pit stops and repairs carried out by the driver himself – and the cars had to run on standard French petrol. Duff immediately decided to enter his 3 Litre, chassis 141, the 1922 Double Twelve record car. There seems initially to have been some favourable response from WO until he read the detailed regulations, at which point he seems to have decided the whole race was madness and nobody would finish. Duff was not a man to be easily deterred, and succeeded in having his car prepared by the Works and in gaining Clement's services as a co-driver, but it was made clear that it was a private entry without official Works backing. The car was prepared in the Experimental Department under Clement's supervision, Saunders and Besant also going to France with the car as the mechanics. On the eve of the race WO had second thoughts, and together with Hillstead caught the night boat to Dieppe and the train to Le Mans. By that evening, with the cars racing and with all lights on, the Grand Prix d'Endurance had got into his blood. The journey, though, by train, was a bit of a nightmare, and as WO put it "Next time we take the car."

That first race was held on the 26th/27th May. The surface was so rough that it had to be sprayed before the start in an attempt to stop it breaking up. Rudimentary acetylene lighting was supplied and operated by the French Army. A mixed bag of cars came to the start line (this was before the classic Le Mans start with the drivers lined up opposite the cars), with the Bentley the sole foreign entry. The absence of front wheel brakes was an obvious disadvantage and Duff and Clement were forced to use the escape road at Mulsanne more than once. The astonishingly rough surface, which did indeed break up rather badly, showed the need for stone-guards. Both headlamps were smashed and the petrol tank holed. Clement had to ride four miles the wrong way around the track on a borrowed bicycle with two bidons of petrol to plug the tank with a wooden bung, this repair lasting long enough to drive back to the pits. A more permanent repair was effected with cork and soap, but a lot of time was lost. The brackets for the Hartford shock absorbers broke as well, first at the back and then at the front.

Clement set the fastest lap at 66.69 mph and the duo finished equal fourth with an average of 50.05 mph. The race was sponsored by Rudge-Whitworth, the wire wheel manufacturers, who put up a Triennial Cup for results over three years based on an Index of Performance. This Index of Performance produced a minimum qualifying distance for any car based on the cubic capacity of the engine. The original idea was that competitors would qualify in the first two years by exceeding the minimum mileage for their car and then compete for the Cup in the third year. This was almost immediately dropped in favour of a Biennial Cup, but right from the very start little attention was paid to the winners of the Rudge-Whitworth Cup and far more to the perceived winners who had covered the greatest mileage. It was not, however, uncommon for the two to be the same. This strong showing by the

ABOVE **Chassis 141 at Le Mans, 1923. The pits were little more than tents. Clement at the wheel, Duff in the passenger seat, Hillstead visible between Duff and Besant, the latter leaning on the back of the car.**

RIGHT AND BELOW **Chassis no.7, ready for the Boillot Cup, in the yard at service and in the lane outside, with Moir at the wheel.**

Bentley in the first race was highly encouraging and paved the way for entries in future years.

Duff continued his Continental racing by taking in the Spanish GP at Guipuscoa, along with a second Bentley entered by Jose Carreras. Repairs had to be effected after Duff hit a low wall in practice and he was easily leading the race until near the end when a stone thrown up by another car smashed his goggles. Duff lost control and ran into a wall, causing considerable damage to the car. The organisers were so impressed that they still awarded him first place in his Class. Duff then rejoined the Bentley Team at Boulogne for the Georges Boillot Cup on 7th September, accompanying Clement in chassis 246 and Moir in chassis 7. Moir's car was fitted with twin Zenith RA48 carburettors, another of the set-ups tried at various times on the 3 Litre in order to get better performance than the 5 jet Smiths. This is not to denigrate the Smiths instrument, because the 45VS and 45BVS (the "streamlined" version with revised porting made for Bentleys) was a very good carburettor. By careful choice of the 5 needles it produced very even performance with no flat spots, unlike the Claudel-Hobson which often had terrific flat spots over certain narrow speed ranges, but the

Smiths throttle openings were too restricted for really high-speed performance.

The team's performance at Boulogne was not a success. Clement retired on the second lap with a burnt out piston: these were a new type made from magnesium alloy, known as "Miralite". Duff was lying fourth after three laps, but retired shortly afterwards, explaining that his car had been on fire five times due to the petrol pipe chafing. Duff's car had to have its front shock-absorbers removed before the race because of a broken bracket. Moir spent a lot of time at the pits fiddling with his car several times during the race, gradually working his way up from eleventh place at three laps to sixth at the finish, one hour and twenty five minutes behind the winning Hispano-Suiza of Garnier. Moir's car had hour-glass pistons, which he said "just rattled". Clement also took over the No. 2 2-seater for the short distance events, finishing second overall. The tests comprised a 3km sprint from a flying start, a 1km standing start sprint, and then a 500m hill-climb.

On the money side, things were looking somewhat healthier. Roughly one hundred and fifty chassis had been sold by the end of 1922 and some two hundred more were sold in 1923. In June 1923 the mortgage of £7,500 from October 1921 and that for £8,750 from January 1923 were paid off.

The first Bentley team to be fielded since the 1922 TT – the three Boillot Cup cars. From the left, Duff in chassis no.141, Clement in chassis no.246, and Moir in chassis no.7. WO can be seen between Nos.25 and 26, with pipe in hand.

The last outstanding loan of £7,500 from April 1921 was also paid off later in the same month, and Prideaux Brune, already a shareholder, became a director. However, this happy debt-free position lasted just three days until the 29th June 1923, when The London Life Association of 81 King William Street lent the Company no less than £40,000. The security put up against the loan would give anyone nightmares:

"1. ALL that piece of land in the Parish of Willesden in the County of Middlesex formerly part of Oxgate Farm and containing 4 acres or thereabouts bounded on the North-East by the Edgware Road on the North-West by land now or lately belonging to Joseph Phillips and on the South-East by Oxgate Lane.

"2. ALL that shop known as 3 Hanover Court in the County of London TOGETHER with the Basement occupied therewith And Together with the shop front and entrance to the shop as now fitted.

"3. ALL and SINGULAR their plant machinery

and fittings utensils of trade designs patents raw and finished materials stampings castings motor cars chassis bodies stock furniture tools jigs and all fixed and moveable chattels and effects which then were or at any time during the continuance of the security should be in or about the above mentioned properties or in the posstesion of all or any of the Company's Contractors and all the assets of the Company and all their undertaking and property whatsoever and wheresoever both present and future.''

There was no need to include the kitchen sink – it had already gone. It is not clear quite what this money was needed for – presumably simply to keep the whole operation going. Hillstead says that the money was needed to speed up production in order to meet a demand for quicker deliveries, but does not say that meant more sales. In view of the way in which Burgess' correspondence with WO suggests

EXP 3 in France in 1923, posed outside the Café Clement with AFC Hillstead (left) and Frank Clement. The car has been fitted with Lucas bell head and sidelamps, as used on many of the early production 3 Litres.

the manner in which production was set up, with axles and possibly the gearbox being delivered to Cricklewood as complete assemblies, it is likely that expansion to bring all the assembly work in house required more expenditure. Again, development work on the six cylinder engine and chassis would have needed a significant level of expenditure.

It is also possible that something rather more sinister was afoot. Hillstead devotes a whole chapter in *Those Bentley Days* to a trip to the French Grand Prix made by himself, WO and Clement in EXP 3. The Grand Prix was held on the 2nd July, and as far as can be deduced from Hillstead's account they left London roughly four days before the event – June 28th. The mortgage detailed above was dated 29th June. Hillstead takes up the story – ''One of WO's first jobs on his return to London [after the Grand Prix] was to circumvent a carefully planned coup by a certain group to secure control of the Company at the expense of the original shareholders. Apart from that hideous embarrassment, we were still woefully short of working capital.''

Frostick implies that the money was necessary

simply to pay for the expansion of the whole enterprise (as suggested above), and that they were in fact simply running to stand still, or even slipping slowly backwards. However, things are not as simple as that. Had sales of the 3 Litre expanded at an adequate rate, the cost of manufacturing would have dropped. In the early 1920s, quantity in production was still equated with reduced unit costs, and there must have been a production rate using the existing facilities that would have allowed the Company to make money. Despite increases that would have been essential to cope with higher production, indirect costs in terms of design staff, office and general administration would not have risen proportionately and at a certain sales rate the Company would have become profitable. As Ian Lloyd observed in *Rolls-Royce – The Years of Endeavour*, "There is a strong similarity between the situation of Bentley Motors and that of the Springfield Company [Rolls-Royce's American subsidiary] which proved equally incapable of solving almost identical problems. In both cases the manufacturing profit on a small turnover was too small to support overhead expenditure."

The only solution was an increased turnover and hence increased manufacturing profit with the same overheads. In the final analysis, sales were inadequate to sustain the Company. By August 1924, they had issued 120,000 shares. Remember, though, that 30,000 were issued for considerations other than cash, that some went free to Keigwin, and that of the rest the vast majority were paid up for less than the £1 nominal value. As far as is known, the 10/- call was never called in, perhaps because the Board were frightened that if they

The four wheel braking system was largely the work of Burgess, and the first installation was fitted to his car EXP 4, ME2431. These photos were taken in September 1923, when the new braking system was unveiled to the press and announced as a standard fitting on the 1924 model. Burgess demonstrates the stopping power of the new system on the rough track opposite the Works, on the other side of the Edgware Road.

called in all the unpaid calls on the shares there would be a revolt!

At Olympia improvements to the 3 Litre for the 1924 season were announced – principally the adoption of four-wheel brakes, on Perrot principles at the front. Earlier experiments with hydraulics had been abandoned, so WO and Burgess modified the Perrot methods at the front to develop the so-called "Stage 1" Perrot shaft, as described earlier. Front wheel brakes were introduced as standard on the 1924 Model. The first 1924 chassis was 352, although some numerically later chassis were delivered as 1923 models.

The long chassis continued to be offered for some time at a reduced price of £875 without front wheel brakes, presumably either in an attempt to sell off some obsolete chassis or to use up stocks of obsolete parts. Only two such chassis were sold (nos. 801 and 802), both to Nosawa & Co., of Tokyo. Stocks of the old shoes with cast iron liners were used up on the handbrake. The cast iron liners themselves were pretty effective in service, with no problems drying out the brakes if they became wet, but the slightest oil spillage reduced braking effectiveness to almost zero. The earlier unbaffled

ABOVE AND ABOVE RIGHT **The special paint brush rig fitted to EXP 4 for measuring stopping distances. The Perrot shaft is basically the Stage 1 pattern, with superficial differences in the frame bracket. The axle is pretty much the production pattern beam used with four wheel brakes, with a considerably stronger beam section than the unbraked axle. The radiator trunnion is presumably an experimental part, as it is quite different from the production pattern.**

RIGHT **EXP 4 with four wheel brakes fitted, after tests on an oil and water covered surface.**

banjos were prone to leakage through the Timken caps, because it was difficult to maintain the correct $0.002'' - 0.004''$ radial clearance between the banjo casing and the acme thread oil flinger on the halfshaft. Wear in the Timken races can also allow the car to drop slightly, enough to allow the oil flingers to touch. The nearside of the car is the more prone to oiled-up linings due to road camber. All the brakes were lined with Ferodo, the lining area of the rear brake shoes also being increased. The original cast iron liners had been cut from solid

rings and covered less of the circumference of the shoes, so the castings for the shoes were altered.

In the quest for silence the camshaft bevel gears were changed from straight-cut to spiral bevel, with unfortunate consequences. Although these gears never experience true load reversal in the same manner as a back axle, they nevertheless have a tendency to climb into mesh on the overrun and for teeth to break off. This of course does not happen with straight-cut gears, and the only time spiral bevel gears were used in the back axle of one of the

Team cars, they failed. A whole string of customers brought their cars back to Service complaining of noise from the overhead gear and great efforts were made to conceal the problem. WO was somewhat elusive about the cause, "It's not the gears of course, it's that aluminium piece THEY put underneath them!" One of the directors walking around at Service picked up a damaged gear, to be told in full ignorance by his guide it was a back axle gear – a marvellous piece of unintended bluff.

This problem seems to have been endemic between about January and June 1924, during which period many of the 16/32 bevel sets were changed for the later 14/28 pattern. Typical examples were chassis 301: "25/1/24 Bevel gear on camshaft had two broken teeth. New top vertical shaft bevel gear fitted." Chassis 635: "21/7/24 Bevels stripped. New top bevels fitted. .004 camshaft [BM2391] complete fitted. Note – camshaft bent owing to bevel stripping." Matthews, who together with Freddie Settrington made up the Progress Department at Service under Bertie Moir, finally lost patience with Dr. Potts, whose 3 Litre chassis 623 went through two sets of bevel gears in the first week of July, 1925. Matthews managed to des-

Bentley leading lights – from the left RS Witchell, works manager, WO, and FT Burgess, chief designer, photographed outside the Works with a 1924 Model 3 Litre Speed Model chassis.

patch the good Doctor to Cricklewood to see the new 6½ Litre, leaving the following poem on Freddie Settrington's desk:

Oh! Doctor Potts
I wonder what's
The matter with your motor.
We must conceal these bits of steel from off your bevel rotor.
Now how would you enjoy a view of our six-chimney Bentley?
While you're away
We then can play
With your top bevel – Gently.

The ultimate solution to this problem was the located type of top bevel in which the gear is restrained from climbing in and out of mesh by a pair of angular contact thrust races assembled back to back, but this did not appear for some considerable time. It is suspected that the cause of failure was that in an attempt to further reduce the gear noise, the gears were consistently meshed too tight at the Works. That valve gear noise was regarded as a problem is shown by a not inconsiderable number of rocker covers that were drilled to fit lead sheet to the undersides of them.

There were problems with back axle gears as well, though. Bentleys complained that ENV were cutting the gears incorrectly before it was realised that there was more gear noise with more people in

ABOVE **And this is what they would have seen under the bonnet of an early 1924 Speed Model – the five-jet Smiths 45 BVS carburettor, here on the polished chassis. The photos of the polished chassis show well the difference in manifolding between the two carburettors, and the four bolt mounting of the later unit. This picture also shows the dynamo drive with fabric Hardy joints.**

LEFT **The Icarus mascot, designed by Gordon Crosby, that appeared in a number of 3 Litre catalogues in 1923/24. As far as is known, very few were made.**

the car. The cause was distortion of the differential casing itself, which could only be solved by putting more metal in it. The thickness of the aluminium nosepiece casting was increased slightly, and the steel banjo had baffles welded in, which by the end of $4\frac{1}{2}$ Litre production had become so extensive that the banjo was virtually double skinned. ENV were able to compensate for these distortion problems by building in compensating factors to the gear cutting process.

All the earlier gears, for the back axle, engine and gearbox, were machined from BND, a steel supplied by Firths which was immensely tough. BND gears tend to suffer from pitting and flaking on the pitch line long before failure, but continue to run for a very long time, albeit noisily. In later years gears were cut from 5% CHMS (case-hardening mild steel), a material which Nobby Clarke distrusted, so all the Team cars were run on BND without any recorded failures due to the gears themselves.

The TT replica model was superseded by the immortal Speed Model – 9'9½" chassis, red enamel to the radiator and petrol tank badges, high compression engine with A box and 3:78 to 1 back axle – the epitome of the 3 Litre. Hillstead suggests that the enamel of the badges was changed from blue to red because of the red colour schemes that he had

WO and Moir in 'The Sun'.

used on his demonstration cars. Twin SU G5 "sloper" carburettors were specified for the first time, early problems with these instruments due to erratic manufacturing standards having been overcome. However, Speed Models were fitted with the Smiths 45 BVS carburettor until mid-1924 after which slopers were invariably fitted. Virtually all the earlier Speed Models were then converted to SUs. Mention "3 Litre" to Bentley enthusiasts and they will invariably think of a Speed Model with Vanden Plas sports 4-seater body.

The Company were also showing a great deal more interest in coachwork. The classic Vanden Plas four-seater was by now well-established and Bentleys offered in their catalogues a range of more or less standard bodies covering the full range from 2-seaters on the Speed Model chassis, up to six-light saloons on the standard chassis. Weymann methods of body-building were now common, and fabric-covering increasingly so as well.

The Weymann method was devised in France by M Weymann, and the company had a subsidiary at Addlestone in the UK. The framework was built up in wood in the usual way, but none of the wooden pieces physically touched – all the joints were made by metal plates. This produced a light, flexible body without any of the rattles and squeaks associated with more conventional practice. Bodies also had to be flexibly mounted to incorporate movement. The finished frame was then panelled in steel or aluminium or fabric covered in rexine. For fabric bodies the curved surfaces had to be panelled first and wadding was used to smooth out the surface. It should be noted that "Weymann" refers to the method of construction of the bodyframe only and not to the surface finish – it is quite wrong to assume that all fabric-covered cars are Weymann construction. Weymann licensed other coachbuilders to manufacture under his patents and for a fee of £10 they could affix a "genuine Weymann" plate to the body. The standard touring body on the long chassis was built by Freestone & Webb, Gurney Nutting, Vanden Plas and Vickers, and are all virtually indistinguishable. The first body to be built to this pattern by Vanden Plas was on chassis 619, registered VS 911 and the VdP body book records that it was built to a design copied at the Bentley Works on 3rd April 1924. HM was particularly interested in coachbuilding and by 1929 was a director of British Flexible Coachwork Ltd., based at 1, Stanhope St., Euston Rd., with fellow directors Arman Woods and Eric A. Woods.

Early in 1924, experimental work on the new six-cylinder chassis came to fruition. WO quoted an engine of 140 mm stroke by 83 or 84 mm bore, giving a capacity of 4545cc or 4655cc respectively. This engine was built up by Plumley and installed in a chassis of about 12' wheelbase. The front end of this chassis was very similar to the 3 Litre, with a cross-tube supporting a fixed starting handle, and single Hartford shock-absorbers. The whole was then clothed with a Freestone & Webb six-light fabric saloon with a disguised radiator, a sort of hexagonal monstrosity. Unfortunately, this body effectively concealed every detail of the mechanical specification of the car. It was then registered MF 7584 and christened "The Sun". It was this car that was shipped to France by WO and Hillstead to

At the back of the pits – 'The Sun', the prototype 6½ Litre, with tarpaulin over the bonnet to deter curious passers-by.

see the 1924 Le Mans race, accompanied by Moir and Witchell. Kensington Moir was transferred from Service to Experimental specifically to work on the six-cylinder prototype, leaving Pike in sole charge of the Service Department. The French customs officials were by now used to the car "Bentley" but were rather uncertain about this car – as it was not a "Bentley!" Eventually they were passed through but not before a rather disgruntled WO had to pay duty on all the English cigarettes he had brought with him.

Hillstead published three photos of "The Sun", in *Those Bentley Days* and *Fifty Years with Motor Cars*, but unfortunately none of these reveal any details of the car – indeed, one shows it with a sheet over the front to protect it from prying eyes. WO, Hillstead, Michaelis (from Hoopers), Witchell and Moir took the same car to the French Grand Prix a couple of weeks after Le Mans. At Tours the previous year Segrave's Sunbeam had proved itself the first successful British Grand Prix car, but in 1924 the Sunbeams retired and Campari won in a P2 Alfa-Romeo. However, one problem with the Bentley, or rather with the tyres, manifested itself. Low pressure or "balloon" well-base tyres had just been introduced and a set had been fitted and spares

'The Sun' in front of the pits, showing the registration MF7584. Note the same fixed starting handle arrangement as the 3 Litre, and the low pressure balloon tyres, which were not very satisfactory. WO with pipe in overcoat, Moir in jacket and cap. The third gentleman is unidentified, but is probably Witchell.

ABOVE **WO and Witchell in playful mood at Le Mans, 1924; Hillstead commented 'A very rare occurrence on the part of the former'.**

CENTRE **The return journey, after the encounter with the Rolls-Royce Phantom prototype. WO, Witchell and Moir at Dieppe.**

BELOW RIGHT **'The Sun' under wraps before boarding at Dieppe.**

BELOW **Hillstead's caption: 'WO and Moir in mid-channel. Moir's attitude speaks for itself'.**

carried. However, tyre life proved to be alarmingly short although the ride was considerably more comfortable, with the Dunlop tyres running at a pressure of 16 lbs on a 2½ ton car. They were on the last set of covers, when approaching a Y junction at speed on the way back to Dieppe, another strange car was seen making for the same junction on the other leg of the Y. As the two cars drew level with the occupants staring at each other, it became clear that each realised the other's identity – the prototype New Phantom Rolls-Royce and the prototype Six Cylinder Bentley. These two continued neck and neck, each a match for the other until one of the Rolls crew's cap blew off and WO was able to ease off to conserve the last set of covers.

Back at Cricklewood a post-mortem was held, the decision being taken at a meeting between WO, Burgess and Moir, to increase the capacity substantially, as it was almost certain Rolls-Royce would do the same thing. An increase to 100x140 mm or 6597 cc gave a useful increase in power and torque, with an RAC rating of 37.2 bhp, and it is interesting to speculate whether it is any coincidence that Hispano-Suiza used exactly the same dimensions in their 6597 cc H6! WO was unhappy with the power of the 4½ Litre engine anyway, which was hardly suprising considering that the new car had only 1½ times the capacity of the 3 Litre, for twice the weight – the Speed Model tourer weighing roughly 26 cwt compared to "The Sun's" 51 cwt. It was to be quite a protracted gestation, as no official photographs of the 6½ Litre were published until June 1925.

June of 1924 was a momentous date. 41 cars came to the start line for the second 24 Heures du Mans, featuring one foreign entry – Duff, Clement and 3 Litre Bentley. This year they were driving a new car, chassis number 582 registered XT 1606, and preparation work had been carried out with greater collaboration from the Works than 1923, even if the entry was still a private one.

For months in advance, details of the car were worked over – as Pennal recalled, Duff was full of good ideas, most of which were incorporated. First off, after the experiences of the previous year, were stoneguards for the headlamps and protection for the petrol tank, consisting of wood laths strapped around the tank by wire mesh and belting. Duff wanted a spare oil can fitted under the scuttle. This was made up with a pipe into the side of the oil filler neck, so that the can could be emptied into the sump by the driver while the car was on the move. This idea was developed on later Works cars into a sophisticated system with an automatic overflow

tap in the sump to prevent over-filling during pit stops. The team began to pay the attention to pit drill that was to become responsible in no small part for the Bentley Team's future successes.

The race organization was more professional this year, and it seems that it was indeed 1924 when the Bentley was caught by the scrutineer over the width of the wings, necessitated the mounting of a strip barely more than 1/16″ all along both sides! There is a photo of the three Team 4½ Litres in 1928 showing Nobby Clarke standing just behind the scrutineer, Nobby's caption reading that he was keeping a good eye on the old basket. Scrutineering was performed in a large hall, which the cars drove around, being checked at different places for compliance with different aspects of the rules. One of these checks involved running the engine with the exhaust over a tray of sand to see whether it kicked up sand or not – doubtless there was some reason. Afterwards the cars were driven out and photographed, and these instantly recognisable post scrutineering shots invariably showing car, driver, surrounding buildings and crowds of locals seem to have survived for every year except 1923.

The night before the race Pennal was still working late into the night on his own when WO wandered in and deeply upset Pennal by questioning the wisdom of the entry. On the day, however, there were no such qualms as the Bentley gradually worked its way up the field, delayed only by a coachbuilder's staple falling into the gear lever gate and fouling it, until an inverted John Duff fished it out with a piece of wire. The staple had punched itself into the aluminium, making selection of third gear difficult. From then on a large hole was drilled in the aluminium outrigger casting below the gate so that any future foreign bodies of that nature would fall through and WO instructed that all the Team cars should use bolted clips not staples. The Hartford shock absorbers broke up; for some reason those fitted to the Bentley were a special set made up in duralumin. Duff later commented that had he realised the terrible state of the road between the Mulsanne corner and the grandstand, he would have used the more normal steel ones.

One of the very fast Peugeots was disqualified for the most petty of reasons – one of the windscreen support brackets fractured, so the driver tore it off and threw it in the back. This contravened the race regulations, and the subsequent disqualification annoyed the partisan French crowd. The regulations were very complicated and the Bentley almost came unstuck as a result of them as well. Not only was refuelling and refilling with oil and

The Bentley at the pits before the 1924 Le Mans race. Frank Clement on the far left, then John Duff, and probably Leslie Pennal with him under the bonnet of the car, XT1606 chassis No.582.

water only allowed at 20-lap intervals, but also a set average speed had to be maintained for the same period. With 1½ hours still to run, Duff came into the pits to change wheels. The fronts were no problem, but the rears were jammed solid and a lot of time was wasted pulling them off. The motoring press reported ''swollen hubs'' – but years later Adlington (Duff's partner in Duff & Adlington) said that when they eventually pulled the wheels off there was a lot of swarf as well from someone filing the hub splines – and there was considerable suspicion that the car had been sabotaged while in France.

Nobby Clarke was rather more specific about the nature of the problem, writing in 1974: ''In regard to the swollen hub on a Bentley. The late Arthur Saunders confided in me, when I took over from him in 1926, that the ''swollen hub'' he experienced on a Rudge wheel was due to some evil-minded foreigner who hated Englishmen and their motorcars, driving a jet reamer, a triangular-sectioned hardened steel needle-shaped instrument used in the old days for reaming out carburettor jets, about the size of a darning needle, into the slackness between the male and female splines of the hub. When one came to withdraw the wheel, the hardened needle-like reamer broke off and acted like a sprag, tearing the splines as the wheel was withdrawn by a puller hence the need for, and institution of, a day and night watch on the cars.''

Clement drove flat out for the rest of the race and

covered another 90 miles to no avail, because the average for the period had dropped and they were not counted. Duff and Clement still won at a lower average than the previous year, by some 10 miles, having covered 1290 miles – but in truth the figure was nearer 1380 miles. The subsequent celebrations should have been ecstatic, but in truth were more funereal, as WO was an intensely shy man at the best of times and after race victories seemed to withdraw even further into himself. The celebratory dinner did not go with a swing.

Back at Cricklewood, the success at Le Mans was used as ammunition by WO in a constant struggle to convince the Board that money spent on motor racing was money well spent. However, WO expressed a degree of regret over the comparative ease with which they succeeded in the early years, as it possibly led to complacency and a feeling that the race was easy. WO was also filling the correspondence columns of the motoring press in a debate that reached considerable (and disproportionate) heights concerning the origins of the 3 Litre design. This stemmed from a letter written by WO criticising Louis Coatalen's (of Sunbeams) belief that racing improved motor cars and hence touring cars. Coatalen's reply was to the effect that the Bentley was basically a 1914 racing car with mudguards, or to be more exact, a linear descendant of the 1914 TT Sunbeam. WO denied that racing and the production of special racing cars had any benefit to ordinary touring cars, which seems a strange stance for WO to take in view of his belief in racing as a way of enhancing the qualities of the road-going cars, unless WO was more interested in upsetting Coatelen and prolonging the debate for reasons of publicity! After all, in the ''Resume of Policy'' put out in 1930 after Bentley's retirement from international racing (see Appendix II) one paragraph reads: ''And now for the reasons for this extensive racing programme...Of paramount importance are the lessons learned from racing...knowledge which is being applied to the production of ever more perfect cars''. This correspondence raged to and fro among Sunbeam and Bentley devotees, but with little effect except possibly in the introduction of the 3 Litre Sunbeam in 1925, aimed directly at the 3 Litre Bentley.

There can be little doubt but that the 3 Litre Bentley was indeed a direct descendant of the pre-Great War racing cars, with a high efficiency and relatively low capacity, high revving power unit. This was clearly implied in the 1919 brochures and in the subsequent reports in the motoring press, and with Burgess as Head Designer previously responsible for the 1914 TT Humbers and Clive Gallop primarily employed because of his Peugeot experience to work on the valve gear, WO's intentions are made clear. Sunbeam's (and Coatalen's) reaction was to put their six cylinder, twin chain driven overhead camshaft engine, similar to the 1923 Grand Prix engine into a rather unsuitable touring chassis, with cantilever rear springs and a remarkably long wheelbase of 10'9¾". This was probably a result of WO's attempt to rile Coatalen by pointing out that a twin-cam racing engine in a short wheelbase chassis with semi-elliptic springs did not correspond with Sunbeam's production cars. Coatalen obviously had ambitions on Le Mans and two cars were entered for 1924 driven by Lee Guinness and Resta. However, these did not materialise. Two cars were entered for 1925 and had they won Coatalen would have had the last laugh on WO. In the event, of the two cars entered, Davis and Chassagne's finished second. However, the race was won by the 3½ litre La Lorraine of de Courcelles and Rossignol – fitted with pushrod operated overhead valves. A 3 Litre Sunbeam also won the 1927 Six Hours race driven by George Duller, but the model never took off and few were built. Its road performance was not significantly better than the 3 Litre Bentley, and the 4½ Litre certainly was a faster and more refined touring car.

In September 1924 Bentley Motors put out a statement to the press to the effect that a number of chassis – 13 as far as is known – which had been damaged in the fire at Gurney Nutting were being "worked over" by unscrupulous members of the motor trade and offered to the public. Potential purchasers were advised to contact Bentleys at Hanover Court to check the chassis numbers of cars before buying. The fire at Gurney Nutting's Oval Road factory in Croydon completely destroyed the works and they moved to Elystan Place in Chelsea. Bentley's warnings to potential purchasers are somewhat ironic in view of the way they themselves handled the matter. Whether the chassis were insured or not is unknown, but whatever the situation in their usual critical position over money, Bentleys could hardly afford to lose anything. Hence it is not surprising to find bits of those 13 cars cropping up later! Of those, 5 chassis returned:-

– 519 was renumbered 517 by an outside firm and fitted with a saloon body, never to be heard of again, so presumably Bentleys had no hand in that one.
– 541's engine 547 was sold, later to be owned by

Model Farms and Dairies Ltd in Dorset in 1934. One can only speculate on what they used it for.
– 550 was sold to a Mr Seward with a Shortts body and a 1927/28 registration number. Bentleys refused to work on the engine of this car in 1930.
– 552's engine 558 appeared many years later bearing the number 9A on the crankcase, the engine number of a Bentley Motors lorry.
– 583 was rebuilt as a Works Breakdown, later sold to McKenzie's Garage.

Apparently, a squad of breakers was despatched to ensure that the less affected parts would not fall into the wrong hands. Depite this, a 3 Litre turned up at Service with no chassis or engine numbers, having been sold to Sir Thomas Beecham as a "real" Bentley by a notorious firm operating from Great Portland Street, despite the fact that the car was fitted with a number of non-Bentley parts. There were suspicions that parts had "walked" from inside Bentley Motors, but nothing was ever proved.

Doubtless other salvageable parts found their way into some of the production cars. The Works had two 3 Litre based "hacks", known as SCRAP 1 and 2. SCRAP 1 had a 4-seater sports body split from end to end and was built out of discarded bits and pieces, and was used for everything from transporting spares to taking the secretary down to the bank on Friday. This car is believed to have ended up wrapped round a lamp-post in Hendon one night, so the wreckage was taken home and SCRAP 2 emerged. Just before the General Strike in 1926 a proper box-type lorry body was fitted, and was used by Freddie Settrington to move about ½ ton of the *Daily Mail* from Lympne to London. Just after the war, a Mr C.B. Cobley joined the Bentley Drivers' Club with a 3 Litre, the details of which were given as chassis 2A, engine 9A, a 1924 10'10" chassis with a lorry body. The registration number of the car was MH 1757, a late 1924 number issued in Middlesex, the same county as that in which the Bentleys registered the 1922 TT cars, and the 1925 and 1926 Le Mans cars. The crankcase 9A also bore the number 558 from chassis 552, supposedly burnt out at Gurney Nuttings in 1923, so it seems likely that 2A was SCRAP 2 built out of fire salvage. As late as November 1926 Bentleys were advising customers to check chassis numbers of Bentleys offered for sale due to "...certain cars purported to be Bentleys...advertised for sale. These particular cars are actually partly or wholly manufactured from scrap obtained from fire salvage."

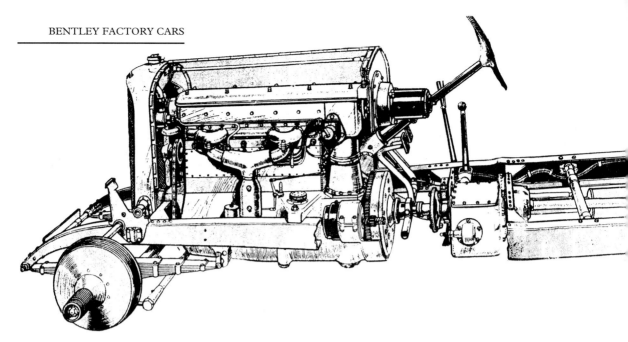

Technical – The 6½ Litre

The 6½ Litre represents the second generation of Bentley design philosophy and is a significant step on from the 3 Litre, but in a rather different direction. While the 6½ Litre was still a sporting car, it was not a sports car in the spirit of the 3 Litre Speed Model. Apart from any other considerations, the 6½ Litre engine was vastly quieter and smoother than the 3 Litre, lacking any sign of the vertical drive periods so often found in the 4 cylinder engines between about 1800 and 2300 RPM. The 8 Litre was a development of the 6½ Litre, albeit representing a greater advance (in terms of the whole chassis, not just the engine) than that between the 3 Litre and the 4½ Litre. The first prototype was on the road by May of 1924. The engine owed much to the overall layout of the 3 Litre, but the shaft drive for the overhead camshaft was superseded by the now well-known three-throw drive, owing much to locomotive practice, but also used by Parry Thomas on the Leyland-Thomas.

Overall dimensions of 83 or 84 mm bore x 140 mm stroke gave a capacity of roughly 4½ Litres, the shortening of the stroke giving more clearance between the con-rods and the crankcase so that the diameter of the crank journals could be increased to 55 mm from the 45 mm of the early 3 Litre crank. The clearance envelope inside the crankcase for the swing of the connecting rod seems to have been excessively tight, and in order to increase the diameter of the 3 Litre crankshaft journals from 45 to 47 mm it was necessary to go from a shell-type rod to a direct-metalled rod, as the deletion of the shell enables the big end of the con-rod to be made smaller for the same weight and strength. At the front of the engine was a fixed starting handle

assembly as fitted to the small sump 3 Litre, but this feature was only used on the first two experimental chassis. All the production chassis used the detachable starting handle arrangement, with the cross-tube deleted.

Twin ML ER6 magnetos were driven by a cross shaft gear positioned in front of the three-throw drive at the back of the engine. The three-throw itself consisted of a steel spur gear on the crankshaft, driving a celeron fibre wheel mounted on the end of a short crankshaft with three-throws. On this crankshaft ran three pairs of sheaves with duraluminium bearings (the two three-throw crankshafts were nitrided), driving a similar three-throw crankshaft mounted on the end of the camshaft. Differential thermal expansion between the cast iron cylinder block and the steel rods was compensated for by spring-loading the sheaves.

The initial design for the drive was covered by patent 212992, referring to a camshaft drive by multiple connecting rods, easily adjustable and damped for compensation movement of centres by springs or fluid dash-pots. The adjustment was by means of nuts on threaded rods, the connecting rods consisting of a pair of rods threaded at each end with the spring-loaded sheaves mounted on them. WO and Burgess presumably considered the option of fluid damping of the sheaves, but went for the simpler and cheaper option of using springs. It is astonishing to find that this patent was applied for as early as December 1922. It was completed by July 1923 and accepted in March 1924.

However, in service these springs proved unreliable and were replaced by packs of washers. The movements of these springs were at one point analysed using a stroboscope. This revised arrangement with the washers was patented, number

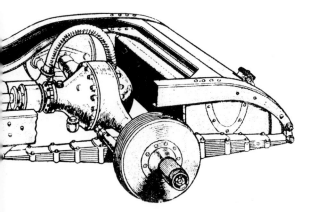

LEFT **The 6½ Litre chassis, *c.* 1925. Notice the cast aluminium headlamp stands/Perrot bearings/wing iron/radiator trunnion mounting. The carburettor is the Type 50 Smith-Bentley. The dynamo is mounted on the bulkhead, as fitted to 1925, 1926 and 1927 Models. The plate clutch can be seen, and the B type gearbox. Drive is via a pot-joint propshaft to the 6½ Litre pattern diff. The petrol tank is the early 19 gallon variety. Note the long, flat springs with rebound leaves.**

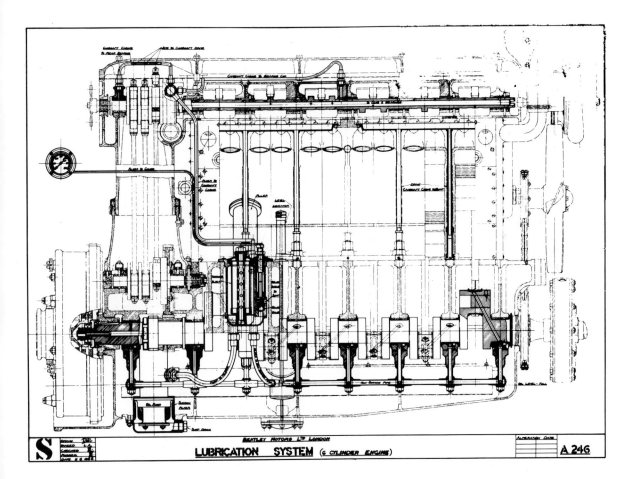

ABOVE **This lubrication chart for the 6½ Litre, drawing A246, shows clearly the layout of the engine of internals. It also helps to explain why the cars were so tall, because with the long stroke and overhead valve gear there is very little wasted space in the vertical plane. Note the dynamo drive coupling on the back of the camshaft. Project code S for six cylinder chassis, drawn by Fred Dewhurst and passed by Burgess.**

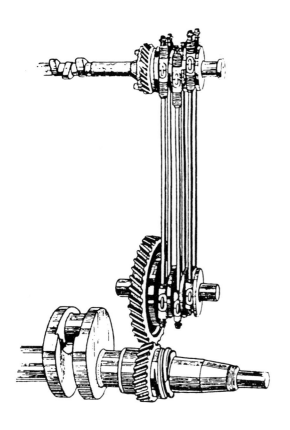

270617, by Bentley Motors, WO and F.T. Burgess, in March 1927. There has been some confusion over whether the washers should be flat or not and many years later WO said that the washers were flat as far as he could remember and any dishing was probably accidental. The patent specification, though, specifically refers to stamping the washers from steel that is either not flat or corrugated, or bending them after cutting, the effect being enhanced by the presence of oil. The camshaft drove a dynamo mounted on the bulkhead as on the 3 Litre and similarly providing some damping to the valve gear.

LEFT **General arrangement of the three-throw drive. The intermediate (larger) gear was made of fibre, Celeron, for quietness. The gear on the camshaft in front of the three-throw is for the magneto cross-shaft. The flywheel mounts on the taper rear section of the crankshaft.**

BELOW **Top of the three-throw drive, c. 1926, showing the springs, twin ML CG6 magnetos and dynamo drive coupling.**

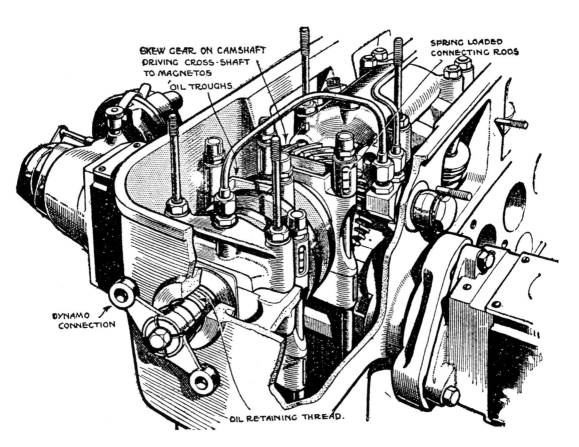

SKEW GEAR ON CAMSHAFT DRIVING CROSS-SHAFT TO MAGNETOS

OIL TROUGHS.

SPRING LOADED CONNECTING RODS

DYNAMO CONNECTION

OIL RETAINING THREAD.

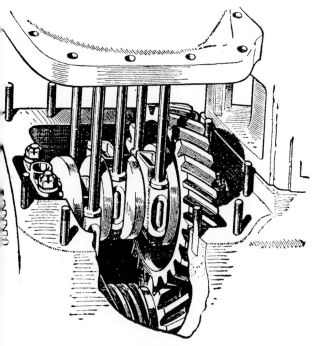

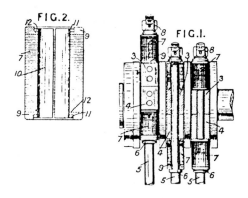

ABOVE **Additional patent specification for the three-throw drive. The patent mentioned dished or corrugated washers, superseding the springs specified earlier.**

LEFT **Bottom portion of the three-throw drive, with the tunnel lifted. The reduction gears can be clearly seen. The gear behind the crankshaft gear drives the oil pump.**

The drive was taken by a completely new plate clutch to a BS type gearbox. Whether the Experimental 6½ Litres used a B box or a BS box is not known, but the BS box was specifically designed for the 6½ Litre and was basically a B box with higher indirect gears, principally third, which was the highest ratio third used on any vintage Bentley box. The clutch on the first experimental car was of the cone type running in oil, a design that was not pursued. One of the problems with the Bentley design of cone clutch is that the spring has to be made stronger, the higher the power output of the engine, to prevent clutch slip. The square section spring used in the early 4½ Litre fitted with cone clutches are often very heavy indeed, making the cars hard work to drive in traffic. The same does not apply to the plate clutch, which uses multiplying levers to enable the use of stronger springs to transmit the power while keeping pedal pressures down. From the gearbox the drive was taken via a pot-joint propshaft to a completely new differential unit. This differential used a double row thrust race to take sideways loads from the crownwheel and was generally more massive and more rigid than that used in the four cylinder cars, with a steel

RIGHT **Steering column mounting on the 6½ Litre chassis. Note the long length of unsupported column above the top mounting. The 8 Litre column top bearings and bulkhead support mounting were positioned much further up the column.**

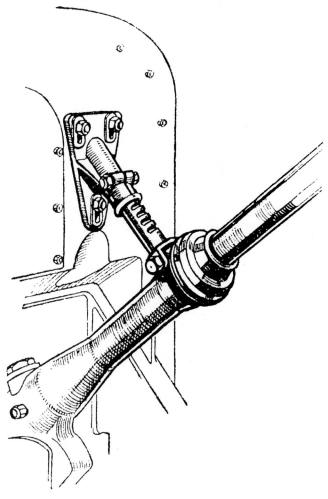

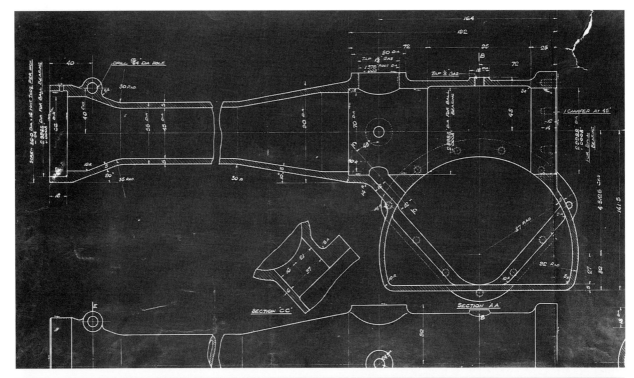

The works drawing of the 6½ Litre steering box, part no. BM5519. Drawn by C.S. Sewell, March 1925, passed by F.T. Burgess.

bridge across the back of the assembly. The brakes on the front initially pulled on via rods (later changed to push-on via tubes), and the Stage 3 pattern Perrot shafts were used running in a one-piece headlamp stand, Perrot bearing, radiator support, and wing stay bracket mounting aluminium casting. Suspension was by semi-elliptic leaf springs backed up by friction shock-absorbers, with rebound leaves fitted above the master leaf – three of these being fitted at the front and four at the back.

Weslake was horrified at WO's achievement in producing only 140 bhp from such a large engine and is on record as saying that the breathing would have been improved merely by reversing the block and using the exhaust ports for the inlets and vice-versa. Rolls-Royce on the other hand, were very complimentary about the high efficiency of the engine! The 7½ litres of the Phantom I, as the New Phantom came to be called, produced only 109 bhp, but there is a fundamental difference in phil-

osophy between Weslake on the one hand and WO and Royce on the other. Weslake was interested in performance, while the others were interested in silence and low speed torque – the ability to pull cleanly in top gear right down to walking speeds (the proverbial single-speed car). In testing a Rolls-Royce to find why it was so quiet, Bentleys found that the valve gear response was practically flat, i.e. there was no significant amount of dwell so the valve was virtually always on the move. Valve gear noise is not so much related to the size of the clearances, but the velocity with which the tappet screw hits the valve. Hence it can be seen that a camshaft with gentle curves will produce less noise from the valve gear – but will suffer correspondingly in terms of power output. It also makes a nonsense of the old concept of reducing the tappet clearances for less noise.

The 6½ Litre engine was effectively strangled by the tortuous manifolding between the single Smiths type 50 BVS carburettor and the inlet ports of the multiport block used on the early cars. Incidentally, several experimental manifolds were made up, but the first was found to be the most satisfactory. The Smiths 5 jet carburettor could be tuned by changing the needles to give a smooth performance throughout the speed range of the engine without flat spots, and it was this smoothness, at the expense of power, which was not really needed anyway, that was required for a town carriage. The SU carburettor was well-known for comparatively poor low-speed performance, which is probably why the SU carburettor was not fitted to the standard 6½ Litre chassis. It is the positioning of the block itself that is more of a mystery.

The reason for reversing the inlet and exhaust sides of the 6½ Litre compared to all the other Cricklewood Bentleys has never been adequately explained. WO says that the manifolding was reversed because the Smith carburettor hung so low in the chassis that it fouled the steering column. Examination of a 6½ Litre chassis shows that this explanation does not make sense. The steering column is mounted on the chassis behind the rear engine mounting and the carburettor hangs down in front of it. Therefore, with the 6½ Litre pattern of steering column, it is impossible for the carburettor to foul. The obvious hypothesis is that the first chassis used a 3 Litre pattern of steering column, which is mounted on the chassis well forward of the rear engine mounting and would indeed make it impossible to fit the Smiths carburettor. This theory is confirmed by the drawings, which show that the 6½ Litre steering column was not designed until mid-1925. Hence they must have used the 3 Litre pattern column on "The Sun". Unfortunately, no offside photos of that car in the period 1924/25 exist to conclusively prove this. By June 1925, when Chas Bowers photographed "The Sun" with its proper Bentley radiator, the usual 6½ Litre pattern column had been fitted.

The design of the 3 Litre steering box is limited in the ratio that can be obtained, because the box is designed to take a full wormwheel, and in the 4½ Litre and late 3 Litre, with the introduction of 5.25 x 21″ wheels with well-base Dunlop tyres in place of the earlier 820 x 120 beaded edge wheels and tyres, the ratio was dropped from 6:1 to 10.3:1, by increasing the diameter of the wormwheel and making the worm with three starts instead of one. With 6.00 x 21″ well base wheels on the 6½ Litres, a very considerable drop in steering ratio was needed. This was achieved by completely redesigning the box to have a segment only of the wormwheel, and provision for adjusting the mesh of the gears was provided by means of an eccentric bush (the cost of manufacture of which must have been pretty high). By this means the steering ratio was

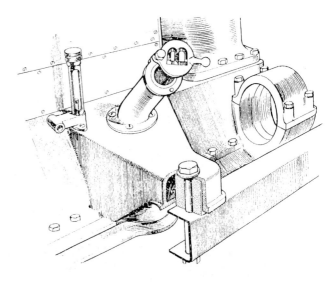

Detail of the rubber engine mounting, as accurately reported in the motoring press.

dropped to a far more satisfactory 12:1. This business of using an eccentrically-machined bush for adjusting the mesh between the steering worm and wheel was a feature of the DFP. The manifolding was reversed again on the 8 Litre. Positioning the carburettor on the nearside greatly complicated the linkages for the hand throttle, mixture control and accelerator, and the first set of controls mounted on the back of the block was made up by Arthur Saunders and then drawn up for production.

Arthur was a real old-school mechanic of the pre-Great War era, and had been largely responsible for preparing the 1919 Show chassis using all the old tricks of colouring steel parts, using various mixtures to dip them in and then heat-treating. He was also the foreman in charge of preparing the TT cars, and was responsible for hand-forming the tubular cross-member for the first 6½ Litre chassis. At that time there was no machine that could produce the required shape without a slight flattening of the tube, but as it took three men a whole evening to make one cross-member it was hardly an economical method of manufacturing! This slight flattening can be seen on all the production chassis. It would have been better if the dip had been made under the prop shaft rather than over it, as it would have made the coachbuilder's job easier when fitting the floor. Many other details of that car differed from the first production cars, but development continued on "The Sun", now fitted with a conventional radiator, and the Number 2 Experimental 6½ Litre fitted with an open touring body up to the 1925 Olympia Show, when the chassis was first shown to the public.

8 The Problems Close In

FROM A COMMERCIAL viewpoint, 1924 was in one important aspect at least, the most successful in Bentley Motors' history. More cars were manufactured (and sold) than in any other year. The implications of this are that the factory had the capacity to produce more cars in later years had it been possible to sell them, which presumably was not the case. Gradual improvements were made to the 3 Litre during its production life, but in general, 1924 seems to have been a fairly quiet year in which about 400 chassis were built and delivered, some at least on the back of Duff and Clement's Le Mans victory (the figures quoted by Ian Lloyd show that 322 chassis were sold, but these figures do not appear to tally with the number of chassis known to have been built). Announcements made at the Olympia Show in October 1924 emphasized that the 3 Litre was to remain virtually unchanged for 1925, with chassis prices of £895 for the standard chassis, and £925 for the Speed Model. A Vanden Plas bodied Speed Model complete sold for £1125 and for an extra £20 5/- 6d, the Bentley owner could have the new balloon tyres, presumably 5.25 x 21″, in place of his high-pressure beaded edge 820 x 120 mm fitted as standard. The so-called "Colonial" model was fitted with 880 x 120 mm beaded edge to give an additional 1¼″ ground clearance. Bentleys did not recommend the low-pressure balloon tyres and at that time were only prepared to fit them at purchaser's risk. This position was to change, of course, 18 months later when 5.25 x 21″ wellbase Dunlops became fitted as standard.

The "Bentley Standard Coachwork" four-door five seat touring body on the standard chassis was available at £1225 complete, and a Freestone &

Webb Weymann saloon specially developed for Bentleys was available for £1325. Twin "Sloper" SU G5 carburettors were fitted as standard to the 1925 Speed Model, described in a road test as being "smoother and quieter than it was". Some of the very early Slopers had the usual two point mounting to the inlet manifold, but at 90° – i.e. with the bolts in line with the manifold rather than perpendicular. The tops of the barrels were circular, with milled tops, rather than the more usual pattern with near-triangular tops secured by four screws. The early 1924 Speed Models were reputedly fitted with 1¾″ single SUs. L.C. McKenzie, well-known Bentley restorer and rebuilder of the 1930s and 1940s, asserted in 1937 that the 1924 Speed Models were fitted with twin single jet Smiths carburettors, but the author has never seen any such instruments. The last Speed Model known to have been fitted with a single Smiths carburettor was chassis 854, a 1925 model Speed Model chassis delivered in January 1925. It is certain that the early 1925 Speed Models were fitted with the 5 jet 45 BVS Smith carburettor until all the Speed Model production was switched to the SU early in 1925.

However, Hillstead recalled 1924 as being the most precarious year from a financial point of view, with the heavy costs of developing the 6½ Litre. HM's comment "Nothing to pay the wages with next Friday – go out and get some money" was often to be heard at Hanover Court. The writing was already on the wall for those who could see it, and no attempts were being made to secure any alternative sources of income. A study of the figures reveals the nature of the problems. A manufacturing profit of £13,529 was turned into a loss of

**Bentley Standard Coachwork. This long
wheelbase touring body was devised by Bentleys,
and made for them by most of the major
coachbuilders. This particular example is by
Freestone & Webb.**

£56,700, the biggest loss made in any year except
1931. Much of this was caused by the heavy costs
of experimental work, the figure quoted being
£15,284, but a balance on the indirect expenditure
of £26,548 remains unexplained (in no other year
was the figure more than £6,787), so some considerable expenditure remains hidden, and one can
but speculate that it was taken up by development
work. It is possible that some of this represented
building work at Oxgate Lane. This difficult position was hardly improved by the withdrawal of the
McKenna Duties, which had imposed an additional tax on imported cars, and, according to
The Times, a slowdown in motor car production
beyond the seasonal average for the middle part of
1924.

The entry for the 1925 Le Mans race was a more
serious affair than before, with a Works car for the
first time – chassis 1138, registered MH 7580 – driven by Bertie Kensington Moir and Benjafield.
WO had obviously noted Benjy's successes at
Brooklands with the Number 2 2-seater and
invited him to become a full team member. In
truth, Benjy was the first of the "Bentley Boys",
that group of rich amateur sportsmen who were largely responsible for putting Bentleys on the map.
Prior to Benjafield's selection to drive at Le Mans
all the Team cars had been driven either by Bentley
employees (WO, Clement and Moir) or pro-
fessionals (Hawkes and Duff). (Although Duff was
not paid to drive, his racing was nevertheless largely motivated by financial reasons unlike the true
"amateurs" who made up most of the Bentley
teams.) The other entry was a private one by that
intrepid duo Duff and Clement, but with full
Works support, in chassis 1040 which always
appeared in contemporary photos registered MD
7187, despite the fact that the licensing authorities

years later denied ever issuing that number. Presumably Duff was up to one of his motor trade tricks!

Both cars were meticulously prepared, 1040 under Clement's supervision at the Works and 1138 under Moir's beady eye at Service and were then wheeled over to Vanden Plas for coachwork. Stoneguards were affixed to the radiators for the first time, subsequently fitted to all the Speed Models as standard, and larger petrol tanks of about 25 gallon capacity. SU "sloper" carburettors were used, begging the question of which carburettors were used in 1924. *The Autocar* of 29th August 1924 implies that "slopers" were used in that year, but the information cannot be confirmed. Gradually the day came and, some two weeks or so before the event, the team left Cricklewood to drive down to the ferry to Dieppe. Doctor Benjafield administered the injections (using an ingredient guaranteed to raise blisters on nitralloy steel sheet!) before sailing, and on arrival in France the cars were driven down to Le Mans where the cars and the mechanics encamped at the Hotel Moderne, the Bentley Le Mans HQ for many years, under the watchful eye of David the proprietor. Years later Nobby Clarke related the tale of David's father, an

ABOVE **For 1925, Bentley Motors entered a works car for the first time, to partner Duff's privately-entered 3 Litre. The works car was chassis No.1138, registered MH7580, prepared under Moir's direction at the service department. The car was basically a standard 1925 Speed model, very carefully built and fitted with mesh guards to the sump and lamps, and a larger petrol tank.**

RIGHT **Duff's car, chassis No.1040, was built at the works, and is seen here in the finished cars shop.**

old man, telling Nobby that "the Boche have come twice in my lifetime – once in 1870 and again in 1914. They will come again in your lifetime."

At Le Mans the daily round of practice sessions commenced, and woe betide any mechanic indulging in "leadswinging in its most virulent form" when early-morning practice was called for – retribution could be dire! The practice sessions were from 5am to 8am, and then from 9pm to 12pm. Understandably, there was considerable good natured rivalry between the Sunbeam and Bentley teams, with a great deal of practical jokes going on. There were problems with the Sunbeams, both of which had to be sent to Paris for repairs, so they were forced to practice in an old Darracq. Certain changes had been made to the regulations since the previous year – hoods had to be raised at the start, and the first 20 laps had to be covered in this manner. The idea was that a proper, 4-seat touring car should have efficient weather equipment. Rapid furling and unfurling of the Vanden Plas hood was practised until the drivers could do it blindfold, and pit routines were rehearsed as well. Finally on the basis of fuel calculations it was decided to put only enough fuel in the tanks to last the first 20 laps, because of the adverse effect on handling of huge quantities of fuel at the extreme rear end of the car. The circuit used between 1923 and 1928 was 10.7 miles long, so 20 laps equated to 214 miles – needing something of the order of 22½ gallons under racing conditions, and therein lay the cause of the first "black" Le Mans (Nobby Clarke estimated a petrol consumption of the order of 9/10 mpg for the 3 Litre cars at Le Mans).

On the fall of the flag, wielded as was customary by the proprietor of Rudge-Whitworth who supplied the major prizes, Kensington Moir in No.10 got off to a terrific start, pursued by the Sunbeam of Segrave. These two drove neck and neck for the first few laps, hoods flapping in the 90 mph gale, but the Bentley fuel consumption calculations had

been performed on cars with the hood down – and had not anticipated Moir's flat-out dice with the Sunbeam. At the end of 18½ laps Moir walked disconsolately back to the pits abandoning a perfectly good Bentley with a bone dry tank. Confronted with this same situation shortly afterwards, Duff was rather more cavalier in his approach. WO refused to give Duff any more petrol on the grounds that it was against the regulations, to which Duff replied, "It's my car and I'll do what I damn well like with it." Armed with petrol in a lemonade bottle, he managed to distract the gendarmes for long enough to pour it in and make his way back to the pits where he continued to work on a "broken" petrol pipe! It is said that Duff poured the petrol into the Autovac, but as his car had air-pressure feed to the tank with no Autovac that cannot be so. Presumably he poured the petrol into the tank and was able to use the air-pressure feed to push it through and prime the system. Other reports suggest that it was the air-pressure pump itself that had failed, which would have given Duff opportunities to introduce more petrol into the system.

This effort was to no avail, however, as one of the carburettor float chambers broke off in the early part of the Sunday morning and set fire to the car.

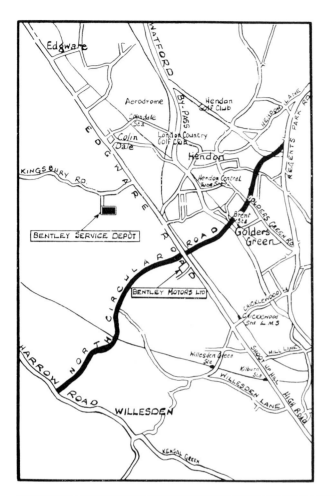

Clement managed to put the flames out, and although it is not known how badly the 3 Litre was damaged, it was certainly hors de combat. Later "sloper" carburettors incorporated a strengthening rib into the casting between the float chamber and the carburettor body. All in all, a very disappointing performance that WO had to justify at more interminable Board meetings covering the financial issues festering below the surface. The Segrave/Duller Sunbeam retired with clutch trouble, but the Davis/Chassagne car finished second, covering 1343 miles, which can't have pleased WO much. The Sunbeam definitely had its problems, though – the back axle bent after hitting a ditch, such that both rear wheels were almost touching the wings, the brakes were failing and Sammy had to hold the dashboard up by hand as it gradually fell off!

This division of interests between Service at the

LEFT **A map showing the location of the new service department, from a 6½ Litre handbook.**

RIGHT **The service department dinner menu, from 3 December 1926. The drawing by Tomlins shows members of the service department – the sketch of WO is immediately recognisable.**

BELOW **The interior of the Oxgate Lane service department, soon after the introduction of the 6½ Litre. The cars are all 3 Litres or Standard Sixes.**

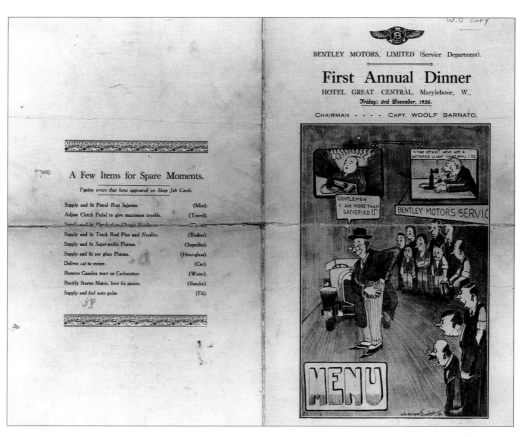

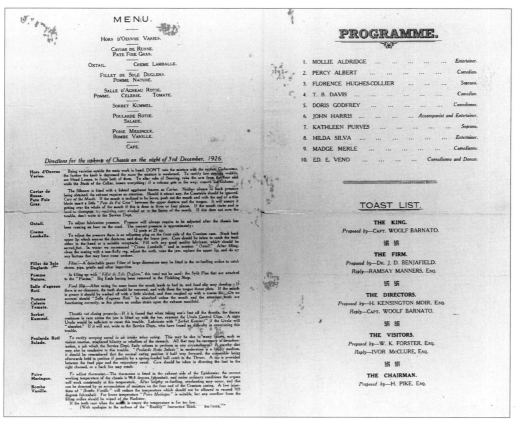

ABOVE **Bentley's other service department was at 112/118 North Street, Glasgow. Plans were afoot in 1931 to establish a Midlands service depot at Leeds, but this was never finalised. Service departments were, though, established in Paris and Cannes. The car is YO133, the second 6½ Litre built, Chassis No. WB2552. Fitted with a saloon body by Hamshaws, WB2552 was delivered to Woolf Barnato in May 1926, and was road tested by *The Autocar* in their issue of 17th September 1926. The photo was taken after July 1927, when**

WB2552 was fitted with the 1928 modifications and sold to S F Menzies of Edinburgh. From left to right, not known, L H Alexander, George Peach, John Sherriff, not known, not known, Ian Cunningham.

BELOW **A slightly earlier view of Glasgow service, with two 3 Litres. George Peach is seated in the middle; behind him is thought to be Effie MacDonald, the clerk.**

top of Oxgate Lane and the Racing Shop in the main Works could not be allowed to continue. WO managed to lease space from Vanden Plas at Kingsbury, where the Service Department was installed, coupled with a management shake-up. Hubert Pike continued as the Service Director, with Kensington Moir moving to the Experimental Shop to work on the 6½ Litre. His place was filled by Nobby Clarke, who had been Works Superintendant. Nobby also took over responsibility for the Racing Shop from Arthur Saunders, Nobby keeping an eye on both until the workload at Service became too much for Pike to handle on his own. The Racing Shop remained at the Works until late 1927, when it moved to a shop leased from Vanden Plas, converted from Kingsbury House stables. In later years the Racing Shop pretty much ran itself, Nobby principally involving himself at Le Mans. Nobby was a great favourite of the customers, some of whom went to Kingsbury just to soak up the "blarney"!

Principal among the usual financial problems was the need to raise large amounts of money to put the 6½ Litre into production. The first extended press descriptions of the new car, with photographs of "The Sun" fitted with a proper Bentley radiator appeared at the end of June 1925, but as with the showing of EXP 1 3 Litre to the press in January 1920, the Company was a long way from debugging the chassis and putting it into series production. This money was simply not available from cash-flow from sales of the 3 Litre and could not be borrowed, as absolutely everything was mortgaged with the London Life Association. Frank Clement was taken off experimental work and sent on a sales drive around the agents with a new Speed Model to try and generate more business, but despite Clement's enthusiasm it made little impact. Some means had to be found to raise the money, and two options were available – the first to produce a quick-selling model, the second to find a major source of capital from some private source.

The first option, supposedly HM's brainchild, resulted in the "Light Tourer", the requirement for which was a complete car to sell for under £1000, made from existing parts, with 4-seat coachwork which should not be too attractive and with significantly less performance than the Speed Model 3 Litre. This requirement was met by using the 9'9½" chassis, with the low compression single Smiths carburettor engine, the B type gearbox and 4.23 back axle of the standard chassis, and shortening the steering column and hence the scuttle by 3" to increase the room in the back of the body. It is

noticeable in the "bathtub" bodied 3 Litre, how much closer to the dashboard the dynamo is because of the short scuttle. Leaflet 8b, issued in August 1925, listed the salient points of the new model. The open touring car with the standard "bathtub" body was offered in three standard colours, grey, maroon or dark blue with leather to match, with hooding in black mohair. An extra charge of £12 was made for any other colour! Some more perceptive customers saw through this and asked awkward questions about the difference between the Light Tourer and their Speed Model – questions that must have been exacerbated when it became clear that the Light Tourer was indeed also available in chassis form, at £795, for specialist coachwork, and that many of them were fitted with A type gearboxes. Although Hillstead was specific about the B box being fitted to the Light Tourer, among the Bentley Motors drawings is a schedule for a Light Tourer with the A type gearbox and many of the chassis were obviously made to that specification.

The majority of the 39 built, however, were fitted with the familiar "bathtub" body, probably conceived by Vanden Plas, who built at least 20 such bodies including the demonstrator. The Light Tourer obviously achieved some of the results needed, selling at £995 complete, but could never have been more than a temporary measure. By October 1925, the Light Tourer was being offered as a tourer at £995, with coupé coachwork at £1175 or with 4 door saloon coachwork at £1195.

WO put it thus: "There wasn't a day without anxiety and when we didn't have to confront the dilemma of having a new model, which we knew we could sell, and at a substantially higher profit than the 3 Litre, but which was so expensive to get out of the experimental and into the production stage that on several occasions I was certain we would go under and down." In other words, the classic cash flow crisis so often faced by manufacturing concerns. WO and Moir even toured all the agents, presumably in "The Sun", persuading them to pay deposits on chassis that would be supplied later at a special discount. Jack Withers & Co had a demonstrator 6½ Litre in October 1925, advertising that "We are the only agents in Great Britain with a 6 Cylinder demonstration car." As there were only two 6½ Litres in existence, the fact that one was available to Withers for demonstration purposes is significant. The car must have been "The Sun" with its proper Bentley radiator. WO and Withers had been friends for many years, Withers having been Leonie Withers' step-brother and responsible

RIGHT **The first publicity photograph of the new 6½ litre in July 1925. The car is 'The Sun', fitted with a standard 6½ litre radiator. The engine is now 6½ Litres, but differs significantly from the production unit. The cylinder block has a water system arrangement similar to the 3 litre, with a pipe to the front of the block and an internal gallery. The crankcase casting presumably does not have the cast-in pocket for the rubber mounting, as the rubber mounting has been effected by bridging the engine bearer with a gearbox bridge-piece and then bolting to the frame through two large lumps of rubber. It is likely that the engine was originally bolted straight to the frame through fibre blocks as on the 3 Litre, and this temporary rubber mounting came later. The carburettor is the Smiths 50BVS.**

for her and WO meeting, which goes some way to explaining Withers' favoured treatment. By these and other means they managed to hold the show on the road for a little bit longer. At £1450 the chassis price of the new model was £525 more than the Speed Model 3 Litre, and according to WO did not cost half as much again to manufacture.

It is pertinent that the 26 cars of the "NR" 3 Litre chassis series sold in 1926 were 1925 cars that had been completed but not sold, rebuilt to 1926 specifications, renumbered, and sold as 1926 cars.

It is likely that the marketing of the Light Tourer was at the expense of the existing models. Nobby Clarke made the following rather cryptic remark in a letter to Stan Sedgwick: "Remember the renumbered chassis when we had to bring in the Light Tourer 3 Litre. Stan this guy would not know the whole story behind this move even then this dodging of numbers had better be forgotten?? If he was to say partly because we could not sell them he would be nearer the mark." Unfortunately the letter referred to cannot be traced. Considering the

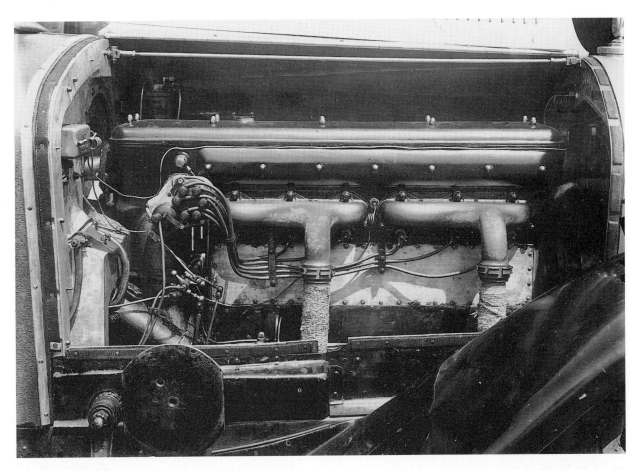

ABOVE AND LEFT **Offside and nearside of the engine of 'The Sun', taken at the same time, but with bonnet and deckboards fitted. On the offside view, note the temperature gauge fitted into the cylinder block between the exhaust manifolds. The steering box casting was later altered, by the addition of a cast-in boss for an oil nipple. It looks as if the chassis was originally fitted with a central accelerator, with the arm showing to the outside of the steering box, with linkage to the controls. The right hand accelerator looks to be a later addition.**

sales position, it is perhaps significant that Jack Withers & Co advertised in *The Autocar* of 9th October 1925 that they had nine new 1926 model 3 Litres in stock. Typical of this problem was the case of chassis 1183 – delivered to Queens & Brighton with a Gurney Nutting body, it was returned to stock in August 1925, going on to Frank Scott before being sold in November.

More or less coincidental with the Light Tourer, was the announcement of the "Supersports" 100 mph 3 Litre on a 9′ wheelbase chassis frame, nomi-

nally as a result of Benjafield's successes in the No. 2 2-seater, which was also built on a 9′ frame. With strict control over the maximum weight of coach-work fitted to the chassis, this model was capable of 100 mph and was guaranteed as such, but there is little doubt that some of the eighteen chassis made were assisted by a certain amount of speedometer tuning! 100 mph on a 3.53 back axle represents more than the maximum permitted 3500 rpm, and it could be understood from the Bentley literature that the chassis was guaranteed to do 100 mph – without a body! 100 mph in top on 21″ wheels with 5.25″ covers represents 3900 rpm according to Bentley's own charts, the difference in inflated diameter of a 5.25x21″ tyre to an 820x120 beaded edge being effectively zero (32.7″ to 32.6″ respect-ively). The first 100 mph chassis was delivered in March 1925 and the bulk of the chassis were deli-vered during the same period as the Light Tourer – Spring of 1925 through to Spring of 1926 – so it is easy to speculate that the motives behind the 100 mph cars were similar to those behind the Light Tourer.

Interestingly, Bentleys seem to have referred to

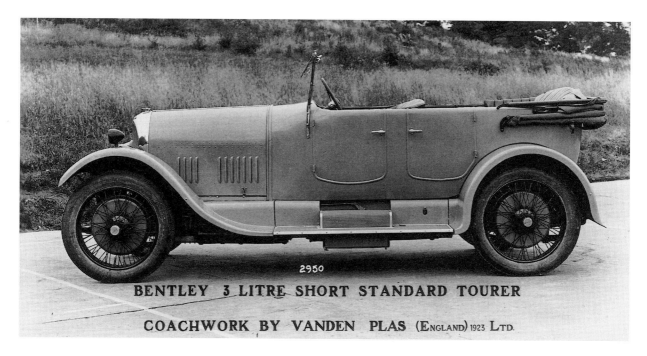

BENTLEY 3 LITRE SHORT STANDARD TOURER

COACHWORK BY VANDEN PLAS (ENGLAND) 1923 LTD.

ABOVE **The Light Tourer, fitted with the standard Vanden Plas 'bathtub' body. This shows off very well the short scuttle, and the double silencer arrangement.**

BELOW **This is almost certainly the first production Supersports chassis, chassis No.1046, outside the double doors at the Oxgate Lane end of the running shop. From the left, Jimmy Jackson, Bob Tomlins, Nobby Clarke, 'Shorty' Hedges (?), Bill Bowie (?), unknown, unknown. The chassis has the tapered Supersports radiator and bulkhead, but has standard length rear springs, with the compensator overlapping the rear spring hanger reinforcing. On the production Supersports chassis the rear spring was shortened by about six inches and a special reinforcing plate fitted over the spring hanger, which was similarly moved backwards. 1046 was fitted with a two seat body, and passed off test in March 1925.**

Two views of the new Supersports chassis, taken
in April 1925. This is almost certainly the second,
chassis No.1059. Note the corded road springs and
the pressure feed to the standard 11 gallon 3 litre
tank, with air feed into the back of the tank neck
and the pump mounted off a temporary bracket
on the bulkhead. The chassis of 1059 was finished
in April 1925, but it was then rebuilt and renumbered as a 1926 model, chassis No. NR513,
and fitted with a Felber two seat body.

THE STORY OF THE RECORD

HE car used was the latest 1926 Speed model, a production chassis, purchased from the Company by Capt. J. F. Duff, and from which the front wheel brakes had been removed.

The car was driven alternately, for periods of three hours each, by the owner, and by Capt. Woolf Barnato. Throughout the 24 hours' run, no mechanical trouble whatsoever was experienced, the only stops being for replenishment of petrol, oil, and water, and to change drivers. The longest stop for this purpose was 1 min. 40 secs., other stops averaging 1 min. 5 secs. On one occasion, in addition to refilling with fuel and changing drivers, two wheels were changed, in a period of 1 min. 5 secs.

The attempt commenced at 6.32 p.m. on Monday, 21st September, 1925, and the average for the first three hours was approximately 97 m.p.h. At the end of the sixth hour a mist developed which gradually became denser until, towards the middle of the night, it was only possible for the drivers to see about 200 yards ahead, and the strain was somewhat severe. The average was, however, not permitted to drop below that agreed

upon at the commencement, although the drivers were solely dependant upon the faint lights on each side of the track for their guidance. Conditions were not improved by the dampness of the track, caused by the mist. With daylight came a wind, which increased in strength towards the end of the run.

The car continued to run smoothly, capturing no less than 22 records in its course. At times the speed was increased to 100 m.p.h. The engine never missed, and the exhaust note was maintained without variation, until, at 6.32 p.m. on Tuesday, 22nd September, the car was pulled up, having covered 2,280 miles in the 24 hours, at an average speed of 95 m.p.h., a world's record for any of engine.

The driving was at all times done at a speed under the car's maximum in order to ensure a reserve, should delay have been caused through any necessity for small adjustments—as it happened, a needless precaution.

The bonnet of the car was not lifted throughout the run, a certificate to this effect having been obtained from The Automobile Club of France.

Perfection in design with the finest workmanship and materials—aided by splendid driving—enabled the Three Litre Bentley to gain this record. *Every Bentley car has the same qualities.*

ABOVE **Woolf Barnato's first Bentley, a 9′ Supersports chassis No.1106 fitted with a racing two seat body by Jarvis, in the yard at Cricklewood. Head tester Bob Tomlins is at the wheel. From the left, the wooden office block, the end of the experimental shop, the chimney/boiler room with the engine test shop visible above the boiler room, and the Running Shop on the right. The yard was still very roughly surfaced.**

LEFT AND OPPOSITE PAGE **The 1925 Montlhéry run, reproduced from the catalogue issued by Bentley Motors to celebrate the record run. The car was John Duff's 1925 Le Mans car, chassis No. 1040, suitably rebuilt and rebodied.**

different models by the engine tune rather than the wheelbase, which explains the anomalies of a Speed Model built on a long chassis and a 100 mph car built on a Speed Model frame. In both cases the engine was correct for the model, even if the chassis frame was not. The customer could indeed have anything he was prepared to pay for. The 100 mph cars differed from the Speed Model specification in terms of the radiator, which tapered to the bottom in similar fashion to the Standard Six, with special bulkhead to match and the back axle ratio of 3.53 to 1. However, many of the eighteen chassis were fitted with the standard Speed Model radiator. It is possible that there is some significance in the illustration of the 100 mph car in Bentley's October 1925 catalogue. The car shown is Barnato's PE 3200, the text reading: "The car illustrated is the property of Capt. Woolf Barnato and has many splendid performances to its credit." A far more realistic means was necessary to finance the Company, and WO took the decision to approach Woolf Barnato and discuss the matter with him; Barnato is a name that will come to assume great significance in the Bentley story.

After the dismal showing at Le Mans, Duff decided to stay on in France to make an attempt on the World's 24-hour record at the Montlhéry circuit near Paris. This circuit was much smoother than Brooklands and was increasingly used for record attempts. The Le Mans car was stripped of its damaged Vanden Plas body, and fitted by Messrs. Weymann with a light, streamlined, single-seat body. Hassan stayed on to prepare the car, from which the front brakes were also removed and twin Solex carburettors fitted. The 3.78:1 back axle was changed for a 2.87:1. In September, Duff and Barnato managed to put up an average of

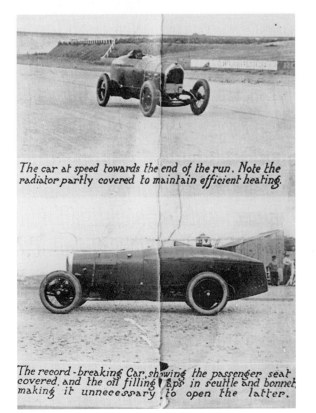

The car at speed towards the end of the run. Note the radiator partly covered to maintain efficient heating.

The record-breaking Car, showing the passenger seat covered, and the oil filling taps in scuttle and bonnet, making it unnecessary to open the latter.

95.03 mph for the 24-hours without mechanical problems, a pretty impressive performance, although it was only a few months later that Garfield and Plessier in the big 45 hp Renault exceeded these figures. It was this strong showing that led WO to believe that the 24-hour record could be pushed over the 100 mph barrier by a 3 litre car, which would bring enormous publicity for the Bentley concern. The Bentley success was not missed by Rolls-Royce and a memo was sent to Wormald, their Works Manager, asking that someone should be nominated to look at the lubrication system of the 3 Litre at Olympia. "The recent extremely dramatic success of the Bentley on the track at Paris, apart from anything else, was an indication of the extremely effective system of lubrication throughout. I have frequently brought to my notice by those who know the Bentley car, the extent to which they have reduced the amount of work necessary to be done by the owner-driver...by making sundry points of the lubrication automatic. It would appear that such automaticity has not resulted in unreliability."

At Olympia of 1925 the Big Six was shown to the public for the first time. The 3 Litre was updated, the "1926 model" being different from a "1925 model". It is very easy now to forget that the cars were indeed updated from year to year, to produce what was to Bentleys, at least, a new model. Integral stoneguards were fitted to the Speed Model chassis radiator, and header tanks fitted to all the radiators. The radiator was also made 1″ taller, which rendered the bonnet line more attractive by making it parallel with the ground. The power of the Speed Model engine was increased by fitting the BM1800 camshaft and lightening the valve gear (although the BM1800 is supposed to give more power than the BM2391 camshaft above 3000 rpm and is always thought of as the Speed Model camshaft, for some reason all chassis from no. 160 up to 1926 were fitted with the BM2391 shaft). All the 1926 Speed Model engines were made to the "Supersports" specification, which included such features as drilled valve spring caps. "Supersports" engines were denoted by "SS" stamped above the engine number on the crankcase, and many of the Speed Model engines were stamped in the same way. A single "S" above the engine number denotes a Service replacement engine, or an engine overhauled by the Service Department.

The 6½ Litre enjoyed a very favourable reception from the motoring press. Two 6½ Litres were shown, a seven seat Harrison saloon with division,

priced at £2250, and a polished 12′ chassis. The saloon must have been "The Sun", whose original radiator had long been replaced by the very tapered radiator used on the early 6½ Litres. This car was registered MF7584, but there is no trace of it after the Motor Show. "The Sun" appeared in all the early press reports of the new model and in Bentley Motor's catalogues. The sales team at the Show consisted of HM, Hillstead, Clement and Longman. The latter became the Sales Manager after HM and Hillstead left.

It would seem that after the Show the polished chassis became EX 2, the second experimental 6½ Litre, registered MH1030 and fitted with an open touring body with central body beam. This car was

THIS PAGE **The No.2 Experimental 6½ litre, MH1030, fitted with open touring coachwork with a central body beam. As with 'The Sun', it has a 3 litre pattern starting handle assembly. The car is seen in Ravensdale Avenue, North Finchley, with F T Burgess and his son Peter.**

RIGHT **The No.2 Experimental 6½ Litre, in Scotland. Burgess took the car on holiday with the Hewitts. HS Hewitt on the left, Burgess on the right, and (from left) Mrs Hewitt, FT Burgess, and Mrs Burgess.**

used extensively in this form by Burgess, and was driven by Hillstead on occasion as a demonstrator. EX 2 was the first 6½ Litre in which Percy Northey, a representative of Rolls-Royce was driven. Northey later reported favourably on the engine and general performance of the car, commenting that "this is the first time I have ever sat behind any engine which could be described as of a Rolls-Royce type." The steering also came in for particular praise. This car had a long career, being fitted with "The Box" saloon body and used by WO as his own car, travelling to Le Mans in 1927, fitted with experimental steam-cooling for Alpine tests, and then the first experimental 8 Litre engine in 1930.

From contemporary papers in surviving Rolls-Royce files it is clear that they were initially worried by the 6½ Litre. Their comment on the high-efficiency of the 6½ Litre engine in producing 140 bhp ("It is, however, a remarkable power output being 20% better than the Hispano, an engine of the same size") seems strange when compared to Weslake's criticisms, and their assessment of the 6½ Litre performance with an all-up weight of 2 tons receives favourable comment (the Phantom engine produced 109 bhp). Strangely, on driving the car, their comments make something of an about face, despite the figures quoted and used in their initial assessments being correct. The first production cars were delivered in March 1926, the second production chassis number WB 2552 with saloon coachwork by Hamshaw, being allocated as a demonstrator. There were design problems with the 6½ Litre, though, similar to the problems also being suffered by other manufacturers of large, fast cars, but the superior performance of the Bentley over most of these exacerbated the problems. Bentleys also had neither the time nor the money to indulge in long, rigorous testing programmes to ensure the absolute integrity of the design before manufacturing. To stop the transmission of torque vibrations, the engine was flexibly-mounted to the chassis frame through thick rubber blocks. This robbed the chassis of any rigidity imparted by means of the crankcase (Hispano-Suiza's solution to this problem was to use a massive aluminium crankcase rigidly mounted to the chassis at four points to stiffen the front end).

The front of the chassis was very weak in torsion and could act in a similar manner to a tuning fork if vibrations of the wrong frequency occurred and were allowed to increase undamped, leading to total loss of steering control and brakes – shimmy or axle tramp. This problem is also connected with the use of a beam axle with heavy wheels, tyres and braking equipment. If one wheel is deflected up and back by road shocks such as a pot-hole, the whole axle can vibrate about the centre point if the springs allow it to. Three different patterns of front springs were used on the 6½ Litre at different times, and on the Team Speed Sixes an additional shock absorber was mounted on the track rod to damp out steering vibrations. Purely by accident, it was noticed that tramp was worse if the chassis was driven without a bonnet, leading to the conclusion that the friction between the bonnet and the bonnet tape acted as a substantial damper to such vibrations. Shortly after this discovery, the ring-pull pattern of bonnet catch with very strong springs was introduced. This weakness of the front end of

LEFT **One of the very first 6½ Litre chassis, photographed in July 1926 after a smash. This shows the rod brakes, expansion chamber for the exhaust system, single Hartford shock absorbers and the early plain bonnet with opening vents.**

ABOVE 6½ **Litre engine, inverted on stand at the works. This shows well the lubrication system, with the oil pump on the far right. In front of the oil pump are the oil pump drive gear (right) and the steel gear for the reduction gearing to the camshaft drive. This is a later engine with crankshaft damper, photographed in November 1926.**

BELOW LEFT **Midships view of the 1927 model 6½ Litre chassis, with the BS box, no servo, and front brake operation by ⁵⁄₁₆″ rods. This shows well the arrangement of the plate clutch and its withdrawal mechanism, and the disc pattern clutch stop used on all models except the 3 Litre and cone clutch 4½ Litres.**

the frame was improved on the 8 Litre chassis by using a more substantial front cross-member bolted into attachments in the dumb-irons, and on the Derby Bentleys built by Rolls-Royce and other cars of the 1930s bumpers were fitted with very substantial spring steel blades to break up and damp out any such vibrations before they built up to a dangerous level. That was after WO lost control of an early 3½ Litre Derby Bentley on a test drive due to tramp and ran into a Wolseley Hornet.

Royce noted in a memo of 13th October 1925 that Elliot, one of Rolls-Royce's designers, reck-oned the engine was impractical because of the six cylinder monobloc construction. Royce speculated that the camshaft oil could get into the cylinders, but there is no evidence of this happening. Royce was right, about the chassis, which was not really very good for coachwork because of lateral and tor-sional flexibility. As Forrest Lycett put it many years later, "Frankly the Speed Six struck me always as a fortuitously successful improvisation, coming as it did of tainted stock – the standard 6½ Litre. Weak frames and cross-members, coupled with harsh suspension, huge wheels, and a high centre of gravity, rendered fast wet road driving and braking an adventure." E.W. Hives, head of Rolls-Royce's Experimental Department, drove a very early 6½ Litre in July 1926, and was very crit-ical in his report. The car was chassis WB 2569, fit-ted with open four seat coachwork by H.J. Mulliner, belonging to Eric Horniman. Hives' comments are interesting in the light of later modifications:

"1. Suspension.
". . .double friction Hartfords front and rear. . .The springs were the Woodhead type with several leaves above the main plate. The riding at high speed was very good – at low speeds it was poor." The comments are similar to those made by *The*

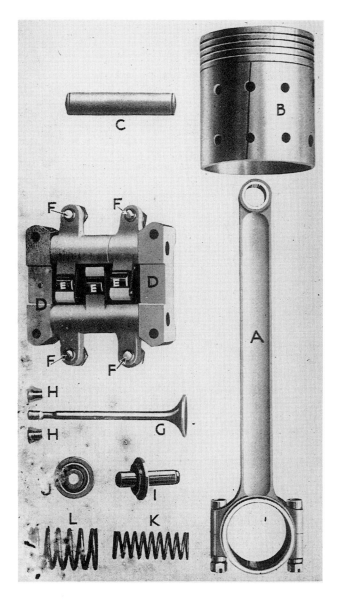

6½ Litre drilled BHB piston, con rod, and valve gear components. This shows well the dual rockers in their rocker box with the camshaft rollers and ball-ended tappet screws.

Autocar about the Speed Model 3 Litre in 1925, and are typical of Hartford friction dampers. If the dampers are adjusted for high speeds, they are harsh at lower speeds. Later 6½s were fitted with B&D friction, later hydraulic, dampers.

"2. Power and Performance.

"...the performance to be similar to the Phantom. The max. speed would be approximately 75 mph..." (Despite the 140 bhp of the Bentley to 109 bhp of the Rolls-Royce.)

"3. Silence

"There was a fair amount of noise from the engine. This came from valve operation, piston knocks, magnetos and noisy dynamo. The gearbox gears were very noisy. The rear axle could not be complained of..." The dynamo, of course, was later moved out in front of the radiator. The other complaints are surprising, because an unworn B, BS or C gearbox is very quiet, and the 6½ engine is very quiet compared to the 3 Litre. However, see later comments on 6½ Litre transmission noise on p.294.

"6. Steering.

"The steering is remarkably light, free from road shocks...On the whole the steering is one of the best features of the car."

"7. Cooling.

"The owner complained of heat in the front seats...the car lost a considerable amount of water and had to be filled up every day...we should anticipate that they will have trouble with over heating." The problems with heat in the driving compartment were principally caused by the positioning of the exhaust system (see p. T214). Clement later recalled "A bit of overheating on the 6½ Litre in the early days" as one of the main teething problems.

"8. Engine vibration

"The engine is very free from torque reaction vibration and crankshaft vibration." As WO put it, "We also had to do something about the general roughness of the engine, which was very marked on the early production models. We cured this (we were too ashamed to confess at the time but I don't see that it matters now) by setting the engine on very carefully camouflaged rubber mountings [these consisted of a block of rubber fitted into a pocket cast into the mountings of the crankcase that sit on the chassis side-rails].

I think that the 6½ Litre must have had the first rubber-mounted engine, and it was a very successful makeshift, giving us silence in a car that already had astonishing flexibility and reliability."

This last feature was indeed not mentioned in the catalogues. Rolls-Royce were obviously not slow in cottoning on, as Royce commented in a memo to Hives in August 1929 "If we could swallow this we could use thicker rubber engine mountings and prevent the engine roughness from reaching the frame" (presumably referring to the Phantom II).

"9. Brakes.

"The brakes are operated direct by foot pressure only: they are very poor." Again, this was known, and early on the brakes were changed from pull-on at the front via $\frac{5}{16}''$ rods to push-on via $1''$ tubes. This latter arrangement was much more satisfactory and was further enhanced by the Dewandre servo in 1928. Later cars were provided with more holes in the levers in the system, so that the balance of the brakes front to back (and side to side if necessary) could be varied by positioning the clevis forks in different holes.

"Conclusion.

"His opinion [Mr. Horniman] is that the Bentley is better than his Silver Ghost. Taking it all round the Bentley is the nearest competitive car we have tried – it is a good car but there is nothing exceptional about it." Royce was well aware of Bentley's intentions – as he said in October 1925, "Regarding the Bentley, the makers are evidently out to capture some of our trade."

There was an amusing follow up to Hives's report from Northey: "Referring to your Hs2/LG23.7.26, in the first paragraph you refer to the ground clearance under the engine as being approximately 7". You will realise that from time to time information gathered from reports of competitive cars is used by Sales in connections with discussions which inevitably arise with enquirers who are trying to decide what car to buy, and it is of immense importance that any figures used in this way shall not only be used with the very greatest tact, but that the statements themselves shall be correct. Mr. Bentley has heard that a reference has been made to the clearance of his six-cylinder car being only 7", and he is very upset at this, as he

ABOVE **The 50 BVS Smith carburettor. E is the slow speed jet, D the four main jets. These come into operation in sequence as the dome, J, rises, uncovering them in turn. M and N are for the water heating arrangements.**

RIGHT **Rear brake arrangement on the 6½ Litre chassis, showing the two pairs of shoes and the tensioning screws. The drum is the steel drum with cast aluminium fins. Note the Hartford shock absorber, and the asbestos-lagged exhaust pipe over the back axle. The rear brake arrangements on the 3, 4½, 6½ and 4½ Litre Supercharged chassis are virtually identical.**

states that the clearance is not less than between 9¼" and 9¾" at the lowest point, and he feels that our statement is very damaging." Some things obviously do not change! One of the worst design features of the 6½ Litre was not picked up by Rolls-Royce, being the relative positioning of the near-side magneto and the Autovac. Fires are not uncommon with these cars, caused by petrol from the Autovac being ignited by sparks from the magneto, situated almost immediately beneath it.

An early 6½ Litre chassis, prepared for a Willesden Carnival, in 1926. The record run referred to is the 1925 Montlhéry record. The engine is the Bentley Rotary, from the office block, where it was normally suspended from the ceiling.

9 The Barnato Era

WO's APPROACH TO Barnato turned out to be a roaring success, dividing the story of Bentley Motors neatly into two eras. Thus it is necessary to examine the man to appreciate his pivotal role in the Company. To say that Woolf Barnato was born with a silver spoon in his mouth would be an understatement. It was a gold one encrusted with diamonds. Barnato's background is a fascinating story. His grandfather, Isaac Isaacs, a Jew from the East End of London, produced two sons, Barnett and Harry, and a daughter, Kate. Barney and Harry formed a comedy double act, changing their name to the Barnato Brothers after a stage manager suggested that Barney should also take a bow at the end of the act, saying "And Barney, too!". Barney Barnato left the East End of London to make his fortune in South Africa in 1873, with £50 and forty boxes of cigars. In this he was prodigiously successful, by the 1880s owning a half-share of the Kimberley diamond mines, being one of the four Life Governors of De Beers, an elected member of the Cape Assembly and a Lieutenant of the City of London. When Barney sailed back to England in 1897 he was a fabulously wealthy man, and no convincing explanation has ever been put forward for his disappearance over the side of his yacht and subsequent drowning, although it was said (in 1924) that he suffered a breakdown in health soon after the Jameson Raid and had committed suicide. It is possible he was being blackmailed by Von Veltheim, a notorious adventurer and con-man, who later murdered his cousin, Woolf Joel.

After various legal battles, young Woolf, only two at the time of Barney's death, was well provided for. Following Charterhouse and Cambridge, Barnato enlisted in the Royal Field Artillery, serving in France and Palestine, being commissioned as a Captain before the end of the War. Afterwards, Barnato enjoyed an active social and sporting life. He raced a Locomobile at Brooklands, competing in EXP 2's second ever race in 1921, and later an Hispano-Suiza. In 1925 he bought his first Bentley, 3 Litre chassis number 1106 fitted with a neat 1½ seat pointed-tail body, designed by A.P. Compton and built by Jarvis on the 9′ 100 mph chassis, registered PE 3200, and raced this car several times at Brooklands. In that same year the legal aspects of the Barnato legacy were finally sorted out and Woolf suddenly became a very rich man indeed, coincidental with the acute poverty of Bentley Motors.

The case concerned the manner in which Woolf Barnato's cousins, Jack Barnato Joel and Solly Barnato Joel (who by then constituted Barnato Brothers following the deaths of Barney and Woolf Barnato Joel) had dealt with Barney's business interests. It took Woolf Barnato's accountant, David Allan, five years to investigate fully the dealings of Barnato Brothers. The case also involved the will of Woolf Barnato Joel, who was shot by Von Veltheim in South Africa on 14th March, 1898. Young Woolf Barnato had broken off his business agreement with the Joels in 1917 and claimed that he was entitled to a share in the profits of the company between 1897 and September 1916, Barnato Brothers being worth £8,885,000 in 1897. Allan's five years of investigation separated the initial hearings in 1919 and the case really got underway in October 1924, and it is clear that it was a bitterly fought battle. The case was settled in a compromise deal on 20th May 1925, Barnato receiving roughly £900,000, with £50,000 legal costs. With

Woolf Barnato, seen here with Tim Birkin and 'Old Number One' Speed Six (just visible) at the 1929 Le Mans race.

similar figures going to Leah Barnato and to Lady Plunket, Jack Barnato's widow, the Joels were out of pocket to the tune of nearly £3,000,000.

The whole case was again brought up with the appointment of an arbitrator to sort out some of the details. Of agreed costs of £50,000, Barnato had to sue the Joels for interest on late payment, and for a further £26,883.8.11 incurred as "disbursements". The Joels haggled over the late payment on the grounds of Barnato's solicitors dissolution of his partnership with another solicitor, Mr. Spink, making it unclear to whom the balance of £40,000 due should be paid. They haggled over the £26,883.8.11 on the grounds that "disbursements" should not include such things as the printing costs for briefs when Barnato's solicitors had a typist who could have typed them! The whole matter came before Mr. Justice Eve in the Chancery Division on the 20th December 1928, at which point it must be assumed that it was somehow sorted out. Barnato probably inherited in total some £1,400,000, so financing Bentley Motors was, in the early days at least, comparatively small change.

Barnato was apparently not at all surprised when WO got in touch with him, and at their subsequent meeting proved to know already a great deal about the Company. WO speculated later that had he not approached Barnato first, it is very likely that Barnato would have approached Bentleys. After the initial approach, apparently delayed because it was thought in the Company that Barnato was not a good enough driver to race Bentleys at Brooklands, moves were put in hand on several fronts. Hillstead went to see Barnato at Ardenrun in the open No.2

Experimental 6½ Litre EX 2, and answered a whole barrage of questions about the sales side. A deal was gradually worked out, but the terms of Barnato's effective acquisition were not announced until March 1926.

Barnato was not the only person WO approached in the quest for finances. In the winter of 1925/26 he also visited William Morris in a very new 6½ Litre (which must have been the Number 2 Experimental car MH 1030), in an effort to persuade Morris to back the Company. The meeting went well, but Morris was non-committal, and declined to get involved. WO's comments on Morris's success are revealing: "What I think really elevated him to millionairedom was his sound financial judgement in avoiding going to the City for his money. . .he succeeded in avoiding taking on big loans from the merchant bankers, insurance companies, etc. It was the most astute move he ever made." It would seem that WO was referring more to his own bitter experiences. Quite how many people WO did approach at this time is not clear, but it is clear that the Company was on the verge of going into liquidation.

The financial position had gone steadily from bad to worse. A lot of probably unfair criticisms were made after the failure of the 1925 Le Mans race by those who had not been involved, implying that the whole thing had been a waste of time and the Company should not have entered. Such comments must have virtually split the Board. Fewer cars were made in 1925 than 1924 and as the only hope of profitability lay in more sales, once again they were faced with the now familiar dilemma which they faced in the way that the Company always faced it – by producing a bigger and more expensive car, subject to the same vagaries of limited sales to rich customers. There was no money to develop the 6½, but it was possible that the new car could solve the ongoing crisis if enough money could be obtained to put it into production (and, of course, if it then generated enough income). By August of 1925 a total of 133,500 shares had been issued, the 13,500 issued in 1924/25 representing another £13,500 in ready cash if they were all fully paid up. The directors remained unchanged, but in November the Company's offices were moved from 3 Hanover Court to the Works at Cricklewood.

With a deal worked out between the Company and Barnato, an Extraordinary General Meeting was held at Hanover Court on 3rd February 1926, at which it was agreed to wind up the Company, and reconstruct it. The liquidators were authorised

This aerial shot of the works is an enlargement of part of a plate taken by Aerofilms in July, 1925. This shows the layout of the works, which had been considerably extended since the 1922 photos. Substantial extensions had been added to make the detail shops, to the north of the old experimental shop, engine erecting shop, and the stores. The finished cars building has also been built, on the other side of the yard. This was the form of the works more or less from 1923/5 to 1928/9. In 1929/30 the works was built up to the layout seen in Chapter 15.

"to consent to the registration of a new Company to be named Bentley Motors, Limited, with a Memorandum and Articles of Association which have already been prepared with the privity and approval of the directors of this Company." The books and accounts of the old Company were not handed over to the new until April of 1929.

The announcement of Barnato's terms for coming into Bentleys were published in March 1926, coming as something of a shock to the existing shareholders. The old £1 shares were cancelled and replaced by 1/- ordinary shares, worth 5% of the original. At a stroke, the holdings of the earlier participants were effectively wiped out. Barnato took on some 109,400 shares of £1 each, as preference shares, the company being capitalised at roughly £112,000. The nominal capital was £175,000,

composed of 162,500 preference shares of £1 each and 250,000 ordinary shares of 1/- each. The Preference shares conferred the right to an 8% dividend, and ranked first in the event of the Company being wound up. It was in effect an almost complete takeover, and a new regime was installed at Bentleys, but there was little choice but to accept the terms or go into liquidation (which, of course, they effectively already had). New showrooms were leased in March, Pollen House in Cork Street (known as "the street of the dead" among retailers), with substantially more floor space than Hanover Court and with offices above. Of the existing sales staff, Hillstead and HM, the former was offered a new job, but declined. Hillstead felt that the family aspect of the Company had gone, and a new commercial era created by men from a different background with different pursuits had arrived. There was really no place for HM in the new regime. Barnato had his own financial advisers and could hardly tolerate HM in that role.

WO had signed an agreement binding himself to his Company on 28th February 1919, but this was not sufficient for Barnato. A new agreement was drawn up between WO and Bentley Motors, in which it was agreed that WO would serve the company as Chief Engineer "including provisions for the transfer of the employment to the successors in business of Bentley Motors Limited." Donald Bastow clearly states that WO was forced to sign this new agreement by Barnato, the contract being dated 10th March 1926. This agreement was far more restrictive than the 1919 agreement, in that under the terms of the new agreement WO was barred from competing with Bentley Motors by involving himself in any motor car design work involving internal combustion engines. His re-muneration as Chief Engineer was reduced from £2,000 pa to £1,000 pa.

HM retained the lease on Hanover Court and set up H.M. Bentley & Partners Ltd, continuing to deal in Bentley cars until well into the 1950s. WO

must have been hurt by the loss of HM and Hill-stead, but makes no reference to his feelings at the time at all. HM had managed the financial arrange-ments of Bentley and Bentley and of Bentley Motors, and his loss created a void that was never really filled. As Hillstead commented of the pre-war days, "WO appeared to live in a world of his own and, on the technical side, had been answer-able to no man. But his racing activities were

Works layout at the time of aerial photo, *c.* July 1925.

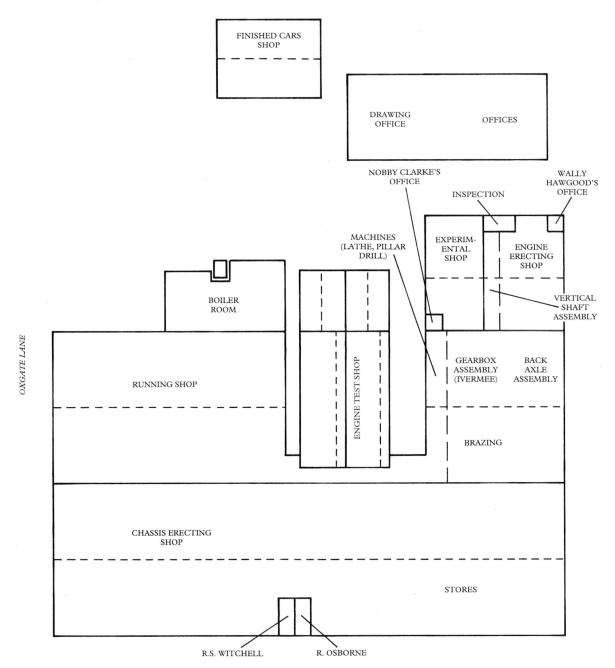

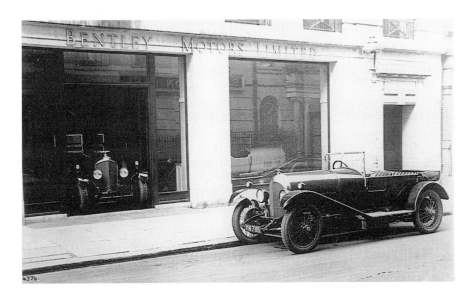

Cork Street may have been known as 'the street of the dead' among retailers, but the new premises were far more spacious than those at Hanover Court could ever have been. These views of the company's new premises in Pollen House were taken in October 1926. It was in the offices above that Barnato installed his financial advisers, and there was a basement as well, used for storing directors' cars and sales cars that were not on display. These views show four 3 Litres (see overleaf) and a 6½ Litre in the showroom, and YM7811 outside, previously Hillstead's demonstrator. Hillstead resigned shortly after the move to Cork Street. Pollen House is still there, and the holes in the façade for the Bentley Motors name can still be seen.

Different angle on the new showroom in Pollen House, 1926. 'Barnato's cronies' are installed upstairs.

governed by how much HM permitted to escape from the kitty." And again (of the early Bentley Motors days) "I became tired of the endless meetings concerned with keeping our heads above water, and how HM remained sane throughout this impossible period I cannot imagine. He had a tremendous ally in his brother, Hardy, and between them they managed in some miraculous way to keep the Company ticking over. This was all the more remarkable as lawyers, [Hardy was the lawyer – HM was an accountant] by the very nature of their profession, invariably have a difficulty for every solution." This guiding, restraining role does not seem to have been subsequently filled by any of Barnato's nominees. Indeed, it is almost indisputable that without HM, Bentley Motors would never have got off the ground, and it is difficult to over-stress the crucial role he filled in the early years.

Under Longman, a completely new sales team was installed at Pollen House in mid-June. As WO put it "Barnato's cronies" moved into the offices above – John Kennedy Carruth and Ramsey Manners, a Jew and a Scotsman, again in WO's words, a very good combination who looked after Barnato's interests very carefully and were deeply suspicious of this venture into the car-making industry. Manners was an ex-Army officer from Glasgow who met Barnato during the Great War, and was known as "Scottie". Carruth was Barnato's Business Manager. Again, a long and involved agreement had to be drawn up between the old and the new Companies, particularly in respect of the positions of the shareholders. It seems that it was not a mat-ter of a straight swap of the old £1 shares for new 1/- shares, as 8900 £1 shares were divided up between the existing shareholders, and in the return for September 1930 these were written down as being worth £4,450 issued for considerations other than cash. However, in the first return after the restructuring dated 18th June 1926, £9,600 worth of shares were issued as commission, for what is not known. The agreement is worded to the effect that the old Company ceased to exist on 30th September 1925 and between that date and the agreement being reached with the new Company in March of the following year had continued to act as agent for the new Company. The agreement recorded that the new Company also took over the full benefits of all arrangements and contracts made by the old Company, including the 1919 agreement between Bentley Motors and WO referred to above.

Barnato came onto the Board as Chairman, and for a while at least WO stayed on as Managing Director. Carruth and Manners each held 10,000 of the ordinary (1/-) shares, which gave them a seat on the Board. The new company was registered on 6th March 1926, with WO, Hubert Pike, Woolf Barnato, Ramsey Manners and J.K. Carruth as directors. Woolf Barnato's address was given as Ardenrun Hall, and other occupations as "none". The Joint Liquidators were W.K. Forster, the Company Secretary, and C.H. McKnight of 4 Union Court, Old Broad Street, EC2. An official statement was made by Forster to the Evening Standard: "It is merely a step in the steady growth and increasing prosperity of the Company. It has been rendered necessary by the fact that a new Company of the same name has been formed to take over the business of Bentley Motors Ltd as a going concern, at the same time introducing fresh capital." Needless to say, the announcements gave no indication of the almost bankrupt state of the old Company or the severe devaluation of the previous shareholders' shares. It is interesting to note that the Company was registered at virtually the same time as the delivery of the first 6½ Litre, so the old Company would have gone under before the 6½ Litre was in production without Barnato's intervention. The true state of affairs is shown in a report of a creditors' meeting held on the Company's premises, (presumably at Cricklewood as Hanover Court had gone to HM) on 10th March. The creditors confirmed the voluntary liquidation of the Company by Forster and McKnight, and were paid off at 20/- in the £. The crunch is the bottom line: a heartstopping £75,000 was needed just to pay them all off.

10 Competition Failures

WO'S AMBITIONS, though, centred on the World's 24-Hour record and the possible publicity from setting this at over 100 mph with a 3 litre car. Benjafield was persuaded to lend the Number 2 2-seater for this and it was carefully prepared with a new engine, number AP 316. Gordon England were then commissioned to build a very light, streamlined single-seat body shell, which was built so low that it sloped down to the driver behind the engine. The Montlhéry track was booked, and in late March, WO with Kensington Moir, Benjafield, Clement and Barnato as drivers went out, the whole team descending on "The Chalet", a castle facing the track.

Transport for the drivers was the Experimental 6½ Litre EX 2 still fitted with touring coachwork. Early runs showed up various problems with the car. The very low steering column had to have a large dent put in the outer aluminium tube to give more footroom and sustained high speeds around the pure oval of the track revealed problems with universal joints in the pot-joint propshaft that did not move and hence failed to get lubricated, becoming very hot in the process. This was aggravated by the very hard suspension. The drivers complained bitterly of how uncomfortable the car was, so WO tried it, easily exceeding the 100 mph mark and returned to the pits to instruct Pennal to let the shock absorbers out one turn each. As Pennal applied the spanner, WO bent down and said "Up one turn, Pennal, up" and so it was done. Benjafield returned from his next run to announce that the car was much better, at which WO fixed Pennal with his most meaningful look! The whole attempt seemed doomed to fail – the target for a 3 litre car was very ambitious and it is clear that the car was very highly and inevitably, stressed.

It is very difficult to work out retrospectively just how many attempts were made – the contemporary press reports contain so many inaccuracies it is difficult to sort out fact from fiction. One source reports a run on the 30th March that lasted for 3 hours. According to *The Autocar*, which seems to have been fairly reliable in its reporting, they were delayed first by problems with the body (probably the steering column as mentioned before) and then by bad weather. The first known run was made on 31st March/1st April, Barnato driving first and averaging 103.47 mph for the first hour. Despite a slight delay because of spark plug problems (which was such a standard excuse it is difficult to know whether to believe it or not!), a World record for 3000 kms was set with a time of 12 hours 23 mins 57.04 secs, an average of 100.23 mph. However, an abrupt end to the proceedings was brought about by a broken valve spring followed by a broken valve after 16 hours 21 mins running. The report concluded by saying that another attempt would be made as soon as another engine was fitted to the chassis. The next report (source unknown) dated 16th April reported a record attempt made on Sunday 11th/Monday 12th April. Duller started at 2.20 pm and averaged 103.09 mph for the first two hours, then handing over to Clement. Two rear tyres were changed and the tank refilled, dropping the average to 100.92 mph at four hours and then 101.32 mph at five hours. Problems then occurred, and a lump of solder blocked the outlet from the petrol pipe causing a delay of 12 minutes, followed by an enforced stop because the car ran out of petrol. At six hours the average had dropped to 96.1 mph. Benjafield took over and the average

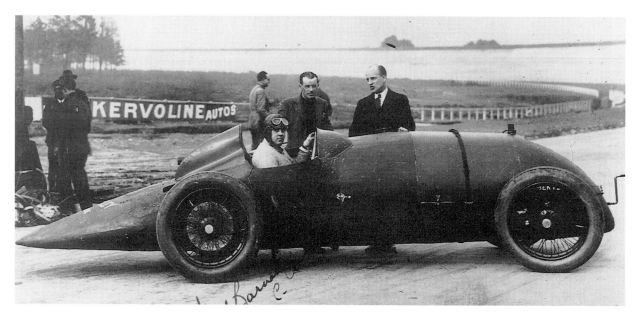

was up to 97.05 mph at seven hours, and it was hoped that if the average crept up to the 100 mph mark towards midnight, the record was still possible. However, in the early hours of Monday a further blockage in the tank occurred and the attempt was abandoned.

Renault, Panhard and Voisin all had their eyes on the 24-Hour record, as did Major Segrave in the Sunbeam, the then-holder of the World Land Speed record. As the Sunbeam had a top speed of nearly 170 mph, Segrave was confident that he could average 110 mph, so there was considerable pressure on Bentleys to achieve. The next run was on Wednesday 28th April, Clement driving for the first 3 hours averaging 104.30 before handing over to Barnato, the average standing at 102.22 at the end of six hours. What happened after that is not recorded. Two engines had been taken out, numbers AP 316 and 1109, the latter from Barnato's own 100 mph Jarvis bodied track car, and with

TOP **Montlhéry, 1926. The No.2 two seater, still owned by Benjafield, as rebuilt for the record attempt, with body by Gordon England. As can be seen, the body sloped down well before the driver, and the steering rake was dramatically reduced. It is easy to see why the car was called 'The Slug'. Barnato is at the wheel, then Frank Clement and Dudley Benjafield. Part of the track can be seen behind them.**

ABOVE **The pits at Montlhéry, the scene of rather too much activity. On the left is the No.2 Experimental 6½ Litre MH1030, 'The Slug', then WO's Morris Chummy on the right, used as transport by the mechanics.**

RIGHT **Engine tests at Montlhéry and Le Mans. Record attempts, rather than racing, 'improving the breed'.**

problems with both of these, the team returned to England to resolve the difficulties. A stroboscope on the valve gear showed that they were hitting a valve spring period at 3250 rpm. The engine speed could have been dropped by using bigger wheels, but there was a primary camshaft period at 3150 rpm. Finally a decision was taken to use smaller wheels, which pushed the engine speed up to 3450 rpm, which it seemed to cope with without any problems.

Time was running out when the team returned to Montlhéry at the end of May. The same drivers as before were available – Clement, Barnato, Duller and Benjafield, with Kensington Moir as the spare driver. The weather was pretty awful when they started at 12 am on the 1st June, but there was little time left. For the first few hours the 3 Litre circulated monotonously at over the 100 mph mark, all on its own on the large oval track. As

24

Horse Power Comparisons

Power curves taken under similar conditions

RPM	GALLOP ENG No. 1165	DAVIS ENG No. LM1341	CLEMENT ENG No. LM1342	24 hour ENG No AP316	24 hour ENG No 1109	1925. ENG No 1127.
3500	82.4	83.9	86	87.5	87.5	83.1
3250	81.25	81.25	84	81.25	82.6	79.9
3000	76.25	78.75	80	77.5	77.5	76.25
2750	69.9	72.2	73.3	69.9	72.1	71
2500	62.5	64.58	66.14	62.5	65.6	65.6
2250	57.2	58.6	60	57.1	58.1	59
2000	51.25	52.08	53.3	51.7	51.25	—
1750	44.5	43.75	44.1	43	43.75	—

LE MANS. 1926   MONTHELERY 1926   LE MANS 1925

R W Clark

darkness fell the red lights around the circuit were switched on, the unlit Bentley circulating as if guided by some external force. All the drivers and WO had returned to their hotel when at about midnight, Duller lost it and spun round, bruising himself against the body cowling. He kept the engine running and drove round to the pits, where he got out very cold, wet and generally fed up. On hearing the engine cut out, Saunders and Pennal had gone round the track in the reverse direction in the Morris to find Duller had recovered the car, and despite being very shaken and rather bruised, said he was alright, so they drove back to the pits. On returning there, they found Duller on his own with Wally Hassan already in the car. What must have seemed a good idea, to keep the car running, albeit at a reduced speed, would have nullified any records gained after that point because the car was driven by an un-nominated person, but driving that short, sensitive and very fast car on the greasy surface almost proved to be Hassan's complete undoing as he lost it to spin through the crash barrier onto the road circuit and roll the car several times before ending the right way up, but with the car severely damaged and Hassan himself unconscious and injured.

Saunders and a Frenchman (called either de Toit or Roger) originally thought Wally was dead: "'E 'as cooked 'is goose!'". They pulled him out of the car and bundled him into the back of the Morris and found a French doctor who proved to be of little use, but the American hospital in Paris took on another patient and in due course Wally made a full recovery.

Pennal went back to the Chateau in the 6½ Litre EX 2, to find Frank Clement and they returned to the track to circulate twice before noticing that the barrier to the road circuit was down, finding the 3 Litre the right way up at a steep angle on the bank considerably damaged with nobody to be seen anywhere. "The Slug" as the car had by then been called, was not as lucky as Wally Hassan – on the return to Cricklewood, WO ordered it cast onto the scrapheap, from which it does not seem to have recovered. Doubtless parts ended up in one or other of the Works runabouts.

Although they did set the World's 12-Hour record at 100.96 mph, it seems unlikely that the 24-Hour record at such speed was really attainable. The margin was very small and the car pretty highly stressed. WO felt that it was possible, as the car could lap at 108 mph and obviously felt this particular failure deeply. It was a failure that was to be compounded by the events at Le Mans later in June

The Bentley Team for Le Mans, 1926, in the yard at the Hotel Moderne. Thistlethwayte's privately entered Supersports chassis No.1179 No.9, 'Old No.7' chassis No. LM1344 and No.8 chassis no. LM1345. From the left, Dusty Miller, the electrician, Harold Easton, Nobby Clarke, not known, Arthur Saunders, Captain Head, Wally Saunders, Billy Rockell, Leslie Pennal.

and there is little doubt that the time and effort put into chasing that elusive World record was at the expense of some of the preparations of the Le Mans cars. With the Le Mans race on the 12th/13th June, by the time all the packing up had been done at Montlhéry it was so close to the race date that Pennal and Saunders went straight to Le Mans without going back to Cricklewood. It can hardly have helped Bentleys much either – having the Managing Director (and hence prime decision-maker) off-site for so long immediately after the delivery of the first 6½ Litre cars was not a healthy position to be in. To add insult to injury, shortly afterwards the 45 hp Renault saloon driven by Garfield, Plessier and Guillon set the record at 108.3 mph, with a fastest lap of over 119 mph, with no mechanical problems at all.

After their disastrous showing at Le Mans in 1925, WO was well aware that they needed to succeed in 1926. Three cars were entered, two new Works cars, Speed Models chassis LM 1344 and LM 1345, registered MK 5206 and MK 5205 respectively, and a private entry on 9' chassis 1179 for "Scrap" Thistlethwayte. T.A.D.C. "Scrap" Thistlethwayte was another wealthy playboy,

better known for racing Mercedes cars in the late 1920s. This last entry caused considerable problems, but Thistlethwayte was adamant that he wanted a 9' car. The regulations for the bodywork were met by Martin Walter who successfully squeezed a full 4-seat body onto the short chassis – even going to the lengths of buying a bare frame from Vanden Plas for £100 to ensure their dimensions would meet the regulations. Gallop was principally responsible for preparing this car and ensuring it would pass the scrutineer. Gallop co-drove the Thistlethwayte car, with the Works cars being driven by Sammy Davis and Benjafield, and Clement and Duller. George Duller was a champion hurdling jockey and a great friend of Barnato.

For the first time ballast to match the weight of

ABOVE AND RIGHT **Burgess with EXP 4, c. 1927, after it had been rebuilt as a 4½ Litre, in Ravensdale Avenue, North Finchley. The 9'9½" chassis was retained, as can be seen from the length of the rear door. It is not known which engine was fitted. The gearbox is a B/BS/C type, distinguishable by the sloping bottom of the casing.**

three passengers had to be carried and in one of those "interpretations of regulations" moves that cause so much controversy in modern motor racing, Bentleys rather cleverly positioned some of the ballast in the form of a lead-filled steel tube between the front dumb-irons immediately in front of the radiator – a position not normally occupied by a passenger. But it did meet the letter if not the spirit of the rules, and usefully balanced up the weight of the large rear petrol tanks and stiffened up the front of the chassis. It was not found to be easy to secure the weight equivalent to three adults in such a way that it did not cause any problems in 24 hours of racing over indifferent surfaces with speeds ranging from 10-15 mph around the Pontlieue hairpin, to nearly 100 mph along the Mulsanne straight.

All three cars got away to a good start, hoods raised for the first 20 laps as in the previous year, but mechanical problems affected Clement and Duller in the form of valve stretch in the early morning. The Gallop/Thistlethwayte short chassis car retired at 9 am with a broken rocker. "Old Number 7", driven by Davis and Benjafield lay in third place for some time, and with barely an hour to go the pits hung out the "faster" sign in the hope

that Davis would catch the second placed Lorraine. Sammy arrived at the Mulsanne corner just in front, only to have the brake pedal go to the floor too late to take the escape road. With just 20 minutes to go "Old No 7" was firmly embedded in the sandbank, a deeply embarrassed Sammy Davis totally unable to extricate her in the last few minutes. The Lorraines swept the board with a 1–2–3 result. WO was bitterly disappointed and decided to withdraw from racing forthwith – it was the second of the "black" Le Mans.

Shortly after the race, Benjafield telephoned the Company to ask if any of the Team cars were for sale. He was informed that the short car was Thistlethwayte's own property and that the other two had already been sold to Henlys. Benjy repaired to Henlys in Great Portland Street, and was soon leaning on "Old Number 7", MK 5206. The salesman initially denied that the car had been caned at Le Mans, but relented when Benjy pointed out that he had been personally responsible for the treatment handed out to the car, which resulted in a suitable adjustment of the asking price! Benjafield then entered "Old No. 7" in the Georges Boillot Cup race at Boulogne on the 29th August. The 3 Litre ran in full road trim, under regulations similar to those for Le Mans. The race was held over 16 laps of the Grand Prix circuit, a distance of 371.52 miles. Thistlethwayte entered as well, but withdrew after Howey's fatal crash in the hill climb, driving a Ballot. From seventh place on lap 1, by the 13th lap the Bentley had climbed to 3rd place, helped somewhat by the fact that only four cars were left in the race. The 3 Litre covered the first twelve laps in 4 hours 25 minutes, only 4 minutes 9 seconds slower than Eyston in the 1½ litre Grand Prix Bugatti the previous day. The pit signalled Benjy to go faster as he could gain second place, (distinct shades of Le Mans), but again the brakes were not up to it. It seems likely that the rods were stretching, and without the single point adjustment in front of the driver's seat introduced for 1927 they had to stop and take up the brakes from underneath the car. Benjy entered a corner some 10–15 miles too fast, and the Bentley skidded and spun round before hitting a tree. Benjafield, his lip badly cut and leaving several teeth firmly embedded in the wheel rim, hurried to the nearest telephone to call up the pits, who were understandably relieved to hear from him as the tannoy had just announced that No. 45 had crashed and the driver been killed. The Chenard-Walckers finished 1-2-3. Moir had been nominated as the spare driver, and Benjy admitted that it had been a mistake to continue when tired with an eminently capable spare driver available. Only the three Chenard-Walckers finished, and the race organisers presented Benjafield with the Cummings Cup as compensation.

1926 was the year of the General Strike, an event that seems to have made remarkably little impact on Bentleys. In fact the strike only lasted for nine days. The shop floor remained at work, and when Tomlins returned from testing a new chassis to find a mob of agitators hanging around the gate he drove straight at them, leaving some at least to pick their way out of the hedge bordering Oxgate Lane.

A 3 Litre engine was set up with a belt drive from the flywheel to the overhead countershaft driving the lathe, so that Bartlett the turner could carry on even if the power was cut. On the last day of publication of the Government's *British Gazette* newspaper no less than 10 Bentleys (and a Rolls-Royce) all in private hands were convoying papers to Cardiff. There was even the "Brooklands Squad" of racing drivers attached to Scotland

Yard, including in its number George Duller, Leslie Callingham, who worked for Shell, and Woolf Barnato. All owned cars capable of between 90 and 100 mph, and Barnato received special mention for covering the 106 miles from London to Birmingham in 2 hrs 11 mins – only 11 mins slower than the express train.

With the financial situation now seemingly assured, development work turned from the 6½ Litre, now in full production, to a successor for the 3 Litre. It was becoming increasingly clear that the 3 Litre was hard pressed in competition and the 3 Litre Sunbeam with its twin overhead cam engine was a significantly faster car, at least in a straight

3 Litre chassis, photographed in the erecting shop circa 1926. The midships view is of a Standard chassis, with B box, strut gear and double silencer. The undershield was a standard fitting on 3 Litre chassis with the small sump engine. The prop shaft is the pot-joint, and the finish inferior to that on most modern restoration projects.

line. WO followed his adage of adding litres to gain performance and maintain reliability, and put the bore and stroke dimensions of the four cylinder up to those of the six cylinder to give 4398 cc from 100 mm x 140 mm bore and stroke. There also seems to have been a certain amount of pressure from customers to produce an enlarged 3 Litre. The RAC rating was 25 hp and initially at least the 4½ Litre was referred to as the 25/100.

At the 1926 Olympia Show considerable detail differences were shown on the 3 Litre chassis – principally the one-piece integral sump casting and duralumin rockers, and other features anticipating the introduction of the new 4½ Litre. It was offered

The rear view is of a Speed chassis, with duplex Hartford shock absorbers and the cast aluminium fish tail. These photos make an interesting comparison with the 6½ Litre photos earlier. There is a strong family resemblance between all the Bentley chassis, although this was starting to go with the 8 and 4 Litre chassis.

in standard, Speed and 100 mph versions at £895, £925 and £1050 respectively. The 6½ Litre remained basically unchanged, in standard form only with single Smiths carburettor, but with a choice of wheelbases from 11′ to 12′6″. A sectioned 6½ Litre engine was exhibited along with a Barker Sedanca, and for the 3 Litre, a Vanden Plas Speed Model tourer and a long chassis with a Mulliner coupé body. The Light Tourer was renamed the Short Standard Model, identical to the standard chassis but a foot shorter (9′9½″ as opposed to 10′10″) with a chassis price of £850. The 6½ Litre chassis was £1450 for the 11′ or 12′ model, or £1500 for the 12′6″ chassis.

Although WO's intention had been to withdraw from racing and concentrate on production, a new team of cars was under preparation for the 1927 season before the end of 1926. With Barnato as Chairman such a move was inevitable and Benjafield had already proposed his 1926 Le Mans car "Old No. 7" as the nucleus of a new team, Sammy Davis immediately agreeing to partner him.

Indeed, it was in late 1926/1927 that the Company really started to take the whole racing business seriously.

Sammy Davis explained the position lucidly in *Motor Racing*: "In 1927, as may well be imagined, Benjafield and myself were determined to retrieve our reputations at all costs, having had some difficulty in facing WO Bentley, and having acquired, too, almost complete knowledge of the exact costs of having prepared a team of three cars in 1926, which gained no return whatsoever. There was not a detail into which WO was not only too anxious to go that we might appreciate the thing the better. Not for the first time we realised that racing is a business for the firm concerned, if a sport for its drivers."

The result of this was a systematic and professional approach. In the autumn before the following year's events, a proper racing programme was set out, with a committee of heads of departments directed by WO. After the events had been chosen the regulations would be studied, previous experience called on and the technical considerations would be sorted out. These would include such things as engine modifications and compression ratios, back axle ratios, and fuel tank design and capacities. With the plan laid out, the accessory manufacturers and the coachbuilders (Vanden Plas, of course!) could be contacted, and the requisite arrangements made.

The chassis required would be allocated for work to start on them by January, to ensure that the cars

ABOVE **This 6½ Litre, YP4205, seen here in France with WO on holiday, is believed to have belonged to Geoffrey Joel, a cousin of Barnato. This is possibly the car that was used for steam cooling experiments, over the Galibier pass.**

BELOW **A pensive WO on holiday in France.**

were ready for the first race of the season. In 1927 and 1928 the Essex Car Club's Six-Hour race was used as a shake-down for Le Mans, in 1929 and 1930 that function being served by the Junior Car Club's Double-Twelve race. For the Le Mans race the body regulations were very tight, and the French scrutineer was, not unnaturally, more inclined to be liberal with his fellow countrymen rather than the British. The 3 Litre engine at 2996 cc was so close to 3 litres that all the engines were inspected by the RAC scrutineers in London, and on the race engines the letters "RAC" can still be found stamped on the cylinder block and the crankshaft. By 1929 the logistics of the Team were such that some five cars, 12 mechanics and two tons of spares were carried to Le Mans. Even though virtually all the Team cars were owned by the drivers, the true cost of the racing as opposed to the "above the line" figures quoted by WO must have been pretty high. But then, as in any large organization it is almost impossible for anyone but an accountant to directly relate the cost of their activities to money when parts and people just seem to appear on request and the effects on other (possibly profitable) activities are never seen.

"Old No. 7" had to be substantially rebuilt after Benjafield's crash in the 1926 Georges Boillot Trophy, and Davis and Benjafield were determined that the car should have at least one race without driver error. Glen Kidston's 3 Litre MK 5205, No. 8 from the previous year was also rebuilt for the 1927 season, and a new car added, chassis ML 1501 registered YF 2503. The trial run for Le Mans was the Essex Car Club's Six Hours Race at Brooklands on the 7th May, run over a 2.616 mile circuit with artificial chicanes with similar regulations to the Le Mans race. Four Bentleys were entered, MK 5205 driven by Clement, MK 5206 driven by Barnato and Benjafield, YE 6029 driven by Tim and Archie Birkin, and YO 3595 driven by Callingham and Harvey. Unfortunately, shortly before the race all the cars were fitted with a new design of duralumin rocker, because of the valve period at about 3000 rpm. There was concern that the steel rockers would fatigue, as the planned race speed corresponded to 3000 rpm. However, Tim Birkin said that such a move at the eleventh hour was madness, and refused to have the new rockers fitted to his 3 Litre – as events would later show, a wise decision.

During the race the duralumin rockers started to break up. This has variously been attributed to a machining error or a design error over a misplaced rivet hole. Certainly the early rockers used a "T"

shaped strap to retain the roller and this was later changed to a simple strip, deleting the top hole at the most highly stressed point of the rocker. Soon Bentleys were circulating with some of their valves out of action, and then as they retired were robbed to keep their team mates moving. But it was all to no avail, as all three of the cars with the new rockers retired, leaving only the Birkin car with the original steel rockers still running. Archie, the younger of the two, was called in because of his wild and erratic driving to be replaced by Frank Clement, and to add insult to injury the gearbox played up, the only solution available being to lift the lid, bash third gear into mesh with a copper mallet, and drive the rest of the race with the rev counter in the red. On the A type gearbox the mainshaft gears slide on two long keys that are located in grooves in the mainshaft and then held by three rivets. In service the keys can fidget and break the rivets, the heads of which then rise up and foul the gears. However, the Birkin car was fitted with a C type box, which like the D box designed to replace the A has a solid multi splined mainshaft, so it is difficult to see what caused the problem, unless it was a broken selector. It was possibly a simple case of abuse; Hillstead commented on Birkin's persistent over-revving in the intermediate gears. They still managed third place behind a Sunbeam and an Alvis driven by erstwhile Bentley Boys George Duller and Sammy Davis respectively.

WO's position as Managing Director, one that he had held more or less since the formation of the company, (HM and WO both held the role of Managing Director at different times, or possibly jointly) changed in April 1927 when the Marquis de Casa Maury became a director. Maury was a banker, of Cuban origin, and was appointed as "Assistant Managing Director". His directorship was confirmed in a return of 17th May. Whether he ever acted as WO's assistant, or was appointed as joint Managing Director (as his post was later described), straight away, is not known. As Barnato had little faith in WO as a businessman (an opinion expressed by WO himself) the appointment of a banker with the attendant financial experience should have helped to make the Company a more viable commercial proposition. In fact it seems to have made little difference and the firm continued much as before. The only evidence of Maury's activities as Joint Managing Director are in the fields of advertising and racing. Dora Steele (secretary to W.K. Forster) commented in a letter to the author: "We saw very little of Casa Maury and the general impression was [that] his name was useful on the

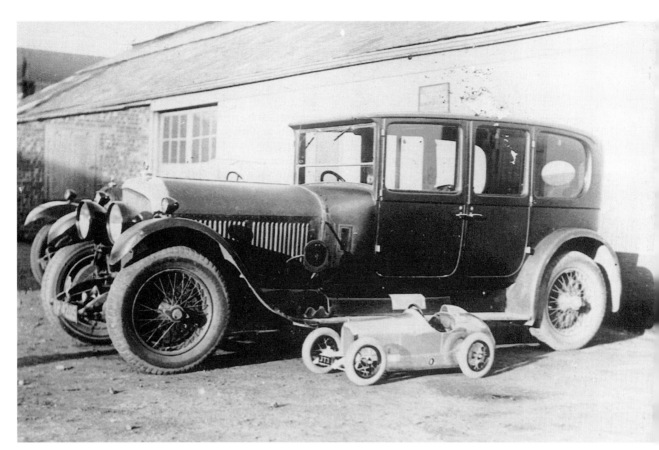

ABOVE **WO with his second wife, Audrey Gore, during the same French holiday. She later introduced him to Margaret, who became his third wife, after their divorce**

LEFT **YP4205 at Cricklewood, outside the running shop, with the model Bentley. This car is thought to be chassis No.WB2566, fitted with a 'Lucifer' saloon by Hamshaws, owned by Geoffrey Joel. In 1929 it was fitted with experimental Smiths lamps and a 12V high frequency horn.**

BELOW LEFT **The exterior of the finished cars shop in 1926. The car is 6½ Litre chassis No. WB2563, fitted with special hunting coachwork for the Nawab of Bhopal by Thrupp & Maberly.**

paper." In WO's words, Maury was no slouch, but seems to have had more interest in coachwork (he drew out the bodies for his own cars), motor racing and women, than in the running of the business. Hillstead adds to WO's comments that Maury was lacking in commercial experience, which certainly seems to be borne out by the fact that his presence seems to have had little, if any, impact on the management policy of the Company in terms of averting the coming crises. Under the name Mones Maury, the Marquis had driven a Brescia Bugatti to third place in the 1500cc category of the 1922 Isle of Man TT race at an average of 49.0 mph, competed in many Continental races and driven in the 200 Miles race at Brooklands. Maury also added a touch of splendour to Bentley's brochures, which tended to be fairly lavish hardback presentations running to 30 pages. Competition among small boys to get a Bentley catalogue or a Bentley tie-pin was fierce!

Developing the 3 Litre into the 4½ Litre.

In general terms the 4½ Litre was more or less a stretched 3 Litre with a host of detail changes and improvements. The engine was fitted with a larger crankshaft, with the same 55 mm mains and journals dimensions as the 6½ Litre. The bigger bores reduced the masking of the valves, which in the 3 Litre are tucked away in the corners of the combustion chambers. Valve gear was the same pattern as the 3 Litre, but duralumin rockers were developed, in separate rocker boxes, these also being adopted on the smaller engine. The new camshaft BM3481 used the same cam profiles as the 3 Litre Speed Model camshaft BM1800, the only real change being the introduction of the one-piece sump with a much improved oil-pump design – but again this had already become a standard fitting on the 3 Litre. Power output went from 85–88 bhp of the smaller engine to 110 bhp, depending in both cases on the state of tune. The chassis frame was the 10'10" 3 Litre standard frame, with a new radiator, bulkhead and a one-piece casting for the Perrot shaft, radiator trunnion and headlamp support, similar in concept to that used on the 6½ Litre. Steering ratio was dropped to 10.3:1 with the fitting of Dunlop wellbase 5.25 x 21" tyres, which had been standard on the 3 Litre since the 1926 Olympia Show. The C type gearbox was used, basically the same as the BS type but with a slightly lower third gear ratio, and a larger 16-gallon tank with a Hobson telegauge was fitted. In virtually all other details the 4½ Litre followed the basic outline of the 3 Litre Speed Model.

The C type gearbox came about to meet a need for a universal gearbox that could be used on all the cars. The A type and its successor, the D type, were only suitable for sporting cars, while the B type was only suited to town carriages. The BS box was not felt to be suitable for the 4 cylinder cars. The solution to this problem was the C box, virtually identical in design to the B box but with a solid, multi splined mainshaft and with the gear lever shaft extended right across the box and supported in a bronze bearing on the nearside. An improved speedo drive unit was adopted, the same as that used on the D box, with a single fibre gear as opposed to the split bronze gear used on the B/BS boxes that used to chew up the steel speedo drive gears because of bronze particles off the split gear meshed with the 3rd/top gear rubbing off during crash changes. The C box ratios were intermediate to those of the A and B boxes, and the C box was used on the 4½ Litre, late 3 Litre Speed Models,

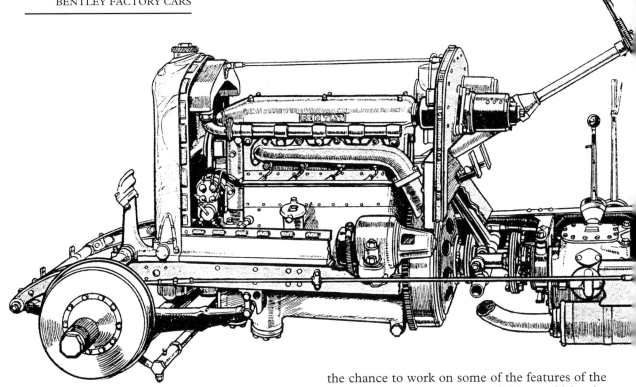

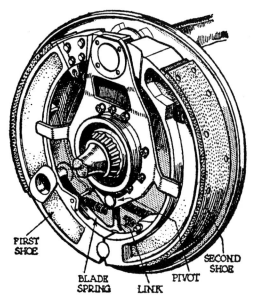

FIRST SHOE
SECOND SHOE
BLADE SPRING
PIVOT
LINK

Detail of the semi-servo 'self-wrapping' brake introduced on the 1929 4½ Litre at the 1928 Motor Show. This brake was used on the 4½ Litre, 4½ Litre Supercharged, 4 Litre and the three Team Speed Sixes.

and the 6½ Litre in both standard and Speed Six form.

It is interesting to compare the 4½ Litre chassis with the 3 Litre. In many areas, the design features are improved, probably because the designers had

the chance to work on some of the features of the earlier chassis. An example of this would be the petrol tank mounting. The 3 Litre tank is mounted at each side by two angle brackets sweated and riveted to the sides of the tank, with two 3/8″ bolts per side, packed out with rubber washers. The last mounting was of similar construction, attached by a single bolt to the rear cross-member. The mountings of the 4½ Litre tank are much better designed. At each side, a casting is sweated and rivetted to the tank machined to a parallel portion on which is mounted a spherical bearing as used on the pedal shaft and the compensator, pinned on one side to provide lateral location for the tank. The bearings are then supported in half castings bolted to the chassis rails, with aluminium packers (hand) fitted to align the tank profile and that of the side rails. The third mounting consists of an angle bracket sweated and rivetted to the rear of the tank, picking up on a stud fixed to the rear tie bar between the dumb iron knuckles. The general arrangement subjects the tank to much less stress from chassis flex.

In other areas, the design is considerably tidied up in that attention was paid to details that are only considered as afterthoughts on the 3 Litre. Examples of this are the bulkhead arrangement. On the 3 Litre, the wiring simply runs round the back of the casting, clipped at intervals. The horn, fuse box and cut-out were then screwed on almost as an afterthought (indeed, many 3 Litres have the fuse-box mounted on the rear face of the bulkhead inside the car). Where the wiring becomes visible

The 4½ Litre chassis circa 1928, an interesting comparison with the 3 Litre chassis shown earlier. Note the cone clutch with fabric Hardy disc, D type gearbox, brake compensator in the rear position, Spicer type propshaft and the 4½ Litre diff casing with forward-facing filler. The latter feature was necessitated by poor access to the rear filler caused by the larger 16 gallon tank. Note the trunnion mounting of the tank, B&D friction shock-absorbers and Bluemel sprung steering wheel.

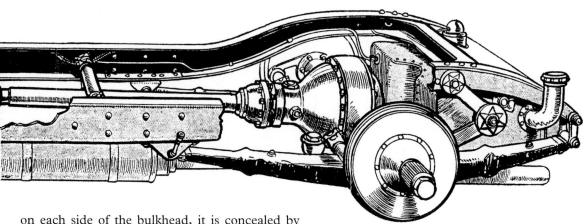

on each side of the bulkhead, it is concealed by pressed covers. On the 4½ Litre the wiring runs inside channels cast into the bulkhead, closed off by aluminium plates. The fuse box, cut-out and horn are mounted on features cast integral with the bulkhead. The general air is that the 4½ Litre chassis was thought through in greater detail than the 3 Litre, which is hardly surprising under the circumstances – it would be surprising if it had not been. Similar comment could be passed about the 6½ Litre, and again about the 8 Litre chassis. The first 4½ Litre engine number EXP 5 was on test in the Engine Shop in February of 1927, and the first chassis ST 3001 on the road soon after.

BELOW **Detail of the cast aluminium chassis bracket acting as headlamp stand, radiator support and Perrot bearing all in one, similar to those introduced on the 6½ Litre.**

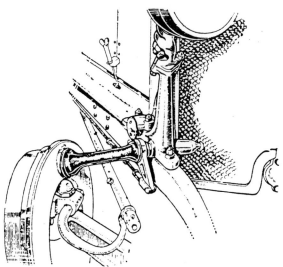

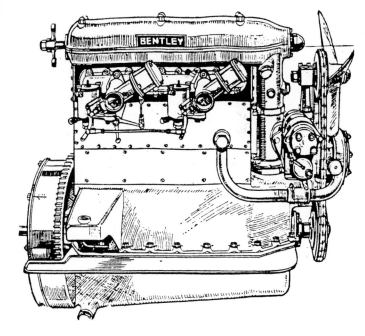

ABOVE **Early 4½ Litre engine with twin 'sloper' carburettors. Note that the 4½ Litre block was square all the way down, unlike the 3 Litre which tapered to the bottom. Hence the 4½ Litre block could be easly fitted with closing plates to tidy it up.**

11 Competition Successes

BY THE TIME Le Mans came round again, Bentleys were badly in need of a good result. Following early successes at the TT in 1922 and at Le Mans in 1923 and 1924, with the exception of the World's 24 Hour Record at Montlhéry in 1925 they had suffered a string of defeats (Le Mans 1925 and 1926, the Montlhéry records saga in 1926, and the fiasco at the Six Hours race), such that WO was almost forced to terminate the racing programme. It is clear from WO's comments that had they failed again at Le Mans in 1927, it would indeed have been the end of the Works Team. While they might have been financially secure, their reputation could not stand further defeats in competition.

Six weeks after the Six Hours race, at Le Mans on the 18th/19th June, "Old Mother Gun" (as the prototype 4½ Litre came to be known) was ready for the race, so Kidston's 3 Litre MK 5205 was dropped from the team. It seems suprising that the 4½ Litre should have been entered so suddenly for a major race, before the model was truly into production and before any long-term evaluation of the new engine was possible, albeit using existing proven components. ST 3001 chassis was passed off on 12th June 1927, six days before the race. The next 4½ Litre chassis was passed off on 25th August – six weeks later. It is likely that the cylinder block, built by Billy Rockell with 2mm oversize valves was fitted to the bottom end of EXP 5 engine for the race. EXP 5 was sent from the Experimental Shop to the Engine Shop to be tested, starting on 11th February 1927. The engine used a BM1800 3 Litre camshaft with rockers and rocker boxes from a 6½ Litre engine. In this form it gave power readings of up to 104 bhp at 3500 rpm, comparing favourably with power outputs of 87.5 bhp from the very hot 3 Litre engines used at Montlhéry in 1926 (which did not have a good reliability record) and 86 bhp from the hottest 3 Litre Le Mans engine (LM1342, No. 8 in the 1926 Le Mans race). The engine ran for varying tests, finally running No. 1 big end. It was then rebuilt and run for a further series of tests, totalling 21hrs 17 mins, before breaking No. 4 rod after running at 3500 rpm. It seems suprising that in view of their run of failures and the design errors resulting in the Six Hours fiasco that they should have exposed their latest offspring to the rigours of Le Mans – unless it was perhaps some form of calculated risk. The implication is made in *The Other Bentley Boys*: "Created in adversity, and thrust into the limelight before it even had its own chassis, the 4½ Litre car...had been an immediate success."

WO wanted Sammy Davis to partner Frank Clement in "Old Mother Gun", which should have greatly increased Sammy's chances of success, but Sammy stood firm and sided with Benjafield to drive "Old No. 7". Callingham partnered Clement and the new 3 Litre (YF 2503) was shared by Duller and d'Erlanger. With three cars competing it was a large group that converged on the Hotel Moderne a full week before the start of the race for practice and preparation, but with an increasingly organized and well-drilled team the Bentley mechanics had plenty of time to tease those panicking and rushing around frantically trying to get their cars to the line. According to *The Autocar*, the Bentley drivers and mechanics were chiefly concerned with setting and focussing headlamps, and working on the carburettors. Memini carburettors were tried and found to give a useful increase in performance, but suffered from flat spots of insuperable

4½ Litre cylinder block. Because of the bigger bores of the larger engine, the valves are not tucked away in the corners of the casting as they are on the 3 Litre.

BELOW The first 4½ Litre 'Old Mother Gun' outside the experimental/racing shop (left hand door) at Cricklewood, circa May/June 1927. This car was basically a Standard 3 Litre 10'10" chassis with the larger engine. Seated in the car is Frank Clement, with behind him from the left Clive Gallop, Dudley Benjafield and Sammy Davis. The 4½ Litre has the 'push-on' brake operation with the Perrot shaft levers reversed, external oil filler neck and no strut gear. Note the single-acting quick release radiator cap and the big tank just visible at the back.

ABOVE **The new No.2 car for the 1927 Le Mans race, YF2503 chassis No.ML1501, outside the experimental/racing shop. Behind is the office block, with the drawing office on the left, more offices, then the door. If one had gone through this door, enquiries was immediately in front, WO's office off to the back right.**

RIGHT **The team cars in the yard at the Hotel Moderne. Just visible in the garage at the back is Callingham's own car, YO3595. Between Nos.2 and 3 Duller, Nobby Clarke and Captain Head of *The Autocar*. Sammy Davis is in the beret between Nos.3 and 1. Note the spare magnetos in the back of No.1, Memini carburettors, and screw pattern petrol tank cap. The oil filler neck can be seen on the offside of the dashboard.**

proportions. *The Autocar* said similar things about the Zita carburettors that were also being worked on, the problems being such that with the Memini carburettors fitted, No. 3 caught fire near Mulsanne, but this was put out before any damage was caused. Frank Clement decided to refit the faithful SUs for the race. Clement in the 4½ Litre was putting in practice laps at over 71 mph, Davis lapping at 69 mph, but even at 5am they were being bothered by carts and cyclists.

WO brought along with him "The Box", the No. 2 Experimental 6½ Litre EX 2 fitted with a very square and rather ugly saloon body, while some of the mechanics had a rather unpleasant night-time run from Dieppe in a lorry driven by a Frenchman who got rather too far into the swing of the racing business. There was every reason for optimism at the start of the race, when the drivers dashed across the track in traditional Le Mans style to raise their hoods before getting away for the first 20 laps. No replenishments of any sort were allowed for the first 20 laps, and the cars were only allowed to start on

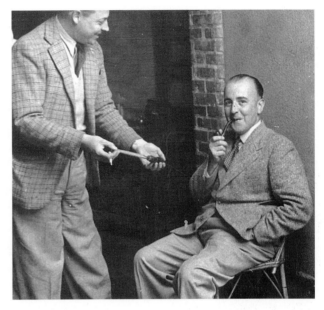

LEFT **Nobby under 'The Box', at Le Mans, 1927.**

ABOVE **A relaxed WO in France. The gentleman with the file is not known.**

the electric starter. The starter also had to be used after pit stops – no push starts were allowed. In those days the Le Mans race was intended as a proper test of all aspects of production touring cars.

Clement put up his usual polished performance to get away first, closely followed by Benjafield in No. 3 and d'Erlanger in No. 2. For some time this trio circulated in front with the 4½ Litre comfortably breaking the lap record in Clement's hands, before handing over to the second drivers after 20 laps. The three time-keepers in the Bentley pits with their six stop watches (and spares) clicked

ABOVE LEFT **'The Box' in the yard at the Hotel Moderne at the time of the 1927 Le Mans race, with two of the Team cars. No.3 is chassis No.LM1344, 'Old No.7' from the previous year. No.1 is the new 4½ Litre 'Old Mother Gun'.**

LEFT **'The Box', the No.2 Experimental 6½ Litre MH1030 fitted with saloon coachwork in place of the earlier touring body. MH1030 was used in this form by WO for several years, and is seen here in France on steam cooling trials. This method of cooling is similar to that used on stationary engines, and the extra bottom tank can just be seen behind the number plate. The body is pretty awful; this car was also sometimes referred to as 'The Coffin'.**

away as the three cars passed in line astern well into the evening, before first Callingham and then all three cars were overdue and the pit members looked at each other anxiously wondering what could possibly have happened.

In fact at White House Corner, Tabourin had crashed his Th. Schneider, ending up across the track and Callingham had run into it blocking the road. Duller had piled into the wreckage next, vaulting over the wheel to safety, but too late to run back and warn Davis of the blocked road in front of him. Some sixth sense, aided by "a scatter of earth, a piece or so of splintered wood" in the road and a headlamp seemingly pointing vertically upwards, warned Sammy, who came around the corner slower than usual, but was unable to avoid running into the mass of cars that seemed to fill the whole of the road. Davis and Duller then set off in search of Callingham, finding him dazed and partially deafened from the continual noise of the 4½ Litre at racing speeds. They managed to extricate "Old No. 7" from the wreckage and finding a way through, Sammy drove gingerly back to the pits. The damage was fairly extensive – the front axle was pushed back, the offside wing buckled and the head and side lamps on that side damaged, the dumb-irons bent and (not found until after the race!) the neck of one of the steering ball joints cracked. But the Bentley was still driveable, so

Victory line-up after the event. The crash damage to the car can be seen, to the offside front. Sammy Davis stands up in the driver's seat, Benjafield with bouquet consoles Chassagne, the failure of his Aries let the Bentley through to win.

Davis lashed down the wing with wire and tied a Smiths Wooton lantern to the scuttle on the off-side, was allowed by the officials to replace the buckled wheel, and drove on. That another of their products was used as a stand-in for the smashed (Smiths) headlight was something that Messrs Smiths Industries were rather proud of. Two mechanics were sent to guard numbers 1 and 2 to prevent any moveable bits being robbed. The brakes operated in firing order on the four wheels and without the offside headlamp some of the corners were very trying in the dark, but "Old No. 7" kept on going and Davis was able to hand over to Benjafield at the allotted time.

Benjy took over with only Davis' advice on the state of the car and gradually increased his lap speeds, but Chassagne and Laly in the 3 litre Aries were still several laps in front and seemed unbeatable. The battery box had to be further lashed up, but by late morning, still several laps down on the Aries, WO and Nobby thought they could hear untoward noises from the Aries' engine, which they thought to be coming from the camshaft drive. The "Faster" sign was hung out for that battered Bentley. And faster they went. Davis was rewarded by seeing the Aries stopped by the road with Chassagne's head under the raised bonnet for two laps running. Finally the Aries was delayed at the pits and retired with a broken camshaft drive, the Bentley sweeping into the lead with less than an hour to go, with Benjafield stopping to give Sammy Davis

the honour of driving the last lap.

At 4 pm on the Sunday they had covered 1472.6 miles at the average of 61.36 mph, some 90 miles more than the regulation minimum distance for a 3 litre car of 1383 miles. Chassagne and Laly must have been bitterly disappointed, but the Bentley team were euphoric, the publicity accruing from the dramatic circumstances of that victory running on for some considerable time. The story of "Old No. 7" at Le Mans has gone down in history as one of the legends of motor racing. Back in London, Sir Edward Iliffe, the proprietor of *The Autocar*, threw a celebratory dinner for the members of the victorious team at the Savoy and during this he proposed a toast to "a lady who should be here". The double doors were then opened and "Old No. 7" wheeled in with her engine running. Both axles had to be removed to haul the car up the stairs, and then refitted in the banqueting room behind a screen. The tables were specially arranged in a horseshoe so the Bentley could sit in the middle, still unwashed after the race and with all the battle scars still showing. Davis and Benjafield were also wined and dined by Bentleys on the 29th June and again by the RAC on 1st July. Woolf Barnato announced during his speech that in 1928 all the drivers would be issued with a gyroscope and a parachute, alluding to George Duller's remarkable leap over the dashboard and out of his car. Duller was often to be seen geeing his car on as if it were a horse.

Sammy Davis was later honoured with the freedom of Le Mans, and a street there was named after him. At Pollen House the concierge maintained press cuttings books, and included in that for 1927 was a cutting from the *Daily Express* of 22nd June 1927 headed "British Motor Triumph – damaged car wins endurance race – new type lorry" but the bit referring to the lorry appears to be missing. Perhaps the apocryphal comment credited to M Bugatti was coined by the *Daily Express*! There was some consolation for the unlucky Frank Clement, who received 1000 francs from each of the Boulogne section of the Automobile Club du Nord de la France and Morris-Leon-Bollee.

In July 1927 Chairman Barnato put his hand in his pocket and lent the Company £35,000 in the form of a debenture. The security was "the Company's undertaking and all its property whatsoever and wheresoever both present and future" (again) subject to the June, 1923 debenture of £40,000 taken out by the old Bentley Motors (the July 1919 firm then in liquidation), with The London Life Association. Barnato's debenture ranked second to that of The London Life, in the event of the Company being wound up.

After Le Mans Clement and Duller entered "Old Mother Gun" for the Grand Prix de Paris in August, a 24 hour race run over a twisting circuit based on the Montlhéry track. The race started at 6 pm on the Monday and the weather was dreadful that evening – driving rain with a violent gusting wind. The 4½ Litre won easily at a very low average of 52.1 mph from an 1100 cc BNC, but not before catching fire twice, the second time losing 45 minutes at the pits. The Bentley travelled 1247.7 miles, 85 more than the BNC. The promoters went bankrupt during the race itself, so for their efforts Clement and Duller received neither cup nor prize money.

"Old Mother Gun" was in action again at the beginning of September in a rather unusual event – the Surbiton MC's fuel race. This was held at Brooklands over 150 miles on an artificial road course, on a handicap basis for fully-equipped sports cars with limitations on the amount of fuel allowed. First off at 2 pm were the 747 cc Austins, allowed 4¾ gallons. Last off at 54 seconds after 3 pm were the scratchmen, Barnato in "Old Mother Gun" and Richard Watney in the Stutz, each

allowed 17¾ gallons. Rubin and Scott entered their 3 Litres, Scott using up his allowance of 12½ gallons after 130 miles. Dingle won in the Austin Seven, at an average of 52.11 mph, followed by Malcolm Campbell's Bugatti at 72.69 mph and Barnato was third averaging 74.28 mph. Barnato's average speed was the highest in the race, but on handicap he started nearly ten minutes behind Campbell. After the race there was some criticism of the seemingly generous allowances made, forgetting the dramatic decreases in consumption brought about by driving under racing conditions (shades of Le Mans 1925). Dingle used 4 of his 4¾ gallons (37½ mpg), Campbell's 2 litre Bugatti only 6.2 gallons of his 8½ gallons, but Barnato used 17.7 of his 17¾ gallons at an average of 8.45 mpg! Scott's 3 Litre averaged 10.4 mpg.

The last major race of the year was the Georges Boillot Cup at Boulogne, but as usual there were no prizes for Bentleys. Callingham took over Scott's entry with MK 5205, but retired with mechanical problems. The oil pump seized, and it took so long to effect a temporary repair that although Callingham got going again, he was stopped by the officials. As Scott was the first owner of YF 2503, the No. 2 Le Mans car, it is possible that he scratched his entry because of the not inconsiderable damage done to the front end of his car in the White House pile-up. He did, however, enter YF 2503 for the Boulogne speed trials and hillclimb, finishing second on aggregate in the 3 Litre class to Tourbier's Panhard.

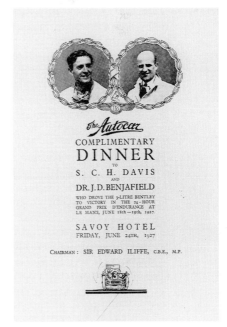

The menu for the dinner chaired by Sir Edward Iliffe at the Savoy.

12 More and More Litres

AT OLYMPIA ON THE Bentley stand no. 126, the new 4½ Litre was officially presented to the public. The 3 Litre stayed in limited production until 1929, but few chassis were sold after 1927, the car being almost entirely superseded by the bigger model. The chassis prices remained unchanged at £895 for the standard chassis and £925 for the Speed Model. The 6½ Litre was considerably revised in that the dynamo was moved from the bulkhead to out in front of the radiator (cutting in at 12 mph and charging at 15 amps), with a new bulkhead with centre casting, new radiator with bottom cut-out, and a new louvred bonnet in place of the earlier plain bonnet with two opening vents on each side. A damper was fitted to the back of the camshaft to make up for the lost damping action of the dynamo. Despite the high cost of these changes, most of the earlier cars were returned to the Works to be updated to the latest model, in much the same way as the introduction of four-wheel brakes had seen many cars returned to Kingsbury for upgrading. However, unlike the four-wheel brake update, these updates to the early 6½ Litres were made at Bentley's expense. This might have been a very decent thing to do, but does seem to have been a rather remarkable move by a commercial concern, the financial implications of which had a considerable impact on profits.

The 4½ Litre was offered as standard with a 10'10" wheelbase, but a short chassis 9'9½" wheelbase variant was available to special order. Only ten of these out of a total production of 667 cars being made, despite which the short 4½s are now reckoned by some to be among the most attractive and pleasant handling of all the vintage Bentleys. The chassis price of the 4½ Litre was £1050, £1295

with a Vanden Plas 4-seater, or £1495 as a Weymann saloon. The 6½ Litre chassis price was £1575 for all the wheelbases offered. The new Bluemel sprung steering wheel was fitted to the 4½ Litre, an interesting example of racing improving the breed. The previous solid wheel, consisting of an aluminium centre and spokes cast directly into a steel rim, was very hard on the driver's wrists and arms under racing conditions. The sprung wheel, consisting of spring steel spokes riveted to a central boss and screwed to a steel rim, was much more comfortable and was used on the first 4½ Litre "Old Mother Gun" at Le Mans in 1927. The disadvantage, of course, of a sprung wheel is that in the event of an accident the wheel rim and spokes merely bend out of the way and the driver is impaled on the end of the column. Benjafield had some reason to be grateful for a solid wheel at the Boillot Cup in 1926, when he only sustained bruises across the chest after hitting a tree. The sprung wheel was also fitted to the 6½ Litre.

The Motor's description of Bentley's programme for 1928, published on 13th September 1927, makes interesting reading. After stating that the new 4½ Litre had been announced some months ago, it then goes on to say that for 1928 a "bold and handsome design of radiator" was fitted and the front end of the chassis cleaned up by the provision of a single aluminium casting on each side, acting as a radiator support, Perrot shaft bearing and headlamp bracket. The implication is that some other design had earlier been used or proposed on the 4½ Litre and this is confirmed by the list of Production Modifications. The new bracket was introduced at chassis ST 3004 and it seems that the radiator design initially proposed was the same as that used

ABOVE AND ABOVE RIGHT **The first publicity photographs of the new 4½ Litre engine, July 1927. The engine is very similar to the late 3 Litre big sump engine, with duralumin rockers in rocker boxes, cone clutch, large water pump, SU G5 'sloper' carburettors, ML ER4 magnetos and fan. The thermostat was later fitted to the 4½ Litre engine and the fan deleted, because the engines were over-cooled.**

BELOW **First publicity photo of the 4½ Litre chassis, July 1927. Again, the chassis is very similar to the late 3 Litre, with rear mounted compensator, solid cast aluminium steering wheel, Hartford shock absorbers, square flange exhaust manifold and bronze pedal shaft trunnions. All these were later changed. The bulkhead is the early pattern with the wiring conduits round the periphery, with Smiths fuse box and cut-out.**

1928 specification Standard Six engine, with the Smith carburettor. Ignition is now by magneto/coil, with ML ER6 magneto on the nearside of the engine and Delco-Remy MRS12 coil unit on the offside.

ABOVE LEFT **Rear three-quarters of the 1928 specification Standard Six chassis; this is chassis no. PR2302, a 12'6" chassis that was the Scottish Show car of 1927 fitted with a limousine body by Hooper. The chassis has the 26 gallon tank with side filler and Hobson telegauge, with piping from the petrol tank 'bump' running down the nearside of the frame to the gauge on the dashboard. The telegauge was an effective instrument, but all too liable to failure of the pipework. The pot-joint prop shaft is still fitted, along with B&D friction shock absorbers, and the exhaust pipe goes over the axle casing. It is noticeable that the inboard, footbrake levers on the back axle are the short pattern. On the later cars, increasing capacity to alter braking leverages front/back and side/side were provided.**

LEFT **Centre view of chassis No. PR2302. This shows the C type gearbox with the special gate, to enable better fitting of the floorboards to prevent heat and fumes getting into the cars. The author knows of only one surviving example of this pattern of gate. Below and behind the lever is the lid of the Dewandre servo. This is an early installation, so the main brake adjuster is still a wing nut arrangement behind the rear gearbox cross-member. Note that this late pattern compensating shaft has five holes in the master lever and three holes in the four footbrake levers; the earlier compensators had two in the master lever and only one in the footbrake levers. This system was introduced to enable more precise setting of the braking system, as the $6\frac{1}{2}$ Litre cars had a much greater spread of wheelbases and completed car weights and weight distributions than the 3 or $4\frac{1}{2}$ Litre cars. Also clearly visible are the 1" diameter tubes to the front brakes and the outside battery trays.**

LEFT **Offside of the 1928 model Standard Six chassis, here seen in the erecting shop at Cricklewood; two further chassis can just be seen in the background, with the doors through into the running shop in the right background. The electrician's cage is in the left background. The pipe clipped to the bulkhead by the steering column is the oiler for the servo. Note the lead seal over the camchest, by the Delco-Remy ignition unit.**

ABOVE AND LEFT **Two views of a 1928 Standard Six chassis on test; the exhaust can just be seen in the offside rear view, showing that the pipe had to have a bend in it and be taken further out to clear the Dewandre servo. This well illustrates the rather crude test equipment. The weighted box over the back wheels was introduced on the 6½ Litre chassis, because of a need to weigh the back down on the long chassis compared to the 3 and 4½ Litres. Each tester had his own wheels, which were evidently used until the tyres were totally bald, and his own batteries. In this case a bit of a swop has gone on, because the battery marked 'H' was George Hawkins' and that marked 'DAB' Bowie's. The batteries are the 6V Young, two of which were used in series.**

ABOVE RIGHT **George Hawkins at the wheel of a late Standard Six chassis at Cricklewood, among the sheds at the north end of the site. These were knocked down later, and the extensions to the detail shops and the machine shop built in their place. This chassis is slightly later, as it has the external single-point brake adjuster, eliminating the need to get out and get under to adjust the brakes. The wing-nut adjuster for this can be seen on the chassis rail behind the handbrake lever.**

on "Old Mother Gun". One such was used on the Works Practice car for Le Mans 1928, 1929 and 1930, registered YW 3774 and generally referred to as EXP 5. That car was fitted with a Gallay radiator, number 18093, 18½″ wide and 2′9″ from filler to bottom. Because of the additional depth the bottom of the radiator was cut away to clear the starting handle nosepiece, in the same way as the production 4½ Litre. Three examples of this special radiator are known to exist, one on a 3 Litre in Canada, one on YW 3774 and that of "Old Mother Gun" on the ex-Fry 3/4½ Litre DLR 8.

The London Life Association were tapped for funds again on the 25th October 1927, for £40,000. This was secured against the Oxgate Lane factory, shop and premises at No. 3 Hanover Court, premises near Kingsbury Road (the Service and Racing Department buildings leased from Vanden Plas), two endowment policies (on HM and WO respectively, taken out with the London Life and amounting between them to roughly £16,000) and a floating charge on the Company's assets. Basically, this loan was used to pay off the £40,000 mortgage taken out on 26th June 1923. The interest had been paid up on the old loan, so The London Life lent Bentleys a further £40,000 to pay off the old mortgage, this money being retained by them and becoming a new mortgage made with the reconstituted March 1926 Bentley Motors. This mortgage had first call on all the Company's assets and property, and would not be called in before the 15th June 1933.

However, should the interest not be paid within 15 days of the due date, the Company cease trading, a receiver be appointed or the security of the Company be at risk, then The London Life could call in the whole amount owing. It would seem that this step was merely to tidy up the financial arrangements of the reconstituted Bentley Motors. It is possible that the insurance policies were those on WO that caused complications when WO rode with Birkin in the Tourist Trophy. It is strange that Barnato's debenture was also secured against the Oxgate Lane factory!

Although the finances were more secure, production still faced problems with the supply of parts. George Hawkins related the following tale: "There was a period, a short one, fortunately, when the firm that supplied Bentleys with the front brake shoes for the 4½ Litre had some difficulty in obtaining the linings. As a result, 4½ Litres were built complete except for the front brake shoes. The brakes on all Bentleys were not hydraulic but operated by rods and levers, so it was decided that we could continue to test the 4½ Litre without front wheel brakes provided we remained conscious of that and not do any speed testing that could be done later when supplies of front brake shoes arrived.

"I was testing one of these one day and returning to the Works at Cricklewood via St. Albans and onto the Watford road when near the Three Hammers pub away ahead of me I saw a milk float drawn by a horse which was cropping the grass at the side of the road. I was a short distance from this milk float when the horse decided that the grass on the other side was better and meandered across the road to try it. I trod on the brake pedal – "Hell no front brakes" – and I hit that milk float with quite a thump. This made the milk bottles rattle which scared the horse which let out a loud whinny and reared up, but fortunately did not bolt. The milk-

man came running down the path from the house where he was delivering to see what the commotion was about and I gave him a good "ticking off" for leaving his horse untethered on a main road.

"The milk float was undamaged but the impact had damaged the radiator of the Bentley and it was leaking water, what I would describe as a fast trickle. I decided that rather than ring the Works for a tow I would try and make it back under my own steam, and steam was the operative word, when I eventually got back to the Works the Bentley was hissing steam like a railway engine. After my fitters had thoroughly checked the engine

LEFT **The outside of the service department, circa 1928. Second on the left is the works' lorry, with the Bentley logo, based on a 3 Litre chassis.**

ABOVE LEFT **Chassis 976, registered OM6832, the racing team hack and spares car in 1927/28, outside the Hippodrome Cafe with 'Old No. 7' at Le Mans in 1927. This car was later sold, and superseded by EXP 5 registered YW3774. It is possible that the original body on YW3774 came from this car.**

ABOVE **The new Vanden Plas works at Kingsbury. Kingsbury Lane is on the right, and Kingsbury House can be seen top right. The long, low building with 'Vanden Plas' on the roof, towards the top of the lane parallel to Kingsbury Lane, was the Vanden Plas paintshop. The topmost portion of this became the Bentley racing shop. The two long shops parallel with the main Vanden Plas building centre left housed Bentley's service department.**

it was established that there was no further damage and that was the end of that little incident."

Before the year was out two new 4½ Litres had been allocated to the new Racing Shop for the 1928 team, chassis number KM 3077 registered YV 7263 and KM 3088 registered YW 2557. As soon as the chassis were completed they were moved over to Vanden Plas to be fitted with a new design of body – the "Bobtail" with a partially faired-in tail with vertical rear spare and a 'D' shaped tank of approximately 25 gallon capacity. For the 1927 Le Mans race an old 3 Litre had been used as the practice car, OM 6832 chassis number 976. This car was a 1925 3 Litre Speed Model with a Carbodies tourer, bought back by the Works and used as a hack/experimental car between 1926 and 1927. This chassis was fitted with engine 974A which was possibly an early (if not the first) experimental 4½ Litre engine, and later engine 525 from Hubert Pike's 3 Litre saloon, which was also used for various experimental trials. OM 6832 was pensioned off in 1927 and a new hack car built to full Team car specification, as it was used as a rolling source of spares at races. The new car, YW 3774, referred to as EXP 5 because it was fitted with that engine was a 10'10" chassis with the taller than standard 3 Litre radiator mentioned earlier, and a lot of Team specification goodies disguised under a tatty old blue touring body off a 9'9½" wheelbase car, possibly 976. This body fell somewhat short of the rear cross-member so the gap was filled with one of the 'D' shaped petrol tanks, with a three point mounting as originally envisaged for that tank. However, the tanks on the Works cars were mounted on a flat board with padding on Wally Hassan's recommendation and the Team car tanks never leaked, while that on EXP 5 suffered from frequent leaks. There were also occasional complaints from owners of standard Bentleys which were passed at high speeds by this disreputable looking heap, an activity much favoured by drivers of EXP 5. The sight of another Bentley on the road was an immediate and more or less irresistible signal for some fun!

The new Racing Shop was based in a building leased from Vanden Plas, more or less round the corner from the main VdP works and the Service Station at Kingsbury. The old Racing Shop at Cricklewood became part of the production shops. The new shop was a long shed with a hard-packed earth floor, to one side of Vanden Plas paint shop. The shop contained about eight mechanics and operated only during the run-up to the racing season. During the rest of the year the mechanics were attached to Service. The Team cars for 1928, 1929 and 1930 were prepared in this shop.

In 1928 again the Essex Six Hours race was used as a dry run for Le Mans, but the second of the two new "Bobtails" YW 2557 was not ready, so Rubin's Team car replica YU 3250, chassis HF 3187 was pressed into service to make up the numbers. Clement and Barnato shared "Old Mother Gun", Birkin drove his own car YV 7263, Benjafield shared YU 3250 with Rubin, and Cook and Scott entered their 3 Litres YV 8585 and YF 2503 respectively as private entries. Rubin's standard 4½ Litre proved to have a distinctly non-standard performance, but the race was run on a handicap formula which gave first place to Ramponi's Alfa-Romeo and second place to Dingle's 747 cc Austin. Birkin finished third, and Barnato set the fastest lap, but the Barnato/Clement car was slowed by braking problems caused by the rods stretching. The three 4½ Litres took the Team prize, Cook finishing 18th and Scott retiring. In the strip reports for the 1929 season Nobby Clarke recommended that the brake rods should either by increased in diameter or the threads rolled to avoid this occurrence (because of the maintenance of the grain in a rolled thread, such components are significantly stronger than the same item with a cut thread. In those days, thread rolling was generally only used by the aircraft industry).

By June the second "Bobtail" car was ready, Rubin's 4½ Litre YU 3250 disappearing, to return later in a much more famous guise as the prototype supercharged 4½ Litre. Barnato and Rubin, as the novices, shared the older and slower (according to Benjafield) car "Old Mother Gun", Birkin and Chassagne YV 7263 and Clement and Benjafield YW 2557. Additional lighting was provided in the

LEFT **The Vanden Plas works, photographed in 1928. The extra lean-to addition to the right hand side of the service department was put up to cope with the increased activity of the department, as more cars were delivered.**

BELOW LEFT **Bentley Boys at service – from left to right, Bill Fruin, Clem Eastman, not known, Pat Smith, not known.**

BELOW **This is the first of the new 1928 season 4½ Litres, Birkin's own car YV7263 chassis No. KM3077, photographed at Kingsbury soon after it was completed in May 1928. Kingsbury House can be seen in the background. This was the first of the two 'bobtail' cars, fitted with a 25 gallon 'D' shaped tank and almost vertical spare wheel recessed into the stub tail of the body. The chassis has the late pattern rear spring hangers with half-time levers, for the forward mounted compensator, 'push-on' brakes, and no strut gear. The exhaust system has a Brooklands silencer, fitted for the Six Hours Race at Brooklands which was this car's first competitive outing on the 12th May, driven by Birkin. It finished third at 72.27 mph.**

form of an extra Marchal headlamp mounted centrally and in front of the two standard Smiths headlamps on a tripod arrangement, giving the cars a curiously cyclopean look. It seems slightly suprising that such a distinctly non-standard arrangement was accepted by the scrutineers.

Although Barnato was the Chairman and source of the Company's major finance, Barnato nevertheless never caused a fuss and obeyed team orders without quibbling. WO, as team manager, was referred to as the headmaster, while the team members were the schoolboys – all part of the games indulged in by those fabulously wealthy sportsmen who populated the Bentley team over the years and were in considerable part responsible for the continued fame of the marque. It was Jack Dunfee who coined the "School" – cars were satchels, and petrol known as ink. WO referred to his relations with "the Bentley Boys" as "part father confessor, part schoolmaster, who was always happy to climb down from the rostrum to join in any classroom foolery." WO was obviously very fond of them and of the prestige and glamour that rubbed off them onto the cars. WO openly admitted to enjoying the aspects of their lifestyle in which he was able to join. Grosvenor Square had its "Bentley Corner", where the Bentley Boys had their London bases, and at a time when a skilled man's wage was barely

TOP LEFT **The three 1928 team 4½ Litres, outside the racing shop. The racing shop is the building on the left; the cars went in through double doors just out of sight to the right. Birkin (left) and Frank Clement with the cars, from the left Birkin's own car YV7263, the first of the two new 1928 cars, with Vanden Plas 'bobtail' body, 'Mother Gun' YH3196, and the second of the two new cars YW2557, also with 'bobtail' body. Note the 'cyclops' lighting set devised for Le Mans that year, 'push-on' brakes, and the absence of strut gear on YW2557.**

CENTRE LEFT **For the 1928 Le Mans, a fairly large party went out from Vanden Plas and Bentleys to watch. Kensington Moir is on the left, possibly Neale of Vanden Plas third from the right, with Edwin Fox of Vanden Plas second from the right.**

BELOW LEFT **The same group with cars; from the left, not known, Halliwell of Vanden Plas, not known, Edwin Fox, Hubert Pike, not known, Frank Fox.**

RIGHT **Bernard Rubin, left, and Tim Birkin, with Barnato's 6½ Litre HJ Mulliner tourer YT8946.**

This car was kept by Barnato from September 1927 until mid-1929, when it was sold to Major Humphrey Butler. It was used on a number of occasions as a demonstrator/road test car, and was written up by 'Migrant' in *Car & Golf*, February 1928. 'Migrant' commented: 'Enviable, indeed, is the lot of he who is fortunate to own one of the splendid cars!' The report concluded: 'But here comes the really important thing. The Bentley Co. are, 'free, gratis and for nothing', bringing all the older 'Six' models up to date . . . The firm deserves the highest credit for doing this handsome thing.'

£5 a week, Barnato was reputed to be spending over £800 a week purely on entertaining (the average industrial wage in 1927 was £4 7¼d per week). The Bentley Boys kept saloon and limousine 6½ Litres for town and country and it was not uncommon for the upper class of the time to be driven from their country residences to the outskirts of London in one car and there be met and carried on into London in another. Birkin was not very fond of driving himself – commenting in 1932 that he drove less than 600 miles per year on public highways, expressing the opinion that the highways were so over-crowded with incompetents as to be really unsafe!

For the first time at Le Mans, the Americans put in a serious challenge in the form of a straight-eight 5 litre Stutz and three 4 litre Chryslers. WO had little time for the American cars of that era – the encounter with the managing director of an American automobile firm late at night in a hotel in Liverpool leaves little doubt about that. This gentleman answered his own question about the role of the engineer in an automobile firm by telling WO that the engineer rated nowhere – only being there to meet the needs of the Sales people. One can well visualize a tired and somewhat fed-up WO

replying "From the standard of your product, sir, I can well believe it", which seems to have gone down very well, the American totally failing to appreciate the true meaning of the answer! There was no doubt, though, but that the Chryslers, and the Stutz in particular, were formidable competition and indeed the Stutz came very close to winning that year. First in trouble was Birkin, with a puncture. It had been found that the cars could safely be driven back to the pits with a flat tyre at reasonable speed, not more than 40 mph, so no jacks were carried. Unfortunately the damaged tyre wrapped itself around the brake drum and mechanism so Birkin had to cut it away, and then proceeded to drive back to the pits at speeds up to 60 mph – hardly reasonable. The wheel collapsed, sending the car sideways into a ditch at Arnage, and Birkin ran back to the pits with his overalls creased and covered in oil. Chassagne didn't need to be told what to do, and with his famous "Maintenant, c'est à moi", set off to run three miles back to the stricken Bentley with a jack under each arm, jacked it up, changed the wheel and drove on. As an old hand Chassagne had the presence of mind to keep the old wheel and hand it over to the pit crew, because otherwise they would have faced certain

ABOVE **The three Bentleys arrive at the circuit on the morning of the Twenty-Four Hour race, 16 June 1928, with the early morning sun casting long shadows. Frank Clement in the new 'bobtail' car YW2557 No. 2, Birkin in No. 3 YV7263, and No. 4 'Old Mother Gun' YH3196. Note the temporary headlamp covers, the third 'cyclops' light for better night vision and the identifying flash on the offside front wing of No. 2. No. 5 is one of the 4 litre Chryslers.**

RIGHT **Nobby Clarke shows the tyre and wheel, wrecked by Birkin in the No. 3 car. Edwin Fox is to the left of Nobby and WO is behind the pit counter on the far left.**

disqualification. But their car was now so far behind with three hours lost that they were effectively out of the running.

Nobby Clarke's recollection of this affair in 1974 was rather different from the usual version: "Despite the statement that Birkin was not carrying a jack on his car when the wheel collapsed after the treatment it had received...I am almost certain that there was a jack on board Birkin's car – on each car, in point of fact. Those were the years when, under the regulations, we were not allowed to assist the drivers, or let them have any tools from the pits. I also remember Chassagne sneaking out of the pits with the Enots hydraulic jack in one hand and the jack handle and the copper hammer in the other, keeping a watchful eye on the Commissaire Sportif (Scrutineer) as he got clear of the pits (by the rear pit door I believe)! Anyhow, if you think of this incident carefully, it must have been a two jack job with the car well and truly in the ditch, a two lift job to get the damaged wheel clear of any obstruction."

Benjafield and Clement were delayed by a

several laps driving to team orders, Brisson wagging the tail of his car to shower the Bentley with stones. As they passed the grandstand on one lap, Barnato pulled alongside to give Brisson the two fingered salute – but by the next lap was back on station, without giving the pit crew any need to hang out signals. Nobby Clarke later said that he could clearly see the words that Barnato was mouthing at Brisson as they passed the pits. WO rated Barnato the best of the Bentley Boys, always appearing to be slower through the corners than the more flamboyant but quicker when measured with a stopwatch. WO was never a great believer in speedometers which in the early 1920s were notoriously inaccurate, (particularly the belt-driven variety used on the very early 3 Litres) and preferred to rely on a rev-counter with speeds in gears inscribed and a stopwatch.

Nobby Clarke noticed first that the frame of the Barnato/Rubin car had gone, drawing WO's attention to the manner in which the door gaps opened up when Barnato got in and out of the car at the pits. Barnato was instructed to push on until the worst happened. With less than an hour to go Barnato came past the pits at reduced speed with thumb down and the car visibly sagging in the middle. The Stutz was not without its problems in that it had stripped top gear, a handicap worsened by the usual American provision of only three forward gears. The regulations specified that up to half an hour was allowed for the last lap, so Barnato cruised round slowly shutting the engine off on the downhills, listening for any evidence of the engine tightening up, at which point he was determined to stop and let it cool off. As it was Barnato crept round to meet the flag, 7.09 miles in front of the Stutz. Any miscalculation by Barnato in arriving at the finish before 4.00 pm on that last lap would have meant a further lap – which would have been disastrous. Birkin put in the fastest lap on his very last, and just covered the minimum distance to qualify for the following year's race. On the way back to Dieppe the chassis frame of that car went as well.

Shortly after the Le Mans race, Barnato lent his 4½ Litre to C.W. Briggs of Rolls-Royce's Sales staff. His report is reproduced in full: "Owing to the kindness of Captain Woolf Barnato I had an opportunity last week-end of giving the 4½ Litre Bentley a very thorough test under practically all the conditions which are possible in England.

"This car has a four cylinder engine, and considering the size of this the flexibility is amazingly good. It will run on top gear down to 8 or 9 mph

fractured oil pipe, but Clement quickly changed over the ends – all the fuel and oil pipes were duplicated with the ends of the spares taped over, so that in the event of failure any pipe could be disconnected and replaced very quickly. More irritating was a tendency for the nearside front door to open of its own accord, symptomatic of a much worse problem. This manifested itself when the top hose parted company from the radiator, with the loss of all the water in the cooling system. As replenishments could only be made every 20 laps, No. 2 was out of the race. It was the reason that was disconcerting – the frame had cracked, caused by a diagonal ridge across the road at White House Corner, which the cars took more or less flat out. The constant flexing of the frames was resulting in metal fatigue at an alarmingly high rate, the frames breaking across the offside lower row of rivets in the front gearbox cross-member. After 1928 the Team car frames were all fitted with strut gear and various reinforcing plates.

With one car down it was only a matter of time before the frames of the other two cars went and although Barnato and Rubin were still leading, the Stutz was not far behind. The Stutz had led from the start until 2am, when the Barnato/Rubin car took the lead. Barnato sat on the tail of the Stutz for

and start and accelerate without jar from this speed. From about 12 to 20 mph there is a heavy torque reaction vibration which disappears very quickly, and the engine is remarkably smooth up to the maximum speed which I attained, which was 90 mph on the speedometer, which from various indications I estimate to have been really 81 or 82.

"From about 36 mph and upwards there is a very high frequency low amplitude vibration from the engine, which is not felt to any extent in the car itself, but on the pedals, steering column, etc., it is quite noticeable. The steering wheel is a spring spoked wheel, and this is very effective in damping it out from the rim. This high frequency vibration does not increase greatly in intensity and is equally marked on the over-run as under power.

"The steering and general controlability of the car are extremely good. It reminds one of nothing so much as a Lancia Lambda car with a much more powerful engine in it. At all speeds the steering is reversible and practically entirely free from road shock. It appears to be slightly higher geared than our present 20, and is absolutely direct in its control. There does not feel to be any elasticity between the steering wheel and the front wheels themselves.

"At the highest speeds it is possible to drive the car within an inch of a desired line or spot.

"Tyres are Dunlop $5\frac{1}{4}''$ medium pressure – unfortunately I was unable to check the pressures at which they were running, but no doubt could obtain this information.

"The suspension was what might be expected in a Sports Car. If the shock absorbers were eased practically completely off, the car was comfortable though not luxuriously sprung up to 35 or 40 mph, but above this bounced altogether too much. A half or three quarters turn of all the shock absorbers made the riding very hard up to 40 mph, and above that very good, but of a hard description.

"The braking was quite adequate without being superlative, and needed considerable pressure to obtain the maximum effect.

"Transmission generally was only fair. The gearbox made quite a considerable amount of noise, and there appeared to be a period in the back axle. The clutch, which is of the cone type, was very

"Capt Barnato :– at the 13th hour 13th lap

"WHERES THAT BLACK STUTZ"

LE MANS VICTORY DINNER

Chairman Captain Woolf Barnato

Lyons Corner House, Coventry Street, W. 1.
7.30 for 7.45 p.m. Friday, July 27th, 1928

Barnato and Rubin (the latter looking distinctly the worse for wear) in the Bentley at the end of the race. The nearside front shock absorber has broken, and that on the offside has completely gone. Stan Ivermee stands to the right of the car, in overalls.

RIGHT 4½ Litre chassis on road test. The chassis is on unlimited trade plates, so this is a 'weekender'. This was a chassis that the tester was allowed to use for a weekend, in a rota with the other three testers. George Hawkins recalls a preference for the 4½ Litre, as his wages did not really run to the petrol consumption of the supercharged or six cylinder chassis.

CENTRE RIGHT Jimmy Jackson in a 4½ Litre test chassis, rounding a bend down near Kings Langley that the testers had a competition on as to who could get round there fastest – the photo was taken by George Hawkins, who was lying in a ditch at the time! As George put it, the 4½ is just about to break away – 'we went down there on test to look at his tyre marks round this curve and there were four, not two – so that chassis must have whipped like an 'S'!'

RIGHT Jackson with another 4½ test chassis, a fairly late chassis with plate clutch and C box.

rough indeed and squeaked badly. It appeared to have only two positions – one full out and the other full in, and it was almost impossible to get the car under way from rest on any gear above second owing to this feature. When the clutch was out there was a very rough grating sound, which the driver appeared to regard as normal.

"This car cannot be considered in any respect as

The exterior of Freestone & Webb's Unity Works in October 1928, with three Bentleys and three Crossleys. The two Bentley cars are 6½ Litre YE2084, left, and 4½ Litre YW7916. The 4½ lacks its nearside front wing and lamp, because it was in for accident repairs at the time. The 4½ Litre chassis in the foreground is a brand-new 1929 chassis, straight from the works, with a very crude delivery seat – the Kigass pump and oil pressure gauge can be seen strapped to the dynamo casing. The chassis has the plate clutch, C box, and forward mounted compensating shaft with half-time levers on the rear spring hangers. The chassis still has the outside battery box, mounted on the nearside.

a competitor of either model of Rolls-Royce car. It is a car which must be definitely placed in the Sports class, and should be compared with the Grand Prix Bugatti. Of its type it is very good indeed."

On Friday 27th July 1928, at 4 o'clock, the faithful gathered at Pollen House in Cork Street for the Annual Ordinary General Meeting to consider the Directors' report to the shareholders and carry out the usual business. Woolf Barnato stood down and offered himself for re-election to the Board, which duly happened. The general tone of the report is rather down. A profit of £1,049.10.4 was recorded, despite the Company making a loss in the first six months of the year and only making a profit overall because of the new 4½ Litre. In the year to 31st March 1927, the Company sold 105 6½ Litres and 228 3 Litres, compared to 219 4½ Litres, 63 3 Litres and 100 6½ Litres in the year to 31st March 1928, emphasizing the point that the 4½ Litre more or less finished off the smaller car. The slight drop in sales of the 6½ Litre is reflected by the comment that "the Company found at first a certain difficulty in entering the very conservative market

for which the 6 cylinder model was designed.'' Converting all the earlier chassis to 1928 specification was an expensive way of sweetening the pill – the £6,378 written off in converting three-quarters of the first 196 chassis represented a significant profit margin. Sales of the 6½ Litre increased to 134 in the next year to 31st March 1929, but that figure was boosted by the introduction of the Speed Six. The comment that the three 4½ Litres finished 1st, 2nd, and 3rd is a surprising bit of misinformation, because they finished 3rd, 6th and 8th overall – but 1st, 2nd and 3rd in class. Note also that the racing expenses were covered by the drivers and suppliers. (A facsimile of the Directors' report is given in Appendix VI.)

M Weymann proposed a challenge race to find the best sports car, Weymann himself propounding the merits of the H6 Hispano-Suiza, but although wagers were laid and Barnato, among others, joined in the correspondence, the race was eventually run as a match race between the Hispano and

ABOVE **The engine of the 1929 model 4½ Litre, in this case engine NX3467 in chassis No. NX3465, as rebuilt by Carl Mueller. This shows the vertical SU HVG5 carburettors, plate clutch flywheel, and the later pattern bulkhead with the wiring conduits moved further in away from the periphery of the bulkhead casting.**

ABOVE RIGHT AND CENTRE RIGHT **Bentley employees at the works, circa 1928/29. These were taken in the area of rough ground to the north end of the works, behind the office block and the end of the detail shops, among the sheds. These were later demolished, along with the wooden office block, to make way for the 1929/30 extensions; the land shown here became the machine shop. The frames stacked up are 6½ Litre, and the tubular cross-member can be seen; this cross-member cannot be fitted once the cross-members have been rivetted, so it has been dropped in by Mechans ready to be lined up, drilled and bolted.**

The Bentley Motor-Cycling Club grass-racing at the Welsh Harp.

the Stutz, the former running away with the race. In the Stutz/Bentley duel, the honours remained in the Bentley camp. Needless to say, the reasons for Clement's retirement and the closeness of the result caused by the imminent demise of the winning Bentley were not revealed to the press, but shortly afterwards the gauge of the chassis frame of all the 4½ Litre cars was increased from 5/32″ to 3/16″. This change was also carried over to the last 3 Litres and the 6½ Litres. The 4½ Litre frame was strengthened several times during the production run, with longer dumb iron knuckles with more rivets, deeper flanges to the centre of the chassis on the bottom rails, and various additional reinforcing plates. The deeper flanges were probably a direct result of the frame failures at Le Mans.

There was no doubt that the performance of the Stutz and the Chryslers was a cause for concern, and in the remaining months of the 1928 season there were few laurels for the 4½ Litres. Birkin entered his own 4½ Litre YV 7263 in the German Grand Prix at the Nurburgring on the 15th July, with Wally Hassan as his riding mechanic, but on a very hot day with the tar melting Birkin could make little impression on the 7 litre supercharged SSK Mercedes and had to be content with 8th place after a terrific drive. The original Vanden Plas "bobtail" body on this car was damaged by fire, so it was rebodied by Vanden Plas (again) with the standard 4-seat body with trunk that is best remembered on the Team cars. In this form Birkin entered the newly-revived Tourist Trophy Race, held near Belfast, along with Humphrey Cook, the latter driving his new 4½ Litre YW 5758, chassis number TX 3246.

The Tourist Trophy series of races had not been run since 1922 and were restarted at the instigation of Harry Ferguson, a local businessman. The race was run over 410 miles on a handicap basis, and on the announcement of the regulations Bentley Motors immediately withdrew the three cars they had entered, amid a public furore. WO was adamant, however. On the twisty circuit the odds were heavily biased in favour of the smaller cars and it was stamina rather than out-and-out speed which brought Bentleys the vast majority of their racing successes. Maury explained their reasons in a long letter to the RAC. In essence, a 4½ litre car had to pass all the 37 entrants at least twice to stand any chance of victory. As usual, WO would not enter a race there was effectively no chance of winning. The 1928 to 1936 series of races were to prove

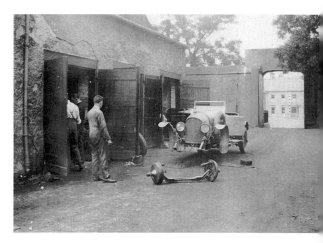

Although their entries for the Tourist Trophy were officially private, Birkin and Cook were ably backed up by Bentleys. These photos show the two cars, No. 54 YW5758, Cook's car, and No. 53 YV7263, Birkin's car, being prepared in Ireland just before the race. The car in the foreground, YW3774, is EXP 5, the team spares car. This car was built up from genuine team car parts as a rolling source of spares, and seems to be performing that function here, with its front axle being grafted onto YV7263. It is noticeable that this car has 'push-on' brakes, corded springs, quick-action radiator filler cap, and, it would seem, a heavy pattern front axle with Stage 3 Perrot shafts. The petrol tank is the 'D' shaped tank used with the 'bobtail' body, and the body is from a 9′9½″ chassis crudely lashed onto a 10′10″ frame. This car went to Le Mans in 1928, 1929 and 1930, in different guises. In the front three-quarters view, Stan Ivermee salutes the photographer, and Wally Hassan can be seen on the far right. The character with the cigarette worked for Cook.

WO's judgement correct, virtually all of them being won on handicap by small-engined cars. Cook and Birkin still entered privately, and doubtless received a degree of support from the Works in doing so. Cook crashed into a low wall in practice, but in the race although Birkin and Cook finished first and second on speed, they could only manage fifth and seventh on handicap. Not to be outdone, Mr. Edward White at the Annual General Meeting of the Royal Irish Automobile Club on the 22nd June 1928, announced the 1929 Phoenix Park race, to be held along the lines of the 1920 Gordon Bennett race.

Birkin also entered his 4½ Litre for the Georges Boillot Trophy on the 8th September. The race was run over 278.6 miles, and it was clear from the start that no matter how hard Birkin drove he had no chance of making up for his handicap in the time available. Birkin drove very hard, averaging 73.16 mph to Ivanowski's 69.73 mph in the 1500cc Alfa-Romeo, but it was the latter who was declared the victor. The Bentley was placed fifth on handicap, the Alfa-Romeos finishing first, third and fourth.

Birkin's fastest time was the all-time record for the Boulogne Trophy race, as it was announced later in 1928 that the series would not be run again. The Alfa-Romeos, with drivers of the calibre of Ivanowski and Campari, were very difficult to beat on handicap, and the SSK Mercedes were very fast from scratch. A previous stretch in engine capacity had staved off the inevitable, but further measures were needed to retain supremacy and Birkin, for one, favoured the new-fangled supercharger.

Problems/Development Work on the 6½ Litre.

The modifications made to the 6½ Litre for the 1928 season were somewhat radical, reflecting a need to make some serious improvements to the chassis. These are detailed on p.192. Further minor changes consisted of the adoption of a Delco-Remy coil ignition unit on the offside in place of one of the ML magnetos, a Jaeger clock to match the speedo and a Hobson type KS Telegauge fuel gauge, connected up to the 25 gallon tank. Previously the tank had been of 19 gallon capacity including a 2 gallon reserve arrangement as fitted to the 3 Litre, without a gauge. Larger Smiths headlamps were fitted and stainless steel used for the gear lever, the gate of which was redesigned to allow better sealing around the floorboards. An extension piece was fitted to the cast aluminium outrigger, carrying a flat steel gate. A

Dewandre servo was fitted, mounted between the gearbox cross-members on the offside, and the operation of the front brakes from pull-on via 5/16″ rods to push-on via 1″ tubes. These changes applied from chassis KD 2122 on, roughly the 200th 6½ Litre built.

In June 1928, Hives and Robotham, (both from Rolls-Royce) drove a 6½ Litre saloon, chassis KD 2112, which incorporated some of the 1928 modifications, but it is perhaps better to skate over their findings! Some of the comments are interesting – "Steering. The Bentley was remarkably interesting. It exhibited practically all the difficulties which we know exist when balloon tyres are fitted to a large car." There was so much flexing in the steering column that they found it very difficult to assess it. The column design and mounting was revised for the 8 Litre, so Bentleys were obviously aware that all was not well. The speedo came in for particular comment: "[the speedometer] appeared to have a characteristic of error increasing rapidly with speed. This confirms the fact that Bentleys are continuing to fit speedometers that are hopelessly optimistic and so maintaining their reputation for speed. This car showed 84 mph by speedometer and actually perhaps did 74 mph." While it is

The 'Other Bentley Boys' with EXP 5 – from left to right, Ivermee, not known, Hassan, Pennal, not known, but worked for Humphrey Cook, Pugh.

Dudley Froy record breaking at Brooklands, September 1928. The car is chassis No.1106, the 100 MPH Supersports chassis built for Woolf Barnato with a Jarvis racing body. The body fitted here is thought to be from an Invicta of Parry Thomas. For the record runs, 1106 was fitted with $4\frac{1}{2}$ Litre engine EXP 5 and a 1922 pattern unbraked front axle. Unfortunately the works records for this car are blank from August 1925 until May 1930. *The Autocar* of 7 September 1928 commented that the 1106 was 'fitted with a $4\frac{1}{2}$ Litre engine taken from a works hack – a machine that had been used as a tender car at Le Mans'. The car referred to is believed to be EXP 5, as illustrated earlier.

certainly known that speedo tuning was indulged in on the 100 mph 3 Litres, these are strong words. It is conceivable that since KD 2112 was fitted with a Hooper body of considerable weight and vast frontal area that the speedo was fundamentally incorrectly set, or that the wrong gears were fitted to the speedo drive box in the gearbox itself. If the car tested was typical, which seems highly improbable, then as they concluded "In production Bentleys seem to be failing badly."

WO's inventive mind was hard at work on the problems of adequately cooling the larger engines. Radiator technology at the time was rather crude, and in hot weather, and particularly when passstorming on the Continent, the $6\frac{1}{2}$ Litre was very susceptible to boiling. This was a condition that most of the large cars of the time, with the considerable thermal mass of the engines and unpressurized cooling systems, suffered from. The position of the

water pump also masked off some of the radiator. Many years later two experts were overheard discussing their cars at a meeting. The Bentley expert pointed out that Bentley had used a proper cast fan like a propeller, while Royce had just used flat plates bolted at an angle onto a hub, to which the Rolls-Royce man riposted that Royce had used six blades while Bentley could only afford four! As has been mentioned before, the 6½ Litre was known to suffer from cooling problems in the early days.

WO's solution to this problem was steam-cooling, a principle often used on stationary engines. This consisted of maintaining a fixed water level in the cylinder block by using overflow pipes set at a certain level, allowing a space above them for steam to be generated. The overflow pipes were then connected to a water tank mounted below the radiator and protruding in front of it, with a filler cap in that position. A pump was then used to draw water from

this tank and feed into the block from above. Baffles in the top of the radiator shell led to a vent in case of excessive pressure build-up. This idea was the subject of patent number 271790 taken out by WO and Bentley Motors, and was fitted to the No. 2 Experimental 6½ Litre MH 1030 with "The Box" body for testing by WO. The scheme proved to be successful, but improvements in radiator design made it unnecessary for the production cars. These changes to the radiator were incorporated for the 1927 Show, which combined with two sets of louvres in each bonnet side in place of the previous three opening vents on each side improved the cooling. In some respects, this last change is an admission of failure on a town car. Louvred bonnets tend to make engine noise more audible at all speeds, bonnets with shutters only when they are open – when needed. They are thus preferable.

It is also clear that exhaust heat getting into the body was a cause of annoyance. The 1928 chassis had more space under the floorboards and the latter were very carefully fitted around the pedals and edges. The gearlever gate was redesigned specifically to allow better sealing and prevent heat getting into the driving compartment. This problem has already been highlighted in Hives's report on driving Eric Horniman's car in July 1926. Despite side vents and a Spinney type ventilator, Horniman complained of heat in the front seats. The car also lost a considerable amount of water and had to be re-filled every day. In March 1929, Major L.W. Cox, of Rolls-Royce, drove a Speed Six with a close coupled saloon body and commented that the front seat became insufferably hot, and that there was a most peculiar oppressiveness about the car which caused him to feel almost ill. Putting the exhaust on the driver's side on the 6½ Litre was undoubtedly forced on Bentleys at the prototype stage and they certainly paid for the consequences before putting it right on the 8 Litre. Tracing cooling problems through the Production Modifications shows that Serck-type tubular block radiators were fitted to the first 20 chassis, followed by Gallay cores, and the redesigned gear lever gate at chassis BR 2367 (delivered in May 1928). The new insulated floorboards were fitted at chassis FA 2504 (July 1928). At chassis LB 2339 (April 1929) asbestos lagging on the downpipe was discontinued on the Speed chassis and from FR 2632 (July 1929) asbestos was deleted from silencers on the Speed Six, so presumably they felt they had sorted out the problems. The improved radiator design and louvred bonnet were incorporated from KD 2122 (November 1927) as chassis past then were "1928 models".

13 The Conception of the Supercharged 4½ Litre

IT HAS BEEN SAID that after 1928 Birkin refused to race a 4½ Litre again unless it was supercharged, and indeed after the 1929 Double Twelve in which he co-drove Holder's privately-entered 4½ Litre UL 4471 he did not do so. It is possible that he only raced with Holder because the supercharged cars were not ready. They were planned for the 1929 Le Mans, and the Double Twelve was only six weeks earlier. It is perhaps significant that Birkin's own 4½ Litre YV 7263 was driven by Cook and Clement. Birkin was supported in his plans by Rubin, but WO's preference was to follow his usual philosophy of adding litres, and developing a competition version of the 6½ Litre.

Birkin approached Villiers to design the installation for the 4½ Litre, and persuaded Barnato to back the project. Presumably Barnato saw this as a means of hedging his bets with two parallel developments taking place, with one at least privately financed out of Birkin's own pocket. WO would have had little say in the whole matter, as by this time he had little executive control of the company that bore his name, and as an engineer he possibly did not want it anyway.

Amherst Villiers had achieved a notable degree of success with the Vauxhall Villiers, a modified 1922 TT Vauxhall team car using a Roots type supercharger with installation devised by Villiers and driven by Raymond Mays. Villiers was given a full set of blueprints of the 4½ Litre engine and was told by WO that the installation would not go under the bonnet, but project out in front of the radiator in the same manner as the dynamo on the six cylinder cars. Villiers then drew up the installation, changing the crankcase, crankshaft, rods, pistons and other details of the engine in doing so.

The first hint of Birkin's project appeared in *The Morning Post* of 1st January 1929, discussing the following year's racing and commenting that "It is more than probable that two leading British firms and one American will have supercharged cars entered for 1929 [for Le Mans], and work is proceeding in secret on these cars at the moment."

Birkin's approach to Villiers resulted in the drawing up of an agreement between Bentley Motors, Birkin and Villiers contained in a letter of 18th October 1928. There were four main parts to the agreement:

1. Villiers was to produce the design for the supercharger and the necessary induction system, supercharger drive and mounting and camshaft timing suitable for a Le Mans type 4½ Litre. Villiers was to manufacture and supply four supercharger units.
2. In the event of Bentleys standardising the supercharger, by fitting six or more superchargers to cars in addition to the first four, Villiers would be paid a further sum in satisfaction of royalty claims.
3. Bentleys were to decide within a reasonable period of time after the first satisfactory demonstration on the road of the supercharger whether or not to standardise it. If the supercharger was standardised, nobody else would be allowed to use Villiers' patents before 30th May 1929. (Unfortunately, a search of patents for the period 1920 to 1931 failed to produce any, either in the name of C.A. Villiers or Amherst Villiers Superchargers Ltd. with respect to superchargers.)

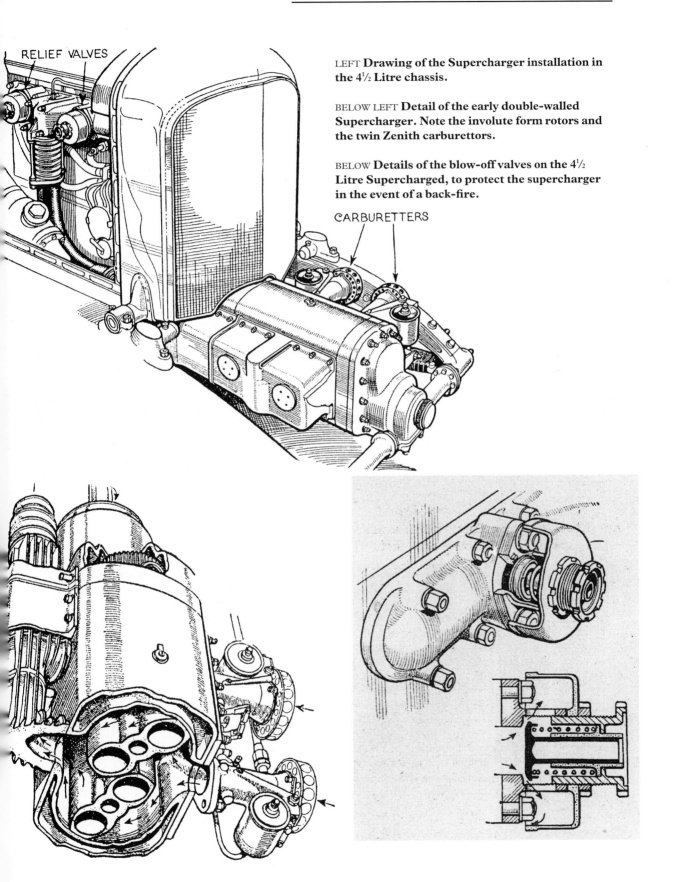

RELIEF VALVES

LEFT **Drawing of the Supercharger installation in the 4½ Litre chassis.**

BELOW LEFT **Detail of the early double-walled Supercharger. Note the involute form rotors and the twin Zenith carburettors.**

BELOW **Details of the blow-off valves on the 4½ Litre Supercharged, to protect the supercharger in the event of a back-fire.**

CARBURETTERS

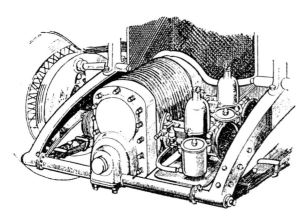

Detail of the ribbed supercharger.

4. If Bentleys standardised the supercharger, all patterns, drawings and tools would become Bentley Motors' property.

In addition to this, Villiers' name was to appear on the superchargers themselves and in the catalogues. "All superchargers made to your design shall carry the name of "Amherst Villiers Supercharger Mark IV" and be so described in the Companys' catalogues of the model to which they may be fitted. In the event, however, of any radical alteration in the design of the supercharger being suggested by Bentley Motors Ltd. and subsequently incorporated in the Villiers design then the name of the supercharger shall be "Villiers Bentley" or some similar name to be agreed." Villiers claimed that the work for Bentleys was done at cost price, in view of the publicity he would receive.

Villiers then set to work on the design of the supercharger, the detail design and drawing work being undertaken by Murray Jamieson. Much of the design work was completed by the end of the year. Birkin set up his own firm, Birkin & Co Ltd., in 1928, moving from his previous workshop in Feltham to premises leased from Vanden Plas in 1929 before moving to Welwyn Garden City. The Welwyn works at 19 Broadwater Road was set up as a proper factory, not just to run the Blower team but also to offer design and development facilities to other companies. Birkin expressed a particular interest in superchargers: "Captain Birkin is especially interested in supercharging and is making a speciality of fitting and obtaining the best results from superchargers on high-grade sports or racing car engines" (*The Motor*, 25th February 1930). This in itself raises an interesting point, in that Birkin & Co. was incorporated in 1928, while Amherst

Villiers Superchargers Ltd. was not incorporated until 1929 – both companies having similar aims in terms of the design, development and installation of superchargers. It does not take a stretch of the imagination to speculate that the two were intended to work closely together and that both were funded by Birkin. Unfortunately the company records for Amherst Villiers Superchargers Ltd. were destroyed by the Public Records Office in 1963.

WO's approach to the need for greater performance was to take the six cylinder engine and develop it into the Speed Six. The cylinder block was revised, with the inlet arrangement of the block cut from three ports to two. The revision of the porting seems to have been Harry Weslake's contribution to the development of the Speed Six. It was for this work that a special test rig was made up, consisting of an airtight box big enough to take a $6\frac{1}{2}$ Litre engine – about six foot high by five feet square. Fitted into this early version of Weslake's gasometer was an instrument to measure airflow.

Weslake later claimed he was retained as a consultant by Bentleys on a retainer of £500 a year, a figure which seems improbably high, but he is acknowledged as having a significant impact on the Speed Six engine. Apparently Weslake also worked on the 1929 $4\frac{1}{2}$ Litre racing engines, revising the porting and fitting a new design of inlet valve. Nobby Clarke's strip reports for the 1929 season identify a new inlet valve number R498/1 (R for racing), recording that all the cars fitted with this valve showed an increase in maximum speed with better acceleration and a generally smoother engine with less pinking and no loss in reliability. Again, presumably a result of Weslake's work.

Although WO is very dismissive of Weslake's contributions in *The Other Bentley Boys*, some credibility is lent to Weslake's account by figures supplied by the late Arthur Watson. Weslake claimed a power output from a modified Speed Six engine of 208 bhp in 1929, using twin Aero Zenith carburettors. These were then changed for 2″ SUs, as Bentleys were using SU (and Smith) carburettors only. The reversion to the SUs resulted in a slight drop in power output. Figures supplied by Arthur Watson read as follows: "1929 $6\frac{1}{2}$ Litre the old Le Mans prototype 206 bhp at 3750 rpm." The three Speed Six engines used in 1930 gave readings of 182, 186 and 189 bhp at 3750 rpm, presumably on SU carburettors.

The production Speed Six was fitted with twin SU HVG 5 carburettors on a box-type inlet manifold, sufficiently close to the block that none of the

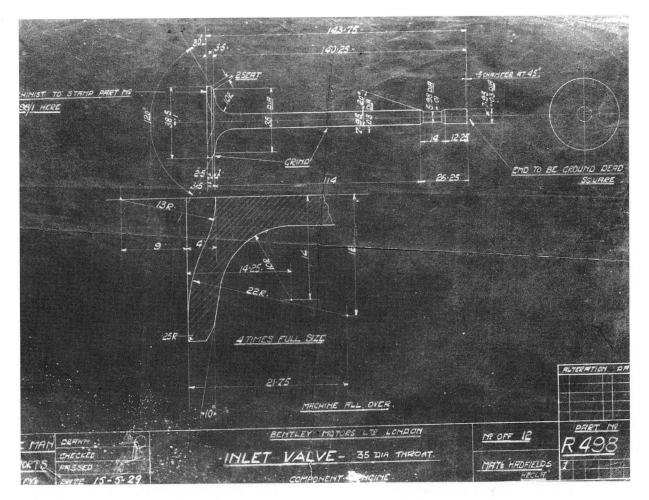

R498, the new racing valve for 1929. Note the project code in the bottom left – "Le Mans Sports".

water jacketing used with the Smith-Bentley carburettor was needed. These modest changes increased the output of the engine from 140 bhp to 160 bhp with no loss of reliability, and fitted to an 11′6″ wheelbase chassis with a parallel-sided radiator and bulkhead the Speed Six was complete. WO is on record as stating that the power output could have been raised to 180 bhp straight away by raising the compression ratio to 5.8 to one, but that this would have resulted in a loss of flexibility.

The first production Speed Six chassis number WT 2265 was fitted with a Gurney Nutting sports saloon body and used as a demonstrator and road test car, while the second, on an 11′ wheelbase, was allocated to the Racing Shop to prepare for the 1929 season. This car, chassis number LB 2332, registered MT 3464, was the most successful of all the Team cars with no less than six wins in major races between 1929 and 1931. For racing, the Speed Six used the front brakes of the 4½ Litre, with pull-on rods running outside the chassis and the newly-introduced servo or "self-wrapping"

brakes, and "Old Number One" as MT 3464 was almost invariably referred to was also fitted with a 4½ Litre differential unit with a specially made-up propshaft. However, it was soon found that the power of the 6½ Litre engine was too much for the differential, and a 6½ Litre unit was later fitted. Weslake recalled the use of 2″ SUs in place of the 1⅝″ units used on the production cars, but no photos of the engine compartment of "Old Number One" seem to exist to confirm or deny this.

The Speed Six was announced at Olympia in October 1928, available on the 11′6″ wheelbase only. The chassis price was £1700 and the Gurney Nutting saloon on show at the stand (no. 137) was priced at £2315 complete. The first press reports appeared shortly afterwards; WO was quoted in *The Morning Post* as saying "The ideal we have kept before us in the design of this car is not to sacrifice one atom of silence and flexibility to speed."

ABOVE **The first publicity photograph of the 1929 specification Speed Six chassis, September 1928. This chassis differs little from the 1928 6½ Litre chassis shown on p.194, with the exception of the Spicer shaft in place of the pot-joint. This was made in several sizes, as a proprietary item; that shown here was used on the 6½ Litre, Supercharged 4½ Litre and 8 Litre chassis. The 4½ and 4 Litre chassis used a shaft with smaller universal joints. The principal difference between the Speed Six and the Standard Six chassis lies in** the parallel-sided radiator and bulkhead, and the twin SU HVG5 carburettors.

BELOW **This early Speed Six chassis No. BA2586 was delivered to Mrs CD Rockingham-Gill, to be fitted with a four seat body by Harrisons. This chassis has the early pattern B&D shock absorbers; the exhaust system has a single silencer, and is curved outwards to clear the Dewandre servo. Note also the heat shield below the toeboard.**

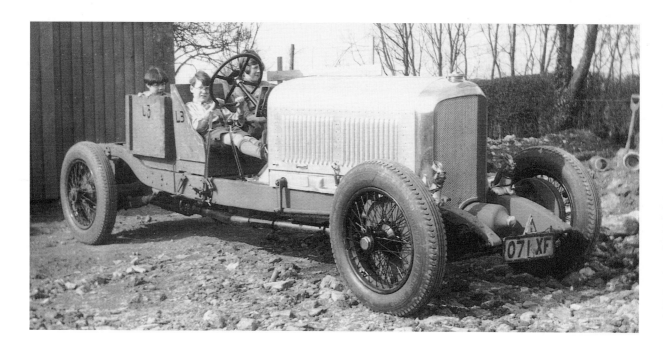

The Speed Six chassis by the petrol pumps at Cricklewood. At the back of the car are Bob Tomlins, left, and Bert Osborne, the works foreman. The building to the right is the experimental/finished cars/car wash building. The building under construction on the left is the new office block.

BELOW A 1929 Speed Six chassis, prepared for the Willesden carnival, at Cricklewood outside the boiler room. On the left is the front of the engine test shop. The model Bentley, JT3, was made by Bob Tomlins, the chief tester, as a replica of Barnato's Supersports 3 Litre PE3200. The chassis has the large Smiths headlamps and the large winged B mascot on the radiator cap.

Maynard Greville, who wrote the piece, concluded "I consider this car to be one of the most remarkable engineering achievements of the century, the balance of speed, silence, and flexibility having been maintained in an unique manner. Regardless of price, this car is the nearest to the ideal road vehicle that I have ever driven." Obviously he was very taken by it. The Standard 6½ Litre continued as before, available on 12' or 12'6" wheelbases. Sales of the Standard Six fell off quite quickly, and very few were made after the first quarter of 1929. The chassis price of the Standard Six remained

LEFT **The model Bentley at Cricklewood!**

BELOW **Clarence Rainbow, fitter in the engine test shop, in a 1929 Speed Six chassis with outside battery box outside the boiler room at the works, after a run with one of the testers. To judge from the height of the driver's seat and Clarence Rainbow's stature, it would be a fair guess that the tester was 'Shorty' Hedges. The engine is interesting in that the carburettor dash-pots are quite different from the usual SU HVG5 pattern; Clarence Rainbow could not remember these, but commented that they were always experimenting with different carburettors in the engine test shop.**

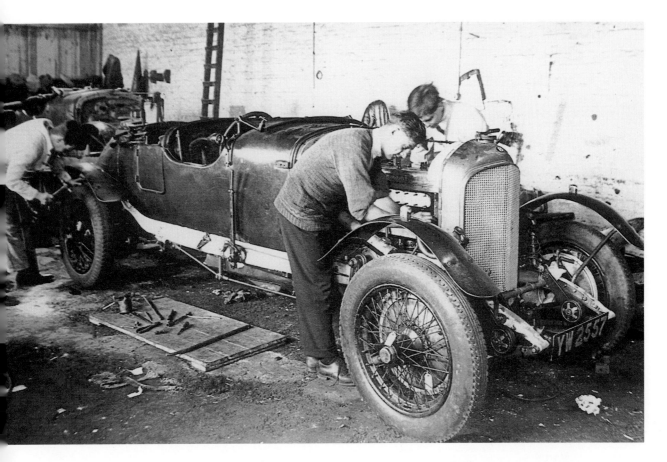

unchanged at £1575, as did that of the 4½ Litre at £1050 for chassis for open coachwork, or £1060 for chassis modified for a closed body. The extra £10 basically paid for a double silencer. The 3 Litre was still catalogued, at £895 for a Standard chassis or £925 for the Speed Model.

Changes were announced to the 4½ Litre, principally with respect to the fitting of "self-wrapping" front brakes. The Team cars had used a form of assisted brake operation since 1927, when the Perrot shafts were reversed with the rods running upwards so that although the brakes still physically pulled on, the action of brake torque causing the front springs to wind up had the effect of pushing the brakes on harder – hence the reference to "push-on" brakes. However, in this form the brakes were almost too good, so by late 1928 the idea was dropped, although the Blower team continued with "push-on" brakes for some time afterwards. The servo brakes were patented by F.T. Burgess and Bentley Motors; patent no. 291978 was applied for on 13th July 1927 and accepted on 14th June 1928. This design was then re-worked by C.S. Sewell, who improved both the cam end of the trailing shoe and the design of the spring link and

Preparation for the 1929 season over the winter of 1928/29. This is the interior of the Kingsbury racing shop, showing the primitive conditions with hard-packed earth floor. The car is the 'bobtail' 4½ Litre YW2557, being rebuilt with a new ³⁄₁₆″ heavy gauge frame with the heavy pattern, bolted strut gear, and four rivet dumb-iron knuckles. The sidelights have been moved from the trailing edge of the bonnet to the scuttle by the screen pillars, and the straight handbrake is for the cable-operated unit used in 1929/30. The team cars were continually changed and modified, throughout their racing careers. Behind YW is another team 4½ Litre.

its adjustment. This new design was granted patent no. 306704, in the names of F.T. Burgess, C.W. Sewell and Bentley Motors, and was accepted on 28th February 1929.

Making brakes that provided effective retardation for the whole of a 24-hour race with fast, heavy cars was something of a problem and the Bentley brakes had to be detuned to last the distance. The brakes were reckoned to last 1700/1800 miles at Le Mans. Other variants were smaller diameter shoes

BENTLEY MOTORS LTD.
POLLEN HOUSE,
CORK STREET, W.1.

26th February 1929.

Dear Sirs,

Re Le Mans 4½ Litre Models

Will you kindly note that the Company are prepared to build to special order a Le Mans type 4½ Litre, and below are given various modifications, any one or all of which can be incorporated at the prices shown.

Various other details which were included on the Le Mans chassis are now Standard specification on all 4½ Litre models, and for this reason we are pleased to be able to point out that the price shows a reduction.

The extras are additional to the normal chassis price of £1050, and the Le Mans type body is £275 retail, the 1927 design being commended.

Back Axle 3.3 ratio in place of 3.53.....................£5. 5. –

"D" type gearbox.......................................25. –. –

Extra pair of B & D Shock Absorbers on back complete with new axle bracket..................................10.10. –

Radiator with stoneguard, and large type Filler Cap........17.10. –

25 gall Petrol tank with provision for a double line of piping both for pressure line and petrol line...........21. –. –

Spring loaded Bonnet Straps fitted with rubber buffers in addition to the ordinary bonnet fasteners..............

Footbrake Adjuster through foot-boards in accessible position to allow driver to make adjustments while running; to be of a type similar to that fitted to the 1929 Six Cylinder...................................15.10. –

Additional Accelerator Pedal, as fitted to the Le Mans Cars...6.10. –

Special Revolution Indicator with large diameter dial......13.10. –

Hour Glass Pistons in place of standard B.H.P..............10.10. –

Petrol Pipe: (a) Standard.............................. No extra charge.
              (b) Single line of Petrol & Pressure
                  Piping to tank rubber covered...........10. –. –
              (c) Double line – do..........................20. –. –

Pressure Fed in place of Autovac.......................... 2. –. –

Split pinning of bolts and nuts as in Le Mans cars.........20. –. –

B.A.R.C. type Exhaust system may be fitted at an extra charge of.............15. – –

Smith Cambridge Thermometer............................... 3. 7. –

Undershield... 16. –. –

25 gall Petrol tank with provision for a double line of piping both for pressure line and petrol line...........21. –. –

Spring loaded Bonnet Straps fitted with rubber buffers in addition to the ordinary bonnet fasteners..............

Footbrake Adjuster through foot-boards in accessible position to allow driver to make adjustments while running; to be of a type similar to that fitted to the 1929 Six Cylinder...................................15.10. –

Additional Accelerator Pedal, as fitted to the Le Mans Cars...6.10. –

Special Revolution Indicator with large diameter dial......13.10. –

Hour Glass Pistons in place of standard B.H.P..............10.10. –

Petrol Pipe: (a) Standard.............................. No extra charge.
              (b) Single line of Petrol & Pressure
                  Piping to tank rubber covered...........10. –. –
              (c) Double line – do..........................20. –. –

Pressure Fed in place of Autovac.......................... 2. –. –

Split pinning of bolts and nuts as in Le Mans cars.........20. –. –

B.A.R.C. type Exhaust system may be fitted at an extra charge of.............15. – –

Smith Cambridge Thermometer............................... 3. 7. –

Undershield... 16. –. –

Asbestos and wire guard to Petrol Tank.................... 7.10. –

Yours faithfully,
BENTLEY MOTORS, LTD.

Letter from Bentleys detailing the list of options on the 4½ Litre, and prices.

with $\frac{5}{16}$" liners instead of the standard $\frac{1}{4}$" liners, coupling the rear brake shoes together, so that operation of the foot brake brought into play all eight rear shoes, and cable operation of both hand and footbrakes. The latter facility was adopted because it allowed the brakes to be taken up easily during the race. In the case of the footbrake the cable ran from the brake pedal to the compensating shaft only, and by means of a large diameter pulley wheel, mounted on a threaded shaft fitted to blocks between the gearbox cross-members, the driver could reach down and screw down the pulley wheel onto the cable, thus taking up all the brakes in one go. Similarly a cable on the handbrake from the lever itself back to the operating lever on the compensator, was provided with a small adjustment on top of the handbrake lever to facilitate taking up wear while on the move. Birkin was in the habit of pulling on the handbrake before a corner, leaving it on and then releasing it on the exit. (This was probably common practice among the Bentley Team.) The "self-wrapping" front brake extended the "push on" design philosophy by using a leading and trailing shoe, the leading shoe contacting the drum first then moving circumferentially to push on the trailing shoe. On releasing the pedal three springs returned the shoes to their original position. The C type gearbox was refitted to the 4½ Litre as standard to replace the D box that was fitted throughout 1928. The lessons learnt in the 1928 Le Mans race were incorporated into the 4½ Litre frame design for the 1929 season, as described earlier.

The Works also offered to the public a Le Mans Replica 4½ Litre, consisting of a standard 4½ Litre chassis with a list of extras available at extra cost, to bring the chassis up to virtually Team Car specification. Double shock absorbers, large diameter rev counter, split pinned bolts throughout and the 25 gallon tank were all offered as extras, and Vanden Plas built a number of Le Mans bodies on customer 4½ Litre chassis with and without the central body beam. Some of these cars were subsequently entered in major races privately – principally Holder's 4½ Litre UL 4471, M.O. de B. Durand's YW 8936, Eddie Hall's UV 3108, and Bummer Scott's UU 5580. Two of these were entered for the first major race of 1929, the Junior Car Club's Double Twelve Hour Race – UL 4471 driven by Birkin and Holder and UU 5580 driven by Bummer and Jill Scott. The official Works entry consisted of "Old Number One" Speed Six driven by Barnato and Benjafield, with Birkin's own 4½ Litre YV 7263 being driven by Clement and Cook and YW 2557

The first competition appearances of the new Speed Six – 'Old Number One' at the 1929 Junior Car Club's Double Twelve race at Brooklands, on the 10th May. Barnato gets to grips with the disintegrated dynamo coupling, with the front portion of the aluminium casing over to the right by the spade. Benjafield, Barnato's co-driver, looks on from the right. The dynamo was removed and thrown in the back, but the Bentley was disqualified because the rules states that the dynamo had to be positively driven. WO can be seen in the pits, wearing glasses.

was driven by Davis and Gunter.

The JCC race was run over the Friday and Saturday from 8 am to 8 pm on each day. Night driving was not allowed at Brooklands because of complaints from the residents of Weybridge and in efforts to appease the locals all cars driven at Brooklands had to be fitted with special silencers. The Brooklands silencer was one of the options offered by Bentleys on the Le Mans Replica 4½ Litres. C.W.F. Hamilton was lent a Brooklands silencer by the Kingsbury Service Department when he raced his 4½ Litre at Brooklands in 1930. The Speed Six quickly proved to be the fastest car in the race, and looked to be a likely winner on the handicap formula.

Unfortunately after roughly nine hours of driving, the dynamo drive coupling broke up, so Benjafield and his mechanic removed it, and doubtless mindful of the strict Le Mans regulations on such matters had the presence of mind to put it in the back of the car. Unfortunately, the race regulations stated that the dynamo had to be positively driven, so the Speed Six was out of the race. Smiths Industries were quick to point out that the problem had been entirely due to the coupling, and that there had been no problem with the dynamo itself.

The organizers asked WO to let the car circulate for the rest of the race "for the spectacle". This went against WO's policy of not revealing the capabilities of his cars and obviously the suggestion didn't please him greatly anyway, and the Speed Six remained stationary. Cook and Clement retired after running a big end, Birkin and Holder also retiring because of back axle failure – due to an oversight in preparation, the standard spiral bevel back axle ratio was not replaced by a straight cut unit. Under racing conditions the load reversal effect on the spiral bevel gears caused them to climb in and out of mesh, leading to rapid failure of the crownwheel and pinion.

Overnight the cars were locked in a compound, the Bentleys with blankets wrapped around the bonnet and radiator, and on the Saturday morning all the sumps were drained and refilled with warm oil and the first few laps covered slowly until oil pressures and running temperatures had returned to normal. Some of the other competitors were still rebuilding their cars and in the immediate period after the 8 o'clock start, there was little activity to convince the casual onlooker that a motor race was in process. During that second day it became increasingly obvious that the handicap race was going to be very close indeed, with Davis and Gunter so close to Ramponi in the Alfa-Romeo that it was virtually impossible to tell at any point who was leading. Both pits hung out the faster signs and peculiarly both cars were called in to strap up battery cases with loose lids. The Bentley was also called in to tighten up a loose bonnet strap, somewhat unnecessary as bonnet catches were fitted as well, but the strap was required by the regulations. Near the very end Ramponi was shown the "all-out" sign by his pit, but no such signal was given by the Bentley pit because the canvas breaker strip could be seen on one of the rear tyres. WO would never have ordered maximum speed under such conditions. Sammy Davis must have been somewhat perplexed, knowing how close the race was but was too professional a driver to ignore pit signals. It was not until the results had been calculated that it was known that Ramponi and the Alfa-Romeo had won on handicap by a margin of 0.003, or less than 200 yards for every hour of the race!

Work on the supercharged cars at Birkin's factory came to fruition just before the 1929 Le Mans race. However, the first run late at night revealed a problem insuperable in the limited time available – the main bearing clearances were too large, causing excessive oil leakage, and a cold oil pressure of 100 psi rapidly dropped to practically zero. Early runs at Brooklands also showed up problems of oiling up the spark plugs. There was no option but to scratch the entry of the Blowers at Le Mans.

It is very interesting to speculate on the background to the supercharged cars: how much of the work was done purely by Birkin and Gallop at Kingsbury and then at Welwyn, and how much was done with the tacit approval of the Works at Cricklewood. Certainly, when *The Autocar* announced the debut of the Supercharged 4½ Litre on 5th July 1929, they hinted that "It is no secret that experiments with Superchargers and Bentleys have been going on for some time." As Birkin had charmed Barnato into supporting the supercharged cars and their entry at Le Mans, the Works were committed to producing a minimum of 50 production chassis. *The Autocar* went on to say that the Supercharged cars were already in production and the chassis price would be about £1500. Bentley Motors had notified Villiers in June 1929 that they were going to standardise the supercharger under the terms of the October 1928 agreement, and Villiers had handed over to Bentleys the requisite casting patterns and drawings. To achieve that level of commitment from Bentleys to build 50 production cars, Bentleys must have had considerable vested interest in the whole thing and as they were lending their name and prestige to the cars, considerable reason to ensure the success of the whole project ("The closest and most friendly co-operation is always maintained between Captain Birkin and Bentley Motors Ltd." *The Motor*, 25th February 1930). Despite WO's comment many years later about the addition of the supercharger being a corruption and perversion of the design of the Bentley engine, even he must have been aware of the commercial possibilities should the car be a success.

It is also highly unlikely that Birkin's works in late 1928/early 1929 would have been sufficiently established to take on a project of such a size. No mention of Birkin's set-up can be found before an article appearing in the *Daily Mail* of 2nd October 1929, albeit by which time the works were fully set up with 16 full-time mechanics plus other staff.

However, the first four superchargers were delivered to Birkin, and it is not known whether the installation was performed wholly at Welwyn by Birkin's people or partly by them and partly by Bentleys at Cricklewood or Kingsbury.

The first supercharger installation was on Rubin's own 4½ Litre YU 3250 and it was that car that appeared in *The Autocar*, subsequent press reports and in Bentley Motors' own catalogues. Birkin bought two new 4½ Litres, HB 3402 and HB 3403 to turn into the first two Team cars (although YU 3250 came first, it is invariably referred to as the No. 3 Team car). Nobby Clarke intimated much later that the "HB" prefix to the chassis series was for Henry Birkin and that that series of cars was to have been supercharged. Presumably because of delays that never happened. The other 23 HB cars were built as standard production 4½ Litres and sold as such, although a certain amount of confusion occurred when YU 3250 had her

YU3250, the first Supercharged car based on Rubin's 1928 4½ Litre, seen in September 1929 in blown form. This shows the first pattern supercharger with the cast-in Villiers logo, supplied by Villiers himself to Birkin, with the wing device cast into the offside trunking casting of the blower installation. Note the number 65 on the wheel hubs, from the 1929 TT race, double Hartfords with the big, square adjusting nuts, light pattern strut gear, push-on brakes, and double-bar pattern quick-release filler cap on the radiator. These illustrations were used by Bentley Motors in their 4½ Litre catalogue, at the time of the 1929 Motor Show.

original chassis number of HF 3187 changed to HB 3404. The "real" HB 3404 was an H.J. Mulliner saloon bodied 4½ Litre delivered to Miss Susan Briggs, so the number of YU 3250 was amended to HB 3404/R, the R for Rubin.

The engine numbers of HB 3402 and HB 3403 were those of the first production cars, SM 3901 and SM 3902 respectively. This has interesting implications, in that Bentley Motors must have tooled up for and laid down the first batch of production engines. All of this would have required the ordering and manufacture of casting patterns and then the ordering, machining and progressing of the castings and other parts before assembly of the engines. This would take a company of the size and structure of Bentley Motors a great deal longer than it would take Villiers' company to order and make the first four superchargers. The installation was largely Villiers' responsibility, and would probably have been done first and drawn afterwards. This would not have presented any problems. However, all of this would be to no avail if Bentleys had not completed the engines. Had the delay come from Bentley Motors, it would explain Birkin co-driving the Speed Six with Barnato in 1929 as a consolation. This hypothesis is substantiated by the agreement between Bentleys, Birkin and Villiers (quoted earlier), which provided only for Villiers designing the supercharger, its drive and mounting, the induction system and supplying data for the camshaft timing – leaving the engine work to be done later by Bentleys.

WO is on record as saying that he "disliked the easy short cut provided by the supercharger" and that "when we wanted higher performance we

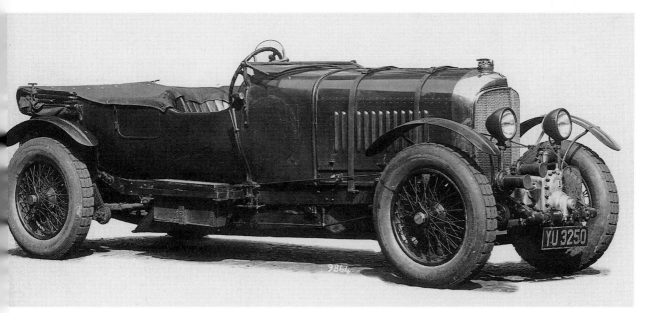

increased the engine size, from 3 to 4½ to 6½, and finally to 8 Litres." It could be argued that the mere addition of litres is as much of a short cut as a well-designed and engineered supercharger installation. Indeed, the supercharger installation proved to be very much less of a short cut than adding litres. It is also arguable that WO's philosophy was quite wrong, as other manufacturers were looking to higher efficiency, small engines rather than larger ones. The success of Alfa-Romeo and Riley, both of whom went the latter route, is perhaps worth bearing in mind. It has also been said that wide-spread interest in forced induction is a sure sign that cars are becoming too heavy. Certainly the increasing upward trend in size and price of Bentley products rendered them totally vulnerable in the event of a depression. WO was obviously aware of these trends in engine design, and was already thinking of the possibilities of a new design. A short-term approach to this problem was the increasing use of Elektron, a magnesium alloy, for castings. This, however, was really an expensive stop-gap which could not really counter the problem of increasing chassis weight with an ageing engine design.

ABOVE **1929 Montlhéry record attempt – Mrs Bruce oversees the refuelling of YV7263, Birkin's 4½ Litre; smoking while refuelling! Wally Saunders is on the left, with Harold Parker, the Shell representative, behind him.**

RIGHT **Closing the circuit before the race, the privilege of the last year's winners – in this case, Barnato and Rubin, in Birkin's Speed Six UL4203. In the passenger seat is the Hon Mrs Victor Bruce, fresh from her successful record run at Montlhéry in Birkin's 4½ Litre YV7263.**

Just before the Le Mans race of that year one of the best-known record attempts made by Bentley cars was enacted at Montlhéry when the Hon. Mrs Victor Bruce drove Birkin's 4½ Litre YV 7263, still wearing No. 5 from the Double Twelve race, single-handed for 24 hours at an average of 89.73 mph. The bare statistics do not detract much from the remarkable nature of that achievement. First of all Mrs Bruce approached WO and after assuaging his initial concern over the whole project, persuaded him to lend her a Team 4½ Litre, along with Jack Sopp and Wally Saunders as mechanics.

Then for twenty-four hours in a big, heavy car, she had to circulate Montlhéry track at high speed. Jack Sopp recalled having to accompany her in the early hours of the morning after one or two disconcerting incidents when she fell asleep at the wheel, the whole affair being conducted in pretty awful weather conditions. To maintain that average for twenty-four hours single-handed, would have been a remarkable achievement for anyone, and Mrs Bruce must have been very proud of the result. Not only did she break the Class C records for 2000 miles, 3000 kms and 24 hours, but also the single-handed record previously held by Thomas Gillette. In recognition, she was invited by the A.C. de l'Ouest to drive one of the Bentleys around the circuit at the start of the 24 Hours race to mark the official closing of the circuit.

When the car was rebuilt after the Le Mans race, it was found to be slightly crab-tracked because of an accident on the way to the circuit – the car was being driven by Parker, the Shell-Mex representative, when he lost it on a fast bend and caught a tree between the dumb-iron and the stub axle. Saunders and Sopp repaired the 4½, fitting another axle shipped over by Nobby Clarke. After the record run, YV 7263 was driven to Le Mans. According to Sopp, "It was WO's wish that we put the car in the race without touching it." Bentley's need to find two further cars to replace the absent Blowers was met by using YV 7263 straight from Montlhéry and pressing YW 2557 into service as well. The day before the race (14th June) *The Autocar*'s somewhat optimistic preview reported that "the super-charged cars. . .are very fast, and they have been tested very thoroughly, but one never knows what a car in its first race will do." In this case, precisely nothing.

The Americans returned to Le Mans in force, with three Chryslers, three Stutzes and a Dupont. Against them there were arrayed five Bentleys. "Old Number One" Speed Six driven by Barnato and Birkin (the latter a strange choice in view of his allegiance to the Supercharged cars, as discussed above. Benjafield was originally nominated to partner Barnato, but stood down because he felt there was more chance of success with Birkin driving), 4½ Litre YW 5758 – Clement and Chassagne, "Old Mother Gun" YH 3196 – Kidston and Dunfee, the "Bobtail" YW 2557 – Benjafield and d'Erlanger and finally Howe and Rubin in YV 7263. Five cars were too many for one pit manager, so Nobby Clarke ran the pits for the first three and Kensington Moir that for the last two.

The choice of Birkin in the No.1 car seems particularly strange in view of WO's comments about his ability to wreck cars and his tendency to play to the gallery. However, Hillstead tells us that Birkin was held strictly to team orders under threat of not being allowed to drive. Humphrey Cook and Chris Staniland were initially nominated to drive, but their places were taken by Kidston and Dunfee – again, perhaps because of the absence of the Supercharged cars. As the Bentley domination of Le Mans strengthened, so the opposition decreased – there were 25 starters in 1929 compared to 33 in 1928, and only 19 cars were entered

in 1930! The circuit itself had changed slightly as well, the A.C.O. building a new link road just before Le Mans, so the cars no longer went into the outskirts of Le Mans itself before turning around the Pontlieue hairpin at speeds reduced to about 15 mph. This resulted in a reduction of the lap distance from 10.726 miles to 10.153 miles. Birkin set the fastest lap ever on both circuits – at 79.289 mph in 1928 and 89.696 mph in 1930, the circuit changing again in 1931.

During the practice sessions, the Speed Six was clocked at 115 mph on the straights, to the surprise of some of the other teams. Right from the start on the 15th June "Old No. 1" pulled away to an early lead which was maintained to the end, at a speed a great deal lower than was achievable because there

Pit work: Birkin fills up the Speed Six with the four gallon churns, under Nobby Clarke's direction, in dark jacket. Note the galvanised funnel for quick filling, and the pit signals painted on the inside of the passenger door. This shows the snap action door handle, fitted to the nearside door on the team cars for rapid entry on flag-fall with the old-fashioned Le Mans start, which entailed the drivers running across the track to their mounts and jumping in before starting up and getting away.

was no reason to stretch the car and possibly reveal the Speed Six's true capabilities. Second place was held by Kidston and Dunfee in "Old Mother Gun" for virtually the whole race. One of the Stutzes managed to move up to third at one point when Clement was delayed at the pits by shifting ballast in the rear footwells bending the brake rods, caused by a nut run down to the end of a thread and not securing the ballast properly. But the American challenge gradually faded until the first four places were all filled by Bentleys. The Howe/Rubin 4½ Litre YV 7263 retired fairly early due to failure of the magneto cross-shaft gear. The car had not been checked over since Montlhéry and one of the magnetos was found to have packed up, having stripped the fibre gear driving the rotor arm (the team 4½ Litres used CG4 magnetos). Howe changed it under direction from the pit counter, as only the driver was allowed to work on the car, but on re-assembly a fibre washer was missed out and the drive seized. It seems likely that it was the paper gasket between the magneto and the cross-shaft gear housing that was omitted, causing the magneto coupling to bolt up directly onto the end of the drive dog which would push the centre cross-shaft gear over and put excessive load on one face, which would result in failure of the gears. Nobby Clarke said that the gears were knife-edged and stripped, but implied that there was something funny about the whole business, without saying quite what. Pennal in *The Other Bentley Boys* simply said: "The ones [cross-shaft gears] on Earl Howe's car were different; they only went through a mistake in fitting – or no prior fitting."

There were problems, though, with the P100 headlamps, which developed a nasty tendency to go out suddenly, particularly on corners. The P100 headlamp was produced by Lucas as a high-quality alternative to the Continental manufacturers, particularly Bosch, Zeiss and Marchal, and was fitted to a number of quality British cars of the era, superseding Smiths lamps on the 4½ and 6½ Litre cars. Unfortunately the bulbs fitted for the Le Mans race took too high a current, which caused the contact spring to lose its temper, and particularly when aggravated by hard cornering, ceasing to make contact. This, at high speed, was rather frightening, to put it mildly. After having all his lights cut out, Jack Dunfee came into the pits and proceeded "to discuss headlights, their short-comings, their types, their little idiosyncracies in terms quite in keeping with his experience of black-out in the forest when cars were jockeying for the 'S' bend and Arnage" (Nobby Clarke's memoirs). Despite Nobby's attempts to hush him up, after blowing off steam Dunfee jumped back into the car and drove on – not realising that Oliver Lucas was in the pits! Fortunately there was a Lucas mechanic

TOP **The Bentley Boys off-duty at Le Mans, 1929. From left to right, not known (worked for Birkin), 'Papa' David of the Hotel Moderne, Chevrollier, Wally Hassan, Jack Selway, Major Williams, Nobby Clarke, Collier, Puddephat, with, seated, Pryke (left) and Howard.**

ABOVE **The Bentley Boys' car – EXP 5, YW3774, fitted with proper touring coachwork in transit for Le Mans 1929. Built out of racing bits, this was certainly a very convenient way to move a large quantity of parts. Nobby Clarke in plus fours leans over; Puddephat in beret stands to the right of the screen.**

LE MANS GRAND PRIX DE DANSE
ARDENRUN
June 29th, 1929
ENTRY FORM

(1) *Name*

(2) *Occupation*

(3) *Cubic Capacity*

(4) *Hobbies*

(5) *Name of passenger*

(6) *Any further information you think it desirable to give*

SIGNATURE :

*No ticket will be sent unless this form is fully completed
and returned to 50, Grosvenor Square, London, W.1*

All entry forms to be sent **by return** to Woolf Barnato, Hon. Secretary,
Hon. Treasurer, Manager, Chairman, Owner, etc., *i.e.*, THE HOST, at
50, Grosvenor Square, London, W.1.
Competitor and Passenger Tickets will be sent on receipt of Entry Forms.
Charabancs for the convenience of those competitors or passengers whose
Cars are ditched or docked will leave 50, Grosvenor Square for Ardenrun,
at 9 p.m. and 11.45 p.m. on Saturday evening.

A GRAND PRIX DE DANSE
will be held at
ARDENRUN, near LINGFIELD
on Saturday, June 29th, 1929

THIS event takes place over a course consisting
of supper, dancing, and fair drinking, commencing
at 10 p.m. on Saturday until 6 a.m. on Sunday.
Any competitors still drinking or dancing after that
hour will be flagged off the course !

The course is a natural one, and of course the only
course open to you is naturally to enter as a matter
of course, (How coarse ! ! !)

An entry form is enclosed for you and a passenger.
No ballast need be carried if you desire to run solo.

in the pits who was able to cobble the lamps together to keep them going. Lucas was introduced to Barnato who simply said "We are doing about 135 mph past these pits, and it's not funny when one suddenly loses one's lights. Do what you can about it, there's a good chap." In fact the Speed Six was less affected than the 4½s. WO was very keen on Bosch electrics but could not fit them to the Works cars as they were not standard equipment and hence not allowed at Le Mans. In 1930 and '31 Bosch electrics were gradually introduced on the 6½ Litre and then on the 8 and 4 Litres.

Shortly before 4 o'clock, the Bentleys slowed down to form up in Team order and drove past the finish line in line astern, an impressive if somewhat theatrical end to an overwhelming victory. The reaction of the press to this 1-2-3-4 victory was as ecstatic as it had been to "Old Number Seven's" dramatic victory in 1927, even if the latter stages of the 1929 race had been pretty boring. Shortly after the race, Bentleys borrowed the four Le Mans victors back from their owners, the 1924 car XT 1606, by then fitted with a saloon body, and lined up XT 1606, "Old Number Seven" MK 5206, "Old Mother Gun" YH 3196 and "Old Number One" MT 3464 in Mount Street Public Gardens with the drivers for a photograph. For some reason John

Duff was absent, perhaps because he was then in Hollywood making a name for himself as a sword-fighter and stunt man. The resulting photograph was used in Bentley's publicity material for 1929 and 1930 and was used as a frontispiece to the 1929 *Hat Trick* Le Mans victory booklet. These booklets were initiated in 1922, with the TT booklet *The Blue Riband of Motor Racing*, consisting of race reports and extracts from the motoring magazines and daily papers, bound up into a souvenir booklet. These were then produced for the first Works Le Mans victory in 1927, *And Again* in 1928, *The Hat Trick* in 1929 and *Plus Four* in 1930, the latter carrying the résumé of policy laying out Bentley's reasons for withdrawing from racing and marking the end of the Works team (reproduced at Appendix II). Barnato celebrated by holding a "Le Mans Grand Prix de Danse" at Ardenrun. The "Entry Form" for this is reproduced here, as it is well worth seeing! Bentleys themselves threw a dinner at the Park Lane Hotel, at which Hubert Pike thanked the drivers for their efforts and presented Mrs Victor Bruce with a silver salver for her achievements at Montlhéry.

After the race, the Bentleys went back to the Finished Cars Shop, over the yard from the Engine Test Shop, so that the engines could be checked

Some driving style analysis at Barnato's Grand Prix de Danse.

and overhauled. It was during this process that the oranges came to light in "Old Number One" Speed Six. Birkin's habit of carrying oranges in his race cars is well-known. WO found the oranges in the Supercharged 4½ Litre that he rode in with Birkin in the 1929 TT roasted from the heat, and photos of Birkin in the 1930 French Grand Prix show him sucking an orange while pressing on through a corner. After Le Mans "Old Number One" was full of oranges that had been bitten into and the juice sucked out, then discarded! The first Blower car UU 5871 was ready for the Essex Car Club's Six-Hour race at Brooklands in July, only being finally assembled on the morning of the race. For some reason Birkin had the first two cars UU 5871 and UU 5872 fitted with 4-seat bodies by Harrisons rather than Vanden Plas. These bodies were built to Harrisons' "BF" or "British Flexible" principles, presumably a variant of Weymann's design, and incorporated at least one poorly thought-out feature in that the screen had to be raised before the lengthened bonnet could be lifted. It is possible that it was due to HM that the Birkin cars were bodied to "BF" patents, as HM

was a director of British Flexible Coachwork Ltd. Coachwork design and construction evolved considerably during the 1920's. Bodies became lighter and were increasingly designed on more scientific lines – as Brian Smith observed in *Vanden Plas – Coachbuilders*: "so far as Bentleys were concerned there was always plenty of power to propel a heavy car and never during racing did the ironwork fail, but Bradley [VdP's Chief Designer] and his forward thinking colleagues appreciated that a reduction in weight would result in better handling and performance". Such sentiments led to changes in design philosophy that reduced the role of the craftsman and increased that of the designer.

The more progressive of the bigger firms developed their own specialities – in particular Gurney Nutting were in the forefront with closed Weymann bodies, the Weymann methods of construction dominating the era. These bodies have limited curvature due to the nature of the framework, tending to be fairly square and upright. Vanden Plas also built bodies described as "British Flexible" but it is not known whether the methods employed were those of Harrisons. Another Vanden Plas variant was the construction of bodies referred to as "Silentbloc" in which the framework was constructed using the silentbloc bushes employed in Hartford shock absorbers. These bushes consist of two steel tubes with the space between them filled with rubber bonded to the two tubes, producing a one-piece bush with a degree of inbuilt flexibility. Quite a few bodies were erected on this principle on Supercharged 4½ Litre and 8 Litre chassis, the idea originating from Vanvooren in Belgium. The products of the Continental coachbuilders tended to be more exotic than their British counterparts, but were hardly in keeping with home-grown tastes for restraint and elegance.

Birkin's car was one of five Bentleys entered for the Six-Hours, there being two Works entries of "Old Number One" Speed Six driven by Barnato and Jack Dunfee and 4½ Litre YW 5758 driven by Cook and Callingham. In addition Holder entered his own 4½ Litre UL 4471 and Scott UU 5580, the latter partnered by Patterson. After the Le Mans style start at 11 am, the Speed Six forged ahead to take the lead after 5 hours to win at an average of 75.88 mph. Alfa-Romeo finished second, with Cook/Callingham third and Scott eighth. Holder retired with engine failure and Birkin retired for reasons that have never been disclosed, but the Blower car appeared to have substantially better performance than the standard 4½ Litre. Birkin did spend some time at the pits "attending to his

ABOVE **Speed Six for Royalty – this Gurney Nutting bodied car was built for HRH Prince George, on chassis No. LB2343, registered YR11, and here seen leaving Buckingham Palace.**

RIGHT **Prince George also owned a 4½ Litre saloon, UV6, a 1929 chassis No. HB3424 with Weyman saloon coachwork by Gurney Nutting, seen here visiting a cinema in London.**

super-heater" (*The Sunday Referee*) and seems to have changed one of the blow-off valve springs. 1929 was the year of peak performance for the Works team with "Old Number One" Speed Six leading the smaller 4½ Litre Team cars to a string of successes, the Blowers shining as a bright hope for 1930, by which time they would be better developed and more reliable and further Speed Sixes for the Works Team would ensure the continued dominance of the Winged B.

The Irish Grand Prix was next on the agenda, held at Phoenix Park near Dublin, over two consecutive days (12th/13th July), on a handicap basis. The first day was for cars up to 1500cc, the second for the larger cars, and as usual the Alfa-Romeos entered on both days posed a considerable threat. Noted playboy "Scrap" Thistlethwayte entered his white Supercharged 7 litre SSK Mercedes, transported over on his own private yacht. No less than seven Bentleys were entered – Birkin driving his Supercharged 4½ Litre UU 5871 and Rubin the No. 2 Blower car UU 5872, both fitted with their unlikely-looking Harrison bodies, Kidston "Old Number One" Speed Six MT 3464, the latter the only Works car, supported by Cook in his 4½ Litre YW 5758, Holder in UL 4471, Scott in UU 5580 and Harcourt-Wood driving Birkin's 4½ Litre YV 7263. As Harcourt-Wood was later to drive for the Blower Team and YV 7263 was invariably used as the practice car (the Birkin équipe never ran to a spare Supercharged car), it would seem he was getting his introduction to Bentley racing. It was clear from the speeds of the first day that the larger cars

would have to be very quick indeed to stand any chance of overall victory on handicap. On a day that was so hot that the melting tar caused some of the field to spin wildly, the Bentleys certainly had their work cut out. Birkin pushed Thistlethwayte so hard that the latter's Mercedes blew a gasket, but Birkin's own car was suffering from overheating and he started to drop back. Ivanowski was going very well in a 2 litre Alfa-Romeo, having set a very fast time on the first day in a 1500cc Alfa, with Kidston in the Speed Six pressing on very close behind. After replacing a wheel on the Speed Six damaged in a skid, the gap of 2½ minutes closed in the last few laps, but not quickly enough, and Ivanowski crossed the line 14 seconds in front of Kidston. Birkin finished third, one of the best results ever achieved by the Blower team, with Harcourt-Wood, Cook, Scott and Rubin 4th, 5th, 7th and 8th respectively. Holder non-started for unknown reasons.

Still in Ireland a month later for the Ulster Tourist Trophy, held at Ards near Belfast on the 17th August, the Blower team was out in force. Birkin drove UU 5871 and Rubin UU 5872, with the No. 3 long chassis car, YU 3250, racing in supercharged form for the first time driven by Harcourt-Wood. Despite WO's dislike of the TT race the Works entered one car, "Old Number One" Speed Six MT 3464 driven by Kidston. The Hon. Richard Norton entered "Old Mother Gun", the first 4½ Litre YH 3196 to be driven by Hayes and Field. Norton had bought "Old Mother Gun" just before the race from Jack Barclays for the princely

TOP **By 1929, the Birkin works were well established at 19/21 Broadwater Road, Welwyn Garden City.**

CENTRE **YV7263 again, inside the Broadwater Road works. This was used as a practice car by the Birkin team. Birkin was also involved in preparing and tuning other cars, some of which can be seen here.**

LEFT **The Six-Hours race at Brooklands marked the first competition appearance of the new Supercharged car. Here, Birkin is seen well up on the banking in the No. 1 Blower, UU5871, fitted with the ungainly Harrison body. They later retired.**

ABOVE **Birkin and WO in UU5871 at the pits. Note the mechanics with fire extinguishers. Seated on the pit counter in a dark suit above the nearside headlamp is Bertie Kensington Moir. The marshal's coat and the standing water show the conditions under which the race was run.**

BELOW LEFT **The Birkin team on the way to the Irish Grand Prix at the Hand Hotel, Llangollen. UL4203, Birkin's Speed Six, then UU5872 the No. 2 car, UU5871 the No. 1 car still wearing race No. 5 from the Six Hours race, and YU3250, Rubin's old car. It is interesting to note that the latter was still fitted with blade wings with Chrysler pattern side-lights, and Marchal lamps on demountable brackets – the latter seems to have been copied on the other Blower cars. Presumably YU3250 suffered some derangement before the race, because it was not entered; Harcourt-Wood drove the practice car, 4½ Litre YV7263. Fifth from the right Rubin, then Birkin, Gallop, and Harcourt-Wood.**

OPPOSITE ABOVE **Post-race celebrations. Jack Dunfee (left) and Barnato in 'Old No. 1', with Wally Hassan side on standing behind the car with Stan Ivermee behind him, with his back to the camera. Nobby Clarke can be seen between them in the dark jacket, leaning on a broom. Note the double Brooklands silencers and the cable-operated handbrake.**

sum of £1000! WO rode in Birkin's car as riding mechanic; Ulster was not a race the riding mechanics looked forward to and WO wanted to set an example by sharing the risks. Bertie Kensington Moir was horrified when WO announced his intention of riding with Birkin and tried to persuade him not to on the grounds of his insurance, so WO arranged a special policy. Birkin says in *Full Throttle* that he asked WO on the off-chance on a night out, and was surprised when he accepted. The insurance agent for the TT arranged the policy for a sum equivalent to WO's other policies. It is possible that it was a personal gesture to Birkin, because the latter must have been well aware of WO's objection both to the supercharger itself and the manner in which it had been forced on his car. While WO's objections might have been professional, there was no reason to extend that to personal acrimony. Either way, WO long remembered

the experience as one of the most terrifying events in his life.

Four SSKs were entered, driven by Caracciola, Merz, Maconochie and Thistlethwayte. Caracciola led from an early stage, pursued by Kidston in the Speed Six, Birkin pursuing Merz. Rubin rolled his Blower at Mill Hill, the car ending up on its back with the wheels in the air. Rubin and his mechanic were both pushed under the scuttle and were unhurt, Rubin later paying tribute to the strength of the Harrison "BF" body which possibly saved his life. Maconochie ran out of fuel and Merz was disqualified for pulling a damaged wing off – in a fit of anger he wrenched it off and threw it away – thus breaking the regulations. Kidston lost the Speed Six in a long skid, narrowly missing a telegraph pole and ending up well and truly ditched with both front wheels off the ground. On handicap, the Alfa-Romeos were really very quick, but

Caracciola drove superbly in the rain to win by a narrow margin from Campari's Alfa-Romeo, virtually the only time during the years 1928–36 when the TT was run at Ards that the event was won overall by a large car. Birkin finished second on speed but 11th overall and the oranges that WO had placed on the floor for refreshment were well and truly roasted. Caracciola "the Reinmeister" was in his element in the wet, WO commenting "thank goodness that by then Tim had decided that Caracciola was beyond the reach of mere

mortals, and had eased up a bit". Harcourt-Wood retired with engine trouble, and "Old Mother Gun" suffered a collapsed wheel but was still moving when flagged off at the end.

On shorter circuits the SSK supercharged Mercedes were very fast and very handy compared to the long wheelbase Bentleys, (WO had thoughts of cutting the Speed Six wheelbase from 11' to 10') but the latter in Speed Six, not Supercharged 4½, form, had more stamina and reliability over longer circuits. However, with their 7 litre supercharged engines, the Mercedes had something of an advantage in speed and power. Motor-racing was fading in importance, though; in WO's words, "the return to London, to a worsening sales situation and a rapidly gathering crisis, was a sobering business". The attitude of Barnato's advisers towards Bentleys had been negative from the start and their opposition to their master sinking his capital into the firm was evidently hardening. Doubtless the signs were already evident to them, as the interest payments on Barnato's £35,000 mortgage were defaulted on by the Company from 29th September 1929.

LEFT **Rubin overdid it somewhat at Mill Hill, rolling right over; fortunately, no-one was hurt. This shows the extra oil pipes used on the Blower cars, and that the Birkin team cars were using the standard 4½ Litre pattern differential in 1929. The production Supercharged cars used the 6½ Litre pattern differential.**

BELOW **WO and Birkin after the race – the faces say it all!**

14 The Stock Market Crash

IN THAT FIRST HALF of 1929, Bentleys reported their first and only significant operating profit of £28,467.19.5, in the financial year ending 31st March 1929. The directors' report for that year makes very interesting reading. The profit mentioned before was after allowances for depreciation and contingencies, and converting all the first production 6½ Litre chassis to 1928 specification. The Service Records do indeed show that the majority of the first 196 chassis were converted in 1928, but why was it done at Bentley's expense? The rest is even more revealing. No dividend was paid on the preference shares, but in order to finance the extensions to the Works a further £25,000 was borrowed from the London Life Association, this extension being made after Barnato himself had offered a further extension to his debenture. Obviously the spending was to continue on an undiminished scale. Carruth stepped down and offered himself for re-election, something which WO probably hoped would not happen.

The London Life's loan consisted of £15,000 and a further £10,000 "with a bonus of ½% in certain events" – with the loan secured yet again against Oxgate Lane, Kingsbury, and basically everything else the Company owned or could get its hands on – this time including the lease on the shop and basement of Pollen House. The £15,000 was advanced on 15th September 1929, the remaining £10,000 in two lots on 15th December 1929 and 15th June 1930. The terms of this loan were basically the same as the first mortgage of October 1927 (see p.197) and, as before, the principal of £25,000 would not be called in until 15th June 1933. This was provided that the interest due was paid within 15 days, the Company did not cease trading or appoint a receiver, or for some reason the security used for the loan became in jeopardy.

That the general trade position was far from well was already known to the Company, as a spokesman had been quoted in *The Star* on the 18th January 1929 as saying "We are doing quite well in spite of bad trade, but the successful flotation of companies proves there is plenty of money, and sometimes after a good week on the Stock Exchange it brings in an order for a car or two". It has often been observed that sales of luxury cars tend to be linked to the health of the Stock Exchange, a situation that must have sent shivers down the Board's spine.

Perhaps a more prophetic statement came from Sir William Sleigh, the former Lord Provost of Edinburgh and a major car dealer in the north of England, at the annual dealers' lunch at the Royal Palace Hotel in Kensington during the week of the 1929 Motor Show. Sir William was quoted in *The Sketch* of 30th October 1929 as saying: "Their [Bentley Motors] racing successes have largely increased their sales up North and I am sure that motorists who favour Bentley products will be glad to hear from Captain Woolf Barnato, the Chairman of the Board of Directors, that Bentleys will race in all the big events in 1930. Previous to this announcement, a rumour had suggested that Bentleys would retire on their laurels. Now I can boldly state that they will take part in the Double Twelve, the Tourist Trophy at Ulster, and the Phoenix Park when that date is settled."

It has been suggested that WO was somewhat bemused by the expansion of the firm and the tremendous successes that had been accrued by his cars, and that he did not have a great deal of control over events. This roller coaster came to an abrupt end as the implications of the Wall Street crash and loss of confidence in the Stock Market brought the Dow Jones Index crashing down, with a consequent "knock-on" effect on the London market, with the disproportionate feeling of helplessness and loss of confidence that accompanies a major Stock Market collapse. In December 1928, President Coolidge told Congress that "no Congress of the United States ever assembled, on surveying the state of the Union, has met with a more pleasing prospect than that which appears at the present time." The *New York Times* average had risen steadily and rapidly since 1927, but the bubble had to burst eventually. It is generally reckoned that the bull market came to an end on the 3rd September 1929. It was not a sudden crash – industrial output indices had been falling since June, but between the 24th and the 29th October the bottom fell out of the market – the depression lasting more or less until the outbreak of the Second World War.

The impact of events was not to be fully felt for some time. At Cork Street the sales organization was overhauled, with Longman and Kevill-Davies briefed to increase sales by any means they thought fit. Again, as in 1925, when Clement had been sent to tour the agents, it was felt that the latter were not pulling their weight sufficiently. Kevill-Davies tackled this by assembling a fleet of demonstrators with which they toured the country, putting on special weeks at each of the agents. At Cricklewood, more extensions were put up for the manufacturing facilities and machine shops were equipped with the latest machinery; the contract for these was placed with Messrs. W.H. Wagstaff & Sons in May. For those whose visions of machine shops of the early part of the century are of banks of machines driven by long belts from overhead countershafts, the neatness of the Bentley shops is something of a revelation. However, the woodblock floor tended to grow humps more or less overnight due to underfloor moisture soaking the blocks, causing them to expand and rise (the Machine Shop was the nearest to the Welsh Harp). It did allow Bentleys to take more control of their detail manufacture and schedules and shifts could be altered as needed. The February 1930 photos of the Machine Shops reproduced in Chapter 15 show work in progress on Supercharged 4½ Litre engine components, but as the crankcases and sumps lying around are evidently fully assembled they are not in the process of being machined. A closer look shows that the components being "machined" are not bolted down to the machine tables, so the

---

DIRECTORS' REPORT TO THE SHAREHOLDERS OF

BENTLEY MOTORS, LIMITED.

FINANCIAL YEAR ENDED 31ST MARCH, 1929.

The financial year now ended may be generally described as one of steady progress. The net profits as shown by the audited accounts amount to £28,467 19s. 5d., after making ample allowance for depreciation and contingencies and after writing off the balance of the cost of converting to the 1928 specification all the first produced six and a half litre chassis.

Your Directors have given earnest consideration to the question of paying a Dividend on the Preference Shares, but in view of the existence on the assets side of the Balance Sheet of underwriting and overriding commission and preliminary expenses amounting to £19,544, they feel that such a course would not be justified, and they strongly recommend that the whole of the balance on Profit and Loss Account be carried forward.

During the past year a new model, the six and a half litre speed chassis, was introduced to the public. Although it is the most expensive chassis produced by the Company it has found a ready market, and the high reputation which it quickly earned has been enhanced by its recent success in the "Grand Prix d'Endurance" at Le Mans.

The Company has been somewhat hampered in the past by lack of its own Machine Shop. All machine work has, of necessity, been contracted out and, although the quality of the work has been of the highest, the procedure has been in some respects uneconomical. The improved financial position of the Company consequent on better trading conditions has enabled your Directors to cope with this disadvantage. A Machine Shop, specially designed to suit the Company's requirements, is now in course of erection at Cricklewood, and it is confidently anticipated that much greater manufacturing efficiency, and consequently increased profits, will be the result of this change of method.

The general offices at Cricklewood have long been inadequate for the Company's requirements and far from consonant with the Company's position. Your Directors have recently authorised the erection of a suitable block of offices and these are due for completion at the end of next August.

In order that the Company's financial resources may not be unduly depleted by this programme of Capital expenditure the London Life Association, who hold a first charge on the Company's assets, have been asked to increase the loan secured by that charge by £25,000, and your Chairman offered to give any extension of the existing loan precedence of his own Debenture to the Company. As a result of the Chairman's offer your Directors are pleased to be able to report that the Association have professed their willingness to advance £25,000 under an extension of the existing Debenture, and this extension will take precedence of the second Debenture to the Chairman.

In view of the great benefits which have been reaped by the Company as a result of our motor racing successes in the 1928 season, your Directors have felt that they are not justified in asking Captain WOOLF BARNATO to bear the whole cost of the racing expenses for that year. The Company have accordingly paid the expenses amounting to £2487, which are debited in the Accounts.

Throughout the past year there has been no change in the constitution of the Board. In accordance with Article 95 of the Company's Articles of Association Mr. J. K. CARRUTH retires by rotation and offers himself for re-election.

Shareholders are invited to write to the Company for further information on any specific point which may be of interest to them.

A copy of the audited accounts will be forwarded to any Shareholder on application to the Secretary.

The 1929 directors' report to the shareholders.

photographs were evidently staged! WO himself says that the only machining work done in-house

was on the 4 Litre and the photos from his album show machining work in progress on 4 Litre sumps, crankcases, blocks and heads. (See Chapter 15, pp.288–9.)

WO was obviously well aware of the prevailing trends in car design and perhaps had deep-seated reservations about supercharging the 4½ Litre engine. It could be argued that supercharging was merely a desperate attempt to prolong the life of an existing design that should have been superseded. In August 1929 WO approached Ricardos at Shoreham with a request for assistance in developing an entirely new engine. This was to be a "much more modern engine design, capable of producing a higher specific output in terms of horsepower per litre and per pound of weight than had been possible with the earlier engines [3, 4½ and 6½ Litre Bentley engines]. Ricardo & Co., in the nineteen-twenties, had considerable success with the development of the well-known side-valve turbulent head, and had been exploring a refinement of this idea, called the "High Power Head". This arrangement involved the use of an overhead inlet valve and a side exhaust valve and it combined the compactness of the turbulent head with the ability to use a comparatively large inlet valve, and consequently had very good volumetric efficiency."

ABOVE **Keville-Davies' travelling Bentley show, here at Maidstone. The idea was to have a special 'Bentley Week' at various agents, to promote sales. From the left, 4½, 4½ with standard Vanden Plas sports four seater, 4½ fixed-head coupé, 6½ saloon, 6½ Sedanca de Ville (probably XV9488 with Thrupp & Maberly body), and another 6½ Litre saloon.**

RIGHT **UW3761, chassis No.SM3903, at the 1929 Motor Show. UW3761 was the production prototype Blower, distinguishable by the light pattern strut gear; all the production cars had the heavy pattern, bolted strut gear. The body was a Vanden Plas standard sports, finished in green over champagne. The brake drums, undersides of the wings, and the body mouldings were all in champagne, the body, upper surfaces of wings, chassis, wheels and running gear were green. The front view shows that the Villiers lettering has been deleted from the supercharger casing, and the Villiers badge has not been affixed to the casting boss (compare views of YU3250, pp.226 and 227).**

(Ricardo & Co., paper DP 6855, 17th October 1962.)

The Supercharged 4½ Litre was almost into

production with the first car (chassis SM 3903, registered UW 3761) ready for the 1929 Olympia Show, fitted with Vanden Plas sports 4-seater coachwork. The Supercharged car was announced at the Paris Salon several weeks before Olympia, but did not appear there. Instead, a 4½ Litre with Vanden Plas 4 seat coachwork and a Speed Six saloon were shown, with Thrupp and Maberly exhibiting a very handsome cabriolet de ville body on a Standard Six chassis.

Although two Supercharged cars were shown at Olympia that year, only one of them was a runner. The following story was recounted to the author by Jack Baker, then an apprentice fitter in the Erecting Shop, in a letter in 1991.

"In 1929, Bentley Motors started manufacturing a series of 50 production 4½ Litre supercharged cars to qualify their entry in the Le Mans endurance race in 1930. I was apprenticed to fitter John Morrison, who was issued with number three of the first chassis series, SM 3903. This vehicle was scheduled to be built as a demonstrator for Bentley Motors. Unfortunately, due to the fact that Amherst Villiers prescribed a modified lubrication system to suit the addition of the supercharger, it was discovered that when the engines went on the test bed, they were totally unsatisfactory and the test staff worked day and night in an effort to over-

come the problems. As a result, a bottleneck of engines developed. By now it was very close to the opening of the Motor Show and the coachbuilders were agitating for their chassis to mount and to complete their coachwork in the short amount of time that was still available. At that stage it became very obvious that the engine lubrication problem required a great deal more time to solve and rectify, so it was decided to install engines into the Show cars for Bentley Motors and, I think, three or four cars for coachbuilder's stands, and to mount dummy supercharger casings with carburettors less the internal rotors, etc., and to make sure that a grub screw was fitted to retain the knurled starting handle cap to prevent inquisitive schoolboys from discerning the omissions. These vehicles were taken to the coachbuilders and after completion of the respective coachwork, they were transported to the Motor Show at Olympia.

"On the Thursday before the Motor Show opening, we had not received an engine to install in chassis no. SM 3903, which was otherwise completed. Needless to say, it was essential for the Company to have a demonstrator, especially as it was a new model being launched and, in desperation, an approach was made to Birkin's stable at Welwyn Garden City and they agreed to lend us the engine removed from the chassis which was

wrecked due to Rubin's crash in the 1929 Tourist Trophy race. Arrangements were made for John Morrison and myself to leave Cricklewood very early on the following day, Friday, in the 3 Litre stores van "scrap three", together with an engine stand.

"We found that the engine we were to collect was on a test bed and being used experimentally. However, after putting the engine in our stand and loading it, we covered it with a white cloth sheet and I was given the job of preventing the sheet from being blown off, and was entrusted with the general safety of the engine in transit. The vehicle was driven by Alf Church and I can well remember that while we were cruising along the Barnet by-pass with little other traffic, Alf closed in on a brand new Renault which was being demonstrated to a young lady by a salesman who promptly accelerated away. Alf increased the speed of the 3 Litre to maintain the gap between the two vehicles. When he had assessed that the Renault was going at full speed, he pulled out and accelerated the Bentley to pass his competitor at a colossal rate. During this episode I found myself struggling to hold the sheet covering the engine and, at the same time, trying to keep my balance and breath against the slipstream, and shouting at Alf to slow down – but it was amusing from my position at the back to see the indignation and fury on the face of the salesman in the Renault, compared with his female passenger who apparently could not help breaking into a broad smile.

"We arrived safely at Oxgate Lane just before lunchtime and promptly unloaded the engine and had a short break, after which we started to prepare the engine for installation into the chassis. John

This underbonnet view shows the engine borrowed from the No.2 Birkin Blower UU5872, and the Autovac originally intended for the Supercharged cars. The petrol filter and Autovac are positioned as per 4½ Litre, but the petrol and suction lines go straight down from the Autovac, across the chassis and out to the supercharger. The suction for the Autovac was taken from the inlet side of the supercharger. It is not difficult to envisage the difficulties not just of maintaining an adequate supply with Autovac, but also of having half a gallon of petrol in the Autovac positioned about three feet above the float levels of the carburettors. The Autovac was deleted before production started, in favour of twin Autopulse pumps mounted off the chassis rail on the nearside.

Morrison and I worked throughout the night and the following day, Saturday, until 4 o'clock in the afternoon, when we were then at the stage that the engine could be run. Incidentally, the tester, Bowie, who was to put the car through it's paces during the remainder of the weekend, had already mounted the test unit consisting of a scuttle and small windscreens and a weighted box-like rear end. The actual starting time was very exciting and certainly the most memorable incident of the whole operation. I can still see the expression of consternation on the faces of those present, including Bert Osborne the workshop manager, Norman Fawkes the charge hand, Bowie, Morrison and myself. We were all shattered by the mechanical noises the engine created – there appeared to be pistons slapping and valve gear chatter, but the most disturbing noise was the unusual whirring noise from the blower rotors and gears – so much so that it was decided to switch off the engine, and a general check was then carried out.

"The engine was then re-started and the external oil sight feed to the blower was checked and the flow slightly increased. When the temperature rose, the noises became less frightening. One had to accept the fact that this engine was built to racing specifications and clearances. Finally Bowie, with a wad of pound notes for petrol, took the car out on the road in preparation for a weekend of testing. On Monday morning he returned with the vehicle, much impressed with it's incredible performance and a list of items to be attended to. These were carried out by the Running Shop and on completion, it was delivered to Vanden Plas, who had a Le Mans type open body ready for mounting. This was done in double quick time and the car was delivered to Bentley Motors on Wednesday morning, the day of the Motor Show opening."

Obviously, the engine fitted to the other show car, the coupé on chassis SM 3902 on Freestone & Webb's stand, must have been a dummy. The engine fitted to SM 3902 was SM 3905. A light-crank engine with an Elektron crankcase, numbered (S?)M 3905, is still in existence fitted to chassis MF 3153. It seems likely that the engine now in MF 3153 was the show engine, later replaced by the real Blower engine no. SM 3905. On the 4th November 1929, SM 3903 was fitted with engine SM 3903, supposedly another Birkin engine fitted to the No. 3 car, YU 3250. According to the Service Records none of the production Supercharged cars ran before April 1930, by which time SM 3903 had been fitted with engine SM 3907. The under-bonnet photo of UW 3761 at the

Show shows that it was fitted with the Autovac, which cannot have been very effective on the Supercharged cars. Even if it could keep up with the petrol consumption, unless the tap on the bottom of the Autovac was turned off every time the car was left, the head of petrol in the Autovac over the carburettors would cause them to flood.

These details of the lubrication problems on the first Supercharged engines have been substantiated by Clarence Rainbow, a fitter in the Engine Test Shop at the time. "Re the oscilloscope [stroboscope]...I was allowed to use it from an instructional point of view, and it was the first Blower that we had on test. I seem to remember the small cover at the front being removed exposing the [oil pump] shaft end, and although running, the shaft appeared to be stationary. I have got this fixed in my mind showing oil leaking along the bottom of the shaft." The Supercharged engine used a different oil pump from the standard $4\frac{1}{2}$ Litre, with longer gears.

The Blower was a major attraction at Olympia, but also the source of a scandal, in connection with designer Amherst Villiers. One of Villiers' agreed spin-offs from the Blower deal was that his name would be on the outside of the casing and mentioned in the catalogues. Villiers took a group of friends on the first day of the Show to show them the supercharger, only to find that the name was missing. Villiers' name was missing from both the Supercharged cars exhibited, the Vanden Plas tourer UW 3761 on Bentley's stand and the fixed head coupé PO 3265 on Freestone & Webb's stand. This broke the terms of the agreement made between Amherst Villiers and Bentley Motors in October, 1928. On the very first superchargers, "Amherst Villiers Supercharger Mark IV" was cast into the front of the supercharger in large letters (see photographs of YU 3250 on pp.226–227). The first four blower units with the lettering were made up for Villiers' firm, with gears and rotors ground by David Brown & Sons. These four superchargers were then delivered by Villiers' firm direct to Birkin. The gears and the involute form of the rotors were ground using the new Swiss MAAG machines, and David Brown were licensed by Villiers to make the production blower units for Bentley Motors to his patents.

Subsequently Bentleys removed the lettering from their superchargers, presumably by altering the casting patterns, and in the catalogues it is obvious that the line "The supercharger is the Amherst Villiers Mark IV" was added as an afterthought. The catalogue reproduces the same photo of YU 3250, but it had been retouched to delete Villiers' name. Conway was instructed to make up name-plates, and someone crept into Olympia to fit them, but by then it was too late and Villiers went ahead with legal action against the Company. Villiers also tore up the special brochures intended to be given away with the cars detailing Villiers' various projects and the general principles of supercharging.

The court case became somewhat involved, and bitter. Villiers visited the Show on the 16th October, observing a cover over the supercharger on the car on the Bentley stand. Villiers' mechanic, Charles Lowe, persuaded Freestone & Webb to remove the aluminium supercharger cover from their car on the 18th, showing that neither of the two Blowers at the Show had the Villiers name or the Villiers logo. Villiers claimed that the lettering "Amherst Villiers Supercharger Mark IV" had been filed off the supercharger casings. It was also noticed that the supercharger was not credited to Amherst Villiers in the catalogues being handed out. Villiers' solicitors wrote to Bentleys on the 18th, pointing out the breaches of the October 1928 agreement and asking for the removal of the unmarked superchargers and the withdrawal of the catalogues. This letter was handed to Arthur Longman, Bentley's Sales Manager, at 7.45 pm on the 18th, on the Bentley stand at Olympia. Longman handed the letter to Purves, the Chief Buyer.

W.K. Forster, the Company Secretary, received a phone call at 8.45 am on the 19th October from Purves, who read Villiers' solicitors' letter to him. At 9.10 am, Longman received a phone call from Forster, telling him to collect up all the catalogues and send them to the printers to be altered, and that the brass plates with the Villiers' logo would be sent down as soon as possible. In the meantime, cards were made up and put on the cars, pointing out that the superchargers were "Amherst Villiers Mark IV". Temporary plates were fitted at 1.30 pm on the 19th, and the final brass plates at 8.30 am on Monday 21st October. The corrected catalogues were received from the printers at 3 pm on the 21st. On the morning of the 19th, Forster sent a copy of Villiers' solicitors' letter to Bentley's solicitors, and Forster and Bentley's solicitor together prepared a letter that was sent to Villiers' solicitors late on the 21st. On the 22nd October, Villiers' solicitors rang Forster, and informed him that they had obtained an injunction against Bentleys. Forster said "I expressed suprise and asked whether they had not received my letter showing that we had complied with the Plaintiff's [Villiers] requirements. They did not deny having received the said letter but added words to the effect that there was more in the case than had yet appeared."

On the 22nd October, Villiers' solicitor C.J. Radcliffe filed the injunction against Bentley Motors in the Chancery Division before Mr. Justice Eve to restrain Bentley Motors from using Villiers' superchargers on their cars unless the superchargers carried the Villiers name. Villiers put it "that unless the Defendant [Bentley Motors] is restrained by this Honourable Court from violating the said agreement by preventing me from getting the said publicity I shall suffer irreparable loss for which it will be impossible later to receive any adequate compensation." The injunction forbade Bentley Motors from fitting, using or displaying cars fitted with a Villiers supercharger, or issuing catalogues describing such cars, after the 25th October if they did not carry suitable acknowledgement. On the 24th, W.K. Forster explained to the court that the omission to comply with the October 1928 agreement was entirely due to an oversight on his part. The lack of the Villiers plates had been noticed two days before the Show, and the plates had been ordered immediately and fitted, as described above. It was Forster's contention that Bentley Motors had done everything to satisfy Villiers' claims, and that the report in *The Times* of 23rd October that Bentleys had ignored the letter of the 18th was untrue. Villiers' solicitor C.J. Radcliffe claimed that when he left his office at 10.10am on the 22nd, he had not received the letter, and it did not come to his attention until after the hearing.

Villiers replied at some length to Forster's affidavit on the 30th October, complaining again about the removal of the lettering from the Supercharger casing, the size and lettering of the cards and plates, and the fact that Bentleys were putting an aluminium cowl over the supercharger that hid the brass plate anyway. Villiers further implied that some of Forster's statements were untrue, concerning the times at which the temporary plates and cards were made up and fitted to the cars at the Show. This was followed by an affidavit from Longman on 1st November, pointing out that Bentleys did not have to retain the lettering, and that their actions had been entirely proper. He concluded categorically: "There has certainly neither in this or in any other respect been any desire or attempt by the Defendants [Bentley Motors] not to fulfill fully all their obligations to the Plaintiff [Amherst Villiers]."

A deal was then arrived at between Bentleys' and Villiers' counsels, by which Villiers received £100. Villiers had been claiming £1,000, and later said that the balance was to follow, and never did. It would seem that he was under a misapprehension

A very late Standard Six chassis outside the running shop, with bolted strut gear and inside battery box, just visible in front of the rear wheel. Rivers Fletcher is on the right, on the weighted test rig on the back of the chassis.

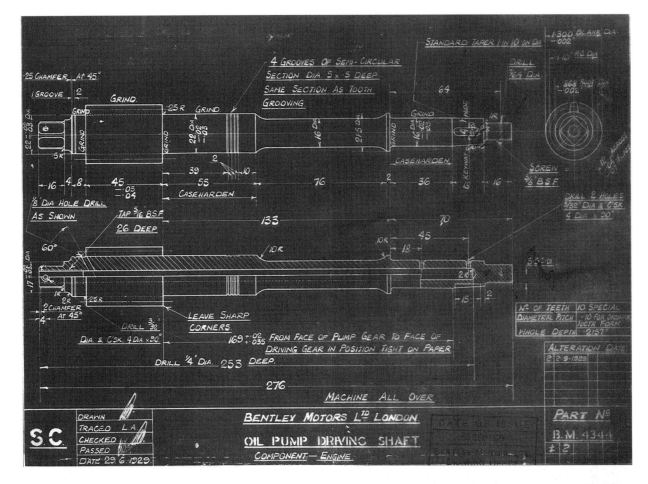

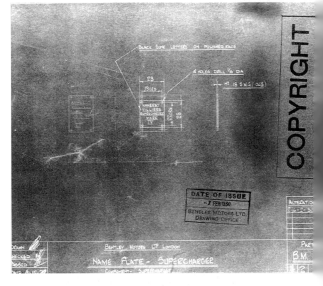

ABOVE **An oil pump drive shaft for the Supercharged engine. This is a good example of a Bentley Motors drawing. Note the alteration date of 2 September 1929, at the time that they were having problems with the oil pump/oil feed mechanism on the Supercharged engine. It was these problems that held up production, and necessitated the loan of an engine from Birkin so that Bentleys could have at least one working car for demonstration purposes. The drawing was passed by CW Sewell.**

RIGHT **BM4863, the Supercharger name-plate. 16th October 1929, checked by Burgess.**

ABOVE RIGHT **BM4152. This is reproduced as an example of a Villiers drawing, simply taken over by Bentleys and given a BM part number.**

FAR RIGHT **BM4124. In this case, the Villiers drawing has been traced by LA (Lillian Atkinson). Drawn by Villiers in April, by Bentleys in October.**

All these drawings are for Supercharged cars only, and bear the project code 'SC'.

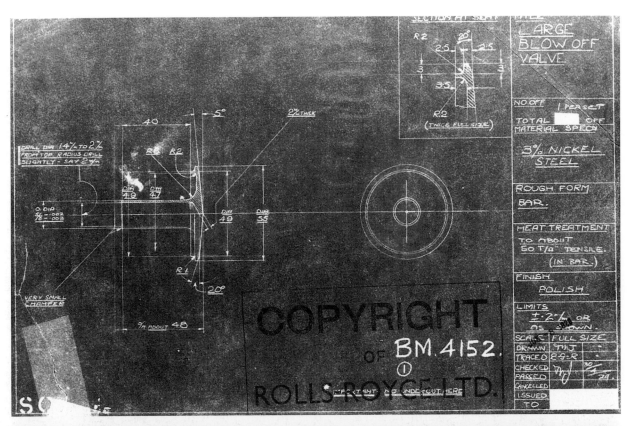

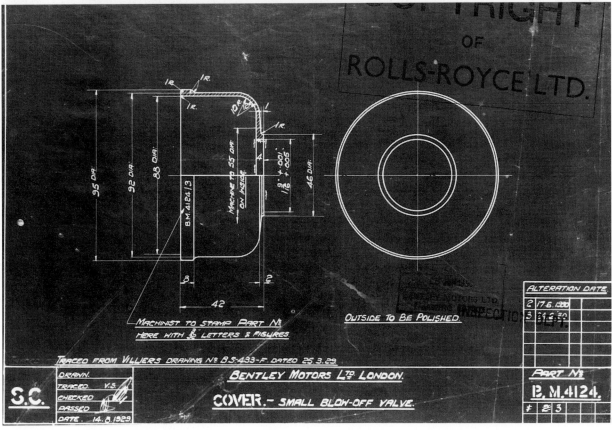

and that the £100 was in full settlement. Certainly Mr. Justice Eve seemed to have got fed up with the whole case, as at the hearing on the 1st November he refused to pass judgement, except to reserve the costs. Basically this was a way of rapping Villiers over the knuckles, because although Bentley Motors had admitted being in the wrong, the costs were not awarded to Villiers as would usually be the case, but were split between the two parties.

Whatever Bentley Motors' reasons, there was one fact that Villiers could not have known – the name plate for the supercharger, part number BM4683, was not drawn until the 16th October, the very day that Villiers was at Olympia – so Bentleys were cutting it pretty fine!

The chassis price of the Supercharged car at £1475 was much higher than the 4½ Litre at £1050, the Speed Six being available for £1800 and the Standard Six for formal coachwork still being offered at £1700. The 3 Litre was no longer catalogued. No fewer than 16 Bentleys appeared at the Show, four on the Bentley Motors stand and twelve on the coachbuilders' stands. Bentleys showed the Supercharged car UW 3761, and a standard 4½ Litre fitted with Gurney Nutting "Prince of Wales" Weymann saloon coachwork. The 6½ Litre was represented by a Thrupp & Maberly cabriolet on the Standard chassis and an H.J. Mulliner Weymann saloon body on the Speed chassis.

Teething problems on the Supercharged cars were relatively minor, after the oil leakage problems had been sorted out and the Autovac superseded by twin Autopulse electric pumps. Problems with pinking and overheating were solved by lowering the compression ratio from 5 to 1 to 4.5 to 1. Nobby Clarke's comment "a Blower eats plugs like a donkey eats hay" is well known, and is probably more of an indictment of the plugs of the time than the car. George Hawkins: "...we started to test 4½ Litre Bentleys fitted with a supercharger. Then we started to burn the spark plugs out so KLG, who supplied the plugs, were contacted and they sent a special set of eight to be tried out. They were handed to me by my foreman with a request to give them a good test. I had them fitted and set off down the Barnet by-pass where there is a straight stretch running by Welham Green towards Jack Oldings where the by-pass joins the A1.

"They stood the test and back at the Works I reported this to my foreman. Good, he said, and by the way what speed did you get up to? Nonchalantly I said 103. Ruddy Hell he said I better see WO about this. I heard that he had gone to WO's office and told him Hawkins had just done 103 down the Barnet by-pass. This shook WO and he said he was not having his testers doing that speed and in future all Supercharged 4½ Litres were to be tested at Brooklands. As you can imagine, this was OK by us, so we went down to Brooklands two at a time in case we blew one up, which we never did, and had a right good time."

On a rather different note, two enthusiasts made an attempt on the American coast-to-coast record sometime in October in a 4½ Litre. J.H. Hanley had just come down from Trinity College, Oxford, and was partnered by R.H. Dutton, also ex-Oxford and a noted oarsman. The record was then held by a Studebaker with a running time of 77hrs 40 mins, or just over 40 mph for the 3200 miles from New York to San Francisco on the Lincoln Highway. The car was supplied by H.M. Bentley & Partners and fitted with a two seat Harrison body with a 25 gallon tank, stoneguards, etc. Very little is known about the run itself, but apparently they were up on the schedule as far as Chicago when two hoods came to their hotel room, and told them at gun point that the existing record still held and they had better forget all about it! In the event they seem to have been stopped by mechanical damage, possibly to the sump, either in Iowa or further west in Wyoming or Colorado.

For 1930 the Speed Six was offered on a 12′8½″ wheelbase as well as the standard 11′8½″ wheelbase, and was fitted with the single port block and 5 gallon sump. The single port block with revised inlet manifold increased power from 160 bhp to 180 bhp. In their deliberations on fitting twin carburettors to the Phantom II Continental, in order to increase its performance to match the Bentley, Rolls-Royce commented that this method of casting part of the manifolding into the block reduced the distance from the carburettors to the cylinders and obviated the need to heat the manifold. They also noted the roughness of the low speed running of the Bentley engine, commenting that it was typical of SU carburettors.

Both the 6½ Litre and the Phantom had their engines insulated from the chassis by means of rubber mountings, a feature that reduces the rigidity of the front end of the frame and aggravates any problems with axle tramp that might occur, a position worsened on the larger cars by the increases in unsprung weight. Indeed, the designer's efforts at the time to reduce axle movement has led to the larger cars of the period being referred to as "springless wonders". It is surprising when looking at the 6½ Litre frame to see the increase in strength

LEFT **Blower and 4½ Litre on test. This is one of the early chassis with the unribbed supercharger, with the early pattern casting over the trunking from the offside of the supercharger up to the inlet manifold.**

BELOW LEFT **George Hawkins with two Supercharged chassis on test at Brooklands. The Blowers were tested at Brooklands because of their performance (see p.250). These are both from the second series of 25, with the ribbed supercharger casings. Note also the short upper arm B&D shock-absorbers, and the heavy pattern front axle beam with integral jacking pads.**

in the centre of the chassis due to the greater depth of both the sidemembers themselves and their flanges, and the additional tubular cross-member between the side-members immediately above the rear spring mountings, and yet to find that the front and rear extremities are no more substantial than those of a 3 or 4½ Litre. WO himself said that "the 6½ Litre chassis was not very rigid and was rather unsuitable for some of the heavy bodies that customers insisted upon". In general the Rolls-Royce papers are very critical of the 6½ Litre, commenting on the lack of rigidity of the chassis, mounting of the steering equipment, poor brakes with the lag of the Dewandre servo, particularly showing up in heavy traffic and an overall impression that the

Bentley was somewhat less refined than the Rolls-Royce as a town car. However, Royce's comment in his memo of 1929 that "We have better staff and better machinery" is not very endearing, and they were to eat their words in barely a year's time over the 8 Litre.

Chassis changes to the Speed Six/6½ Litre were limited to fitting Bentley and Draper hydraulic shock absorbers to the rear, the electrical system benefitting from a new Smith's dynamo, and a Weston ammeter. The new Lucas P100 DB lamps were fitted as standard. Elektron was introduced for some of the engine castings, reducing the chassis weight by roughly 1½ cwt. In terms of coachwork, fabric covered bodies still held sway, either

ABOVE LEFT **Hanley and Dutton with their special Harrison-bodied 4½ Litre, before shipping the car to America for the coast-to-coast record run. Note the long bonnet with team car pattern spring-loaded strap, Lucas P100 lamps with stoneguards, stoneguard radiator, and double rear shock-absorbers. This also shows well the four-rivet pattern dumb-iron knuckle used on the 1929 model 4½ Litre and 6½ Litre frames.**

LEFT **Bowie in a 1930 model Speed Six chassis tows Malcolm Campbell's 'Bluebird' in the Lord Mayor's Show. George Hawkins was sent down to Brooklands to arrange the towing mechanism, and 'Bluebird' was temporarily resident in the finished cars/experimental shop at Cricklewood while the tow-bar arrangement was being made up.**

THIS PAGE **Centre section of the 1930 model 4½ Litre chassis, with the plate clutch, C type gearbox with stainless steel lever, Spicer shaft, and standard 4½ Litre pattern back axle. The 1930 chassis has the inside battery box, slung from bars across the chassis. The frame itself has the deep bottom flanges and light pattern reinforcing brackets over the strut gear uprights. The Blower frame is similar, but has heavier pattern reinforcing brackets above the strut gear legs.**

built as flexible bodies or along conventional coachbuilding lines and merely fabric covered. Indeed, 1929 marked the zenith of the fabric body. Innovations were taking place, and notably Thrupp and Maberly were using light alloy castings for various parts of the bodyshell, including door frames, door pillars and wheel arches. Other firms were using similar castings for windscreen pillars. The 4½ Litre was soon to be out of production, with the last two chassis for the Works Team cars chassis HM 2868 and HM 2869 delivered to Vanden Plas in February 1930 to make up the 1930 team with "Old Number One" Speed Six.

Incidentally, despite the similarities in name, there is no known connection between Bentley and Draper Ltd. and Bentley Motors Ltd. beyond the fact that Bentley Motors fitted the former's shock-absorbers. Bentley and Draper was set up by Percy Holman Bentley and Alick Darby Draper, the company having its base at 4 Fenchurch Avenue, London, specialising in the design and manufacture of friction and hydraulic shock-absorbers.

The 4½ Litre underwent substantial changes for 1930, in that the heavy crank engine as used in the Supercharged cars was fitted as standard. While there were good reasons from the manufacturing point of view in terms of interchangeability of parts, inventory, stores handling, space and ordering for converting the 4½ Litre production over to the heavy crank and rods with the reinforced crankcase, block and sump of the Supercharged car, there were also good reasons for not doing so. Perhaps as WO was, on the surface, so against the Supercharged cars he would not have put the Blower engine in if he had had any say in the matter, but it is unlikely that he had. Before the Supercharged car was marketed, the firm had no way of knowing it would be a commercial flop, and if it had been a success it would have been very easy to have switched over all 4½ Litre production from unsupercharged to Supercharged. WO himself refers to "the Blower engine with the drive spigot cut off the front of the crankshaft". The chassis specification, with the exception of the 6½ Litre pattern differential casing (used in the Supercharged cars only), of the 1930/31 4½ Litres is virtually the same for the Blown/unblown cars. Fitting the Blower engine with its 80 mm diameter main bearings compared to the 55 mm of the standard shaft (and a crankshaft weight of 75 lbs compared to 47 lbs) reduced both the acceleration and maximum speed of the late cars. As the late Harry Charnock commented years later the standard excuse for one's 4½ Litre being slower off the mark than anyone else's was "Blower crank and rods, old boy". Bastow adds to this, by stating that the increase in main bearing diameter reduced the engine output by some 5–10%. The first supercharged car to be delivered to a customer was EU 919, chassis SM 3905, to Captain D'Arcy Hall in April 1930 fitted with a Maythorn coupé body. Some of the pundits were quick to observe that the chassis of the Supercharged car was £425 more than the standard 4½ Litre, but that the performance of the latter was more than adequate for the normal motorist. This is borne out by The Autocar's "Performance Chart for Representative 1929 Cars" contained in their issue of 11th October 1929, in which the standard 4½ Litre had the highest top speed of any car tested at 92 mph.

There was one last major race in 1929, the British Racing Drivers' Club 500 Miles Race held at Brooklands on the 12th October. The BRDC was formed out of informal meetings of Brooklands' habitués held at Dudley Benjafield's house, and the 500 was for some years the fastest motor race in the world – faster even than the Indianapolis "500". Held over 181 laps of the outer circuit the race was for cars without road equipment on a handicap basis. Two Works cars were entered – firstly, "Old Number One" Speed Six fitted with a Vanden Plas 2-seater body driven by Sammy Davis and Clive Dunfee. The Speed Six in this form was so fast that most of the available drivers were frightened off, and Sammy was approached virtually at the last moment. Secondly, Jack Barclay and Frank Clement shared 4½ Litre YW 5758, fitted for the occasion with a long, tapering tail. Jack Dunfee entered the Hon. Richard Norton's 4½ Litre "Old Mother Gun" YH 3196, Rose-Richards and Fiennes Scott's 4½ Litre UU 5580, and Birkin entered the No. 1 Blower UU 5871, partnered by Harcourt-Wood, also fitted with a very light Vanden Plas shell body, a sort of 1½-seater. This body was of rather novel construction in that a minimum number of wooden hoops were used to form the shape and between these were laid spring steel strips in an overlapping lattice pattern, with the intersections held together by special clips. The whole was then covered with wadding and fabric to produce a very light body of vaguely streamlined shape. Birkin and Harcourt-Wood had a very uncomfortable race, with oil blowing back into the cockpit and even getting onto the steering wheel. Unfortunately the outside exhaust system was not well thought out, and WO told Birkin before the race that it would break up. In the race it did indeed start to disintegrate and set fire to the body, causing

LEFT **Birkin at speed on the banking in the 500. The car is the No.1 Blower UU5871, fitted with a special 11/2 seat track racing body by Vanden Plas. Despite the heat shield between the body and the outside exhaust, the system broke up and set fire to the body, putting Birkin out of the race.**

BELOW **'Old Number One' Speed Six in the pits, with obvious tyre trouble. The car was so fast that it frightened off most of the available drivers, and Sammy Davies was drafted in at virtually the last moment to partner Clive Dunfee. The Speed Six was easily capable of lapping at over 125 mph. The track body was built by Vanden Plas, with two interchangeable tails. In the 1929 500, the stub tail was fitted. For the 1931 500, the long tailed version was used.**

Birkin to retire. UU 5871 was then fitted with the famous Thomson and Taylor single-seat body and the Vanden Plas shell transferred to a road car, built up from parts in stock registered UR 9155. Many years later scorch marks could still reputedly be seen on the timbers of the frame.

Jack Dunfee also retired in "Old Mother Gun", making her last appearance in a major race, when the crankshaft sheared immediately in front of the flywheel, an expensive and not unknown occurrence on the 4½ Litre and (rarely) on the 3 Litre as well. The 4 litre Sunbeams "Tiger" and "Tigress" proved to be very fast, but "Tiger" retired with a broken rear spring and "Tigress" cruised round (at 102 mph!) with a cracked chassis frame. Eyston in the 2 litre Sunbeam "Cub" led for a while before the Clement/Barclay 4½ Litre moved into the leading position. Jack Barclay lost control on the banking and disappeared under the scuttle as the big 4½ Litre slid backwards down the banking, handing over to the somewhat steadier Frank Clement.

For the 500 miles Clement and Barclay averaged 107.32 mph to win on handicap, a remarkable performance for an unlightened 4½ Litre on a 10'10" chassis with a a full 4-seat body. The Speed Six finished second on handicap averaging 109.4 mph, the highest average for the race and setting the fastest lap (Dunfee) at 126.09 mph. Barclay and Clement won £250 presented by Humphrey Cook, the Sir Charles Wakefield Cup, £100 presented by Joseph Lucas Ltd. and the Vanden Plas Cup. At the end of the year Bentleys took out a full page colour advert on the front cover of *The Motor* to celebrate the successes of the racing team, and a pretty remarkable record it was too. A Speed Six was intended to go to Montlhéry before the end of the year to tilt at various records, but that project did not come to fruition.

The effects of the Wall Street crash were spreading across the Atlantic and the position of a relatively small company making large and expensive cars relying on a small clientele of well-heeled customers, precisely those most vulnerable to a collapse in the market, was not an enviable one. Up to 1930 motoring was still a pastime for the relatively well off; the 1930s were the decade in which motoring truly started to spread to the masses. The result of this was a fundamental change in manufacturing away from hand building to mass production and the statistics bear this trend out. In 1930 40% of all cars sold had engines of 15 hp or more against 36% in the 8-10 hp area, while in 1939 63% of sales were in the 8-10 hp bracket and only 13% in the 15 hp plus bracket.

WO with Jack Barclay. Barclay needed the cushion because of his diminutive stature.

Leading into 1930 the Works developed the first 8 Litre engine, the Racing Department at Kingsbury prepared the two new Speed Sixes and "Old Number One" for the forthcoming year and the Birkin team at Welwyn Garden City prepared their team of Supercharged cars for their first full year of racing. The Birkin cars were the first entries for the 1930 Le Mans to be received by the A.C. de l'Ouest. The design of the Speed Six engine gradually evolved into that of the 8 Litre with the single port block, 5 gallon sump and on the last engine series, a co-axial Bosch starter motor with modified ring gear on the flywheel. Unlike the Smiths where full load was applied from the start, the Bosch pinion slid into mesh before turning the flywheel, substantially reducing wear on both the ring gear and the starter and eliminating the grinding and crashing noises of an engaging Smiths. The later Speed Sixes were also fitted with a taller radiator and bulkhead, roughly 1" higher than before. The two new Team Speed Sixes were fitted with identical Vanden Plas 4-seat bodies, registered GF 8507 and GF 8511.

The first 8 Litre engine was based on a 6½ Litre crankcase number HM 2870 and was on test in the

ABOVE 'The Box', the No.2 experimental 6½ Litre, in final form with 8 Litre engine. It is possible to distinguish a ring of bolts for the tubular cross-members used with the F type box behind the brake adjuster, and the bracket for the 8 litre pattern Dewandre servo. This car was kept by Bentleys until the Rolls-Royce takeover, when it went to HM Bentleys, who fitted it with a Vanden Plas open body. Note the 8 Litre type 80 hubs to the front, but the smaller type 62 hubs used on all the other Bentleys to the rear.

LEFT Bentley Motors' chief designer for the 8 and 4 Litre cars; Colonel Thomas Barwell Barrington.

Engine Shop by February 1930. It was then fitted to WO's own saloon "The Box" MH 1030, the No. 2 Experimental 6½ Litre car whose long and distinguished experimental career entered a new phase. "The Box" was also fitted with an F type gearbox. Twin SU H08 carburettors with 2″ throats were fitted, producing for the first time a full 6/7 seat production saloon car capable of over 100 mph in near silence and with considerable refinement. WO drove "The Box" in this form for many miles and was obviously pleased with the result. Some of the refinements intended for the 8 Litre appeared on the Speed Six team cars, with

TOP **Centre section of the 8 Litre chassis, showing the F type gearbox. The number 8004 to the top left of the casing indicates that this chassis is YF5002, later fitted with an H J Mulliner saloon body for WO, registered GK706. This shows the tubular cross-members, fabricated in sections, and the Dewandre L3 servo. Visible just behind the rear cross-member is the late pattern reinforcing plate fitted to chassis with the bolted strut gear above the upright.**

ABOVE LEFT **8 Litre pattern hypoid bevel back axle. This shows the single shoe rear brakes, with the single operating lever with twin holes for hand and foot brakes. The mountings are further out for the outrigged springs, and the larger Type 80 hubs are fitted.**

ABOVE **Birkin at Brooklands with his sponsor, the Hon Dorothy Paget, well wrapped up for fast motoring.**

Bentley and Draper hydraulic shock absorbers fitted to the front of the chassis and 8 Litre pattern steering columns.

The 8 Litre was the first Bentley not to be completely designed under Burgess' direction as Chief Designer, as he was away from the Company for long periods due to ill-health, finally succumbing on the 30th November. He was succeeded by Lt. Col. T.B. Barrington, whose appointment as Chief Designer to the Company was announced on the 7th December and made retrospective to the 1st November. Barrington was retained on a three-year contract from 1st November 1929 at £1,500 per annum. Barrington was responsible for the F type gearbox used on the 8 and 4 Litres, and probably much of the chassis work. It is perhaps significant that the 8 Litre chassis moved away from earlier Bentley design philosophy with the double-dropped frame, outrigged rear springs, F type box

and hypoid back axle under a new Chief Designer. Development work continued on the 4 Litre engine, single-cylinder prototypes running on test from early 1930 onwards. The first public hint of the new car appeared in the *Daily Herald* of 22nd May 1930, their Motoring Correspondent noting that "Rumours have been prevalent that a new Bentley 4½ Litre six-cylinder model was to be made, but I can definitely say that this is not so."

The Birkin Works at 19 Broadwater Road, Welwyn Garden City had evolved into a small factory in its own right with a staff of some 30 people, with machine shop and engine test facilities. All this cost a great deal of money and although the Birkin family were making money out of Nottingham lace, it was not enough to cover the ambitions of the son and heir. Of the Birkin brothers, Tim was the last survivor, the first having been killed in the Great War and Archie during a practice session for the

Preparation at Welwyn for the 1930 season – Birkin with the single-seater, the No.1 car UU5871 with special track racing body by Thomson & Taylor. This shows the special bulkhead and Bosch magneto, and the special controls for the carburettors and magnetos used on the Birkin cars. The supercharger is the plain case variety, but is a Bentley Motors unit, not the original Villiers unit. This is indicated by the deletion of the Villiers lettering from the casing.

1927 Motorcycle TT on the Isle of Man. Tim persuaded the Hon. Dorothy Paget, Lord Queenborough's daughter, to put up the money for the team – apparently she had become interested in motor racing after watching the 1929 500 Miles Race, and was in receipt of a private income of some £100,000 a year. The immediate result of this was the rebuilding of UU 5871, the No. 1 car, with the famous Thomson and Taylor single-seat body, and UU 5872, the No. 2 car, was rebuilt on a new 9'9½" frame along with a completely new sister car UR 6571, the No. 4 car.

These new 9'9½" frames were made by Mechans to the same specification as the late 3 Litre Speed Model frames, of ³⁄₁₆" gauge. Bentleys allocated the Birkin Works a series of chassis numbers, HR 3976 to HR 3400, and the new car UR 6571 was numbered HR 3976. The No. 3 car YU 3250 remained basically unaltered. Clive Gallop was appointed Works Manager and Superintendent of Production at Welwyn early in 1929, and it was he who negotiated with Mechans to put the frames through and was responsible for much of the detail work on the cars.

Gallop and Villiers seem to have fallen out in a big way, and Villiers was very scathing about Gallop's abilities and his contributions to the cars. Villiers described Gallop as "a hot air technician". Presumably disillusioned by the legal proceedings and the way he felt he had been treated, Villiers seems to have had very little to do with the cars after the change from the plain to the ribbed casings. The chassis were assembled with fitted bolts made in-house on a Ward capstan and by various means a lot of weight was added to the chassis. The production Supercharged cars had a rated power output of 175 bhp, of which some 35 bhp was needed to drive the Supercharger itself, and the Team cars were obviously more powerful than that. In later years the single-seater was giving power outputs of 240 bhp, but reliability was always the Blower's Achilles heel, due in part at least to a lack of development work. Gallop later admitted that a number of mistakes were made in the process of producing and developing the team cars.

There is rather a sad footnote to the Supercharger story. It is possible that Gallop's views were coloured by perspective and by the fact tht he was evidently not a well man when he wrote the

Work in progress on the No.2 Blower UU5872. These photos graphically illustrate the details of the Birkin cars.

following to WO in March 1958, following on the publication of WO's autobiography containing the oft-quoted "corruption and perversion" comment on the Supercharged cars. Gallop: "Looking back over the years I appreciate how absolutely right you were, at the time, to dislike, for such very good reasons as you have set out, the major error of "blowing" the 4½ Litre. In fact I do not recall any advice or opinion of yours that did not turn out to be absolutely right."

The supercharger itself had to be redesigned, because of problems with the casing. This was designed as a double-wall plain casting, but because the shape of the casting was not even along its length, when heated up it did not expand completely linearly. The resulting distortion caused the rotors to touch. The clearances between the two rotors with their involute form ground on was critical, and the drive gears also had to be made to very high standards. Villiers designed a new single-wall casing with cast-in cooling ribs and it was later found possible to re-use the damaged rotors. The

reason for this double-wall design was that it was originally intended to water-cool the Supercharger, and the plain-case Supercharger has tappings in the top and bottom of the casing for this purpose. The later drawings refer to "Supercharger Non-Intercooled".

The supercharger drew mixture through twin SU HVG5 carburettors (although it is thought that HVG3s with 1⅜" throttles were originally planned for the production cars), and then blew through a long trunking with a blow-off valve near the supercharger up to a special inlet manifold, also fitted with twin blow-off valves. The matrix of the radiator was deepened considerably to compensate for the reduced cooling area, but overheating was a problem on both the Birkin cars and the production cars. The Birkin cars in particular suffered from overheating of the engine oil. The supercharger itself masked off the sump reducing any air cooling effect, so new sumps were made for the racing cars with cast-in cooling ribs, and in later years the single-seater was converted to dry sump lubrication with a pump driven off the front of the supercharger. This finally solved all the lubrication problems experienced, but two years too late. An inevitable problem with the use of a larger diameter crankshaft is increased bearing temperatures because at the same rpm the rubbing speeds are higher due to the increase in diameter. It is virtually as important to get the oil away from a bearing as to it in the first place, so oil pump capacity (flow, not pressure) becomes critically important. Because of this the Blowers used deeper, higher capacity oil pumps than the standard 4½ Litres.

The new car UR 6571 was fitted with a small badge in the form of the Paget crest on the radiator, just above the Bentley badge in recognition of the team sponsor. The Hon. Dorothy was considerably more interested in racehorses than motor cars, so Tim Birkin must have exercised considerable persuasion in bringing her into the fold. Her reputed preference was also for girls rather than men, so it would seem unlikely that Birkin's persuasion was conducted in the boudoir.

Early in March 1930 the single-seater was complete and on test at Brooklands. It soon proved to be very fast, winning the Kent Long Handicap in the opening Brooklands meeting of the year over 12 miles at an average of 119.13 mph. At the Easter Monday meeting, Birkin created a sensation by breaking Kaye Don's lap record of 134.23 mph no less than four times, leaving it at 135.33 mph. *The Sphere* commented that "Captain Birkin's achievement is a wonderful testimonial to his skill as a driver. It is, of course, in addition a real high feather in the cap of Bentley Motors Limited". In May Birkin lapped the Montlhéry track at 138 mph, but was unable to make an attempt on the hour record because all the French time-keepers were in Monaco for the Grand Prix. Interestingly Birkin and Barnato entered a Supercharged 4½ Litre in the 1930 Mille Miglia, with the team to be run by Kensington Moir, but it never came off. Birkin withdrew supposedly because of the logistical problems of driving such a long race so far from home. Kidston, though, did drive in the Monte Carlo Rally in a Speed Six coupé, but crashed under icy conditions near Inverness and arrived at Glasgow minus both front brake drums!

Barnato added his name to the list of sporting activities undertaken by the Bentley Boys by beating the "Blue Train" from Saint Raphael to Calais on the 13th/14th March 1930. Partnered by Dale Bourne, described by Barnato as "a well known amateur golfer," he averaged 43.43 mph from Cannes to Boulogne and arrived in London before the train reached Calais. This feat was undertaken as a private wager, but the French authorities did not see it in the same light. As Barnato was Bentley's Chairman it was seen as an advertising stunt,

TOP **This photo shows the Birkin single-seater UU5871 at the Montlhéry track, at the time of the record attempts in later April/early May 1930.**

ABOVE **Barnato and Clement in the new Speed Six GF8507, before the 1930 Double Twelve, at Brooklands. In the model Bentley is Master Carruth, Jack Carruth's son. This shows the details of the Speed Sixes raced in 1930. The chassis has double B&D hydraulic shock-absorbers to the front, with the heavy pattern front axle beam with integral jacking pads. Just**

visible along the front axle is some form of steering damper. The front brakes are 4½ Litre self-wrapping, with rods outside the frame and special headlamp support bracket castings with bearings for the Perrot shafts. Note the condition of the nearside tyre!

RIGHT **Out at the track, Bertie Kensington Moir (left) and his partner Straker, in front of the latter's 6½ Litre saloon UL4168. Lord March is to the right of Straker. March's Speed Six demonstrator GU6449 is behind Straker's saloon.**

the direct result of this being that Bentleys did not exhibit at the Paris Salon. As part of the agreement reached in 1929, when Bentleys exhibited at the Salon for the first (and last) time, the Company signed an agreement that they would not enter into any competition unless it was backed officially by the French Motor Manufacturers' Association. The Association asked Bentleys for an explanation, but refused to accept that Barnato had acted as a private individual, imposing a fine of £160 or expulsion from the Salon. Bentleys replied that they would neither pay the fine nor exhibit at the Salon, and threatened to withdraw from the Le Mans race as well. However, "the Directors of the firm have closely considered the matter and have now decided that Bentley cars will participate", (unidentified press cutting), *The Daily Herald* of 24th May, under the sub-heading "As regards the 'Official' team, a first-class sensation has developed, and the French authorities may prevent the team from starting", appeared the following: "I discussed the matter with a director of Bentley Motors, and he told me that he thought it likely that the cars would run. He added that the firm felt that they had been rather harshly treated, as Captain Barnato had undertaken the run not as Chairman of the Company but purely as a private

owner, and the firm had not advertised the 'record' in any way." Finally, in the *The Bystander* of 18th June, "Bentley people have officially denied the rumours that recently appeared in the press that they might withdraw their entry for the Le Mans race owing to a difference of opinion with the Paris Salon authorities" – plus ça change!

All three Supercharged road cars were entered for the Junior Car Club's Double Twelve race on 9th/10th May 1930. Birkin and Chassagne were to drive UU 5872, Benjafield and d'Erlanger the new car UR 6571, Kidston and Jack Dunfee driving the long chassis No. 3 car YU 3250. The Works entered the two new Speed Sixes, GF 8507 driven by Barnato and Clement and GF 8511 by Sammy Davis and Clive Dunfee. M.O. de B. Durand entered his 4½ Litre YW 8936 privately, partnered by T.K. Williams. Jack Dunfee got away to a good start and Birkin was leading at the end of the first hour, followed by Jack Dunfee, Barnato and Davis. Jack Dunfee was also the first Bentley retirement from the race at 3.37 pm with a broken valve, followed by Birkin at 4.30 with a cracked chassis frame. It is thought that this was caused by hitting the concrete kerb on the finishing straight, near the point where the two team Talbots collided with fatal results. One of the cars ended up inverted on the track, the other going through the railings and into the crowd. The third Talbot was withdrawn at the end of the first day's racing.

Overnight the cars were locked up in the enclosure, and the following morning the Bentleys cruised slowly for the first few laps after the 8 am start until oil pressure and water temperatures were back to normal. The Speed Sixes were leading by such a margin that they could afford to slow down, the Williams/Durand 4½ Litre catching fire at the pits and later retiring with back axle failure. The pinion race failed, probably due to the use of spiral bevels on a race car. Benjafield's car caught fire at the pits after changing a valve spring, and eventually retired due to back axle failure. Benjy did not give up easily – he pushed the car round to the pits and dismantled the unit to withdraw the differential assembly, only to find the pinion was damaged and the car was hors de combat. *The Motor*'s account of the race included the following: "It was rather amusing hearing the driver and pit manager discussing what should be given out to the Press as the reason for the stop. Kensington Moir had a brainwave. 'Tell them we are changing the back axle ratio to get more speed!'"

Davis and Dunfee had problems with valve springs on the second Speed Six and had to change

LEFT **Famous personalities in the pits – WO and HRH Prince George, talking to Woolf Barnato and Sammy Davis. Behind WO is Wing Commander Butler, Prince George's equerry. Note the stopwatches and record books on the pit counter. Second from the right is the team electrician, Frank Mills, from the service department, with his assistant Garrat on the far right.**

BELOW **The successful Speed Sixes after the race – Barnato and Clement in the winning Speed Six GF8507 (right), and Sammy Davis in beret in No.3 GF8511 on the left.**

two, but the two cars thundered around to take first and second places at speeds of 86.58 and 85.58 mph respectively and on merit with indices of 1.462 and 1.452. There was also *The Autocar* Price Handicap Award, an index relating performance to chassis price. Not suprisingly, this was won by the supercharged Austin, but the Speed Sixes still finished 2nd and 6th despite the huge price tag of the Speed Six chassis. The dress rehearsal for Le Mans for the Works Speed Sixes was an unqualified success, but that for the Blowers was to become an all too familiar tale. Clement and Barnato took home between them the SMMT Challenge Trophy, the Rudge Whitworth Trophy, 150 guineas and £100 from Castrol's Lord Wakefield. With Barnato's vast personal fortune but noted parsimony it would be very interesting to know just where the money ended up!

To celebrate his victory, Barnato threw a dinner at the Kit-Kat Club. The following printed instructions were to be found on each of the guest's plates:

"The race will be run over a number of courses a list of which will be found on your starting places. After having been scrutineered at the Bar, chassis will be parked as the Clerk of the Course directs. As soon as the starting-gun is fired Drivers can let go their clutches and open their throttles wide.
N.B. (1) Always think of the safety of your passenger.
(2) How your mechanic looks behind.
(3) All spanners, jacks, knives, forks, etc. are the firm's property and must not be taken away.
(4) Drivers may only carry one passenger at a time. Passengers may, however, be changed during the Race.
(5) Overtaking – A faster car may overtake a slower one, but only if the passenger waves him on.

Before your chassis is parked at its original starting point, each Driver must see that his carburettor is flooded with at least two cocktails. Passengers, mechanics, spare parts and wives are requested to co-operate, and in any event see that this rule is strictly adhered to.

Map of the Course.

Through the Bar to the Dining-room. After two hours' running in the Dining-room, contestants may try out other chassis in the Ball-room (but only if their own has broken down, seized up or died on them). Refuelling arrangements will by this time have been made downstairs. Contestants are specially requested not to sit about in other firms' pits.

By Order of the Clerk of the Course.
Woolf Barnato."

'By such happy means did Barnato manage to get through eight or nine hundred pounds a week" – *WO – An Autobiography.*

For the Le Mans race on 21st/22nd June 1930, an announcement was made that a single 7 litre supercharged SSK Mercedes was to be entered, driven by the "Reinmeister", Rudi Caracciola, partnered by Christian Werner. Barnato was looking for his hat-trick and came back from America in a fever pitch of excitement. Three Works cars were entered, the new Speed Sixes from the Double Twelve with Clement and Watney sharing GF 8507 and Sammy Davis and Clive Dunfee GF 8511, with Barnato and Kidston sharing "Old Number One" MT 3464. Three Supercharged

WO and Moir in lighter mood at Le Mans, in Sammy Davis' Leon-Bollée.

cars entered as well, UU 5872 driven by Birkin and Chassagne, UR 6571 by Ramponi and Benjafield, and the long chassis car YU 3250 driven by Jack Dunfee and Beris Harcourt-Wood. The Works Team were well prepared for such events and had a comparatively easy time in the week before the race, based at the Hotel Moderne as always, but the Birkin team had their troubles. The low grade French fuel supplied in 2 gallon drums caused the supercharged cars to overheat, so just before the race a decision was taken to run on pure benzole. This necessitated raising the compression, achieved by removing a compression plate between the block and the crankcase. Even though the Work's Team mechanics assisted Birkin's people, there was only time to do this on the two short chassis cars, so YU 3250 non-started.

If it really had only been a matter of removing a compression plate on each car, then there was no conceivable reason why the long chassis car non-started. However, in reality, soon after practice had got underway, it had run its bearings, followed by throwing the flywheel off on the fourth night, so there was no way that car could have been repaired in time. Indeed, it was only through the co-operation of the Leon-Bollée works that the two short chassis cars started at all. Years later Nobby Clarke commented that it was obvious the Birkin people were not up to all the tricks, because they lifted the block right off to remove the compression plate, whereas the Works Team just lifted it far enough to cut up the plate and take it out in bits. The Birkin Team even put out the story that the car was being saved for the Spa Grand Prix to be held the following Saturday, the 5th July.

While all this activity was going on below stairs, the drivers were also enjoying the run up to the race. Dinners were laid on at which great discussions of tactics were held, Benjafield recalling one suggestion that one of the Bentleys should pass the Mercedes at high speed on the approach to the Mulsanne straight and then disappear down the escape road, in an attempt to fool Caracciola and Werner into entering the bend too fast and ditching the car! It is legend that Birkin hounded the Mercedes and ensured the premature demise of his car for the sake of the team, but whether the race was organised as such or whether Birkin was just driving in his usual fashion is a matter for speculation. Birkin had considerable vested interest in seeing a Supercharged 4½ Litre win the Le Mans race, and it seems unlikely he would deliberately sacrifice his car to allow the Works Speed Sixes to take all the glory.

The Mercedes was first away on the flag fall, but was soon pursued by all five Bentleys. Only nineteen cars entered in 1930, one of the lowest entries ever, and of those eighteen started. WO himself was quite happy to win races by scaring off the opposition and indeed by 1930 they had succeeded in doing just that. 26 cars entered in 1931. Gabriel Voisin complained bitterly about the British domination of the French race, and the way in which other manufacturers had been frightened off by the dominance of the Bentleys. Caracciola and Werner had calculated from the 1929 lap times that they shouldn't have too much trouble and would not have to push their car too hard. They were somewhat mistaken because of the manner in which WO always slowed down his cars when victory was certain so as not to reveal their true potential to the opposition or the handicappers. Pretty soon Caracciola's rear view mirror was full of Bentleys, and Birkin passed him on the run into Mulsanne with two wheels on the grass, the rear bald from a stripped tread. To the Germans' astonishment Birkin remained in front for a whole lap before pulling into the pits for a new tyre. The Dunlops used that year suffered from a tendency to lose treads, which would part company from the casing explosively and bend wings before disappearing – one such flying tread hit Benjafield on the shoulder. This

ABOVE **The Bentley Boys and staff at Le Mans – from left to right, Dudley Froy, 'Nibs' Howard (front), Sam Hood (back), Dusty Miller (electrician), not known (service dept.), Puddephat, Hassan, Ginger Pryke.**

BELOW **Customs documents for WO's car, 'The Box', for Le Mans 1930. Note the chassis and engine numbers EX 2 for the No.2 Experimental 6½ Litre MH 1030, apparently still fitted with an experimental 6½ Litre engine (37.2hp).**

The start of the race. Caracciola sprints for No.1, the Mercedes, with the three Speed Sixes waiting in echelon. Note the white identifying patches on the offside front wing of No.2 and nearside wing of No.3.

problem seemed to afflict the Supercharged cars more than the Speed Sixes, with their greater weight and higher centre of gravity. Although Mercedes were using the same tyres they did not appear to suffer from this problem at all.

Sammy Davis then took up the pursuit, despite splinters in the eye from cracked goggles broken by a flying stone, and handed over to Clive Dunfee after 20 laps. Dunfee unfortunately was too keen and not very familiar with the circuit, and put the car into the sandbank at Pontlieue on his first lap. He dug for ages, and Sammy also dug away with a spare headlamp glass, but eventually a bent front axle and a pair of damaged wheels were revealed, putting No. 3 out of the race.

The Mercedes was hard pressed by the Bentleys and the whine of the clutch-engaged supercharger, with a noise akin to a stuck pig, could be heard all over the circuit. The supercharger was only sup-

posed to be engaged for short periods and there was no doubt but that continual use of it would eventually blow the engine. Sure enough, on the 82nd lap the Mercedes was stationary at the pits with, according to WO, water pouring from the front from a blown gasket, and shortly afterwards the retirement of the Mercedes was announced due to "the battery being completely discharged". This version has to be taken with a very large pinch of salt as it was commonplace to blame the manufacturer of a proprietary part for failures to distract attention from a failure with the car. Faced with a terminally-sick Bugatti, Ettore was known to remove the radiator cap "accidentally" breaking the regulations to ensure the sympathy of the crowd for the retirement of the car due to an infringement of a minor regulation! Shortly before this, Eaton, in one of the Talbots, had noticed how powerful the lights of the Mercedes were and *The Motor* recorded that it had been suffering from weakening brakes and flickering oil pressure. WO hurried round to commiserate with the Germans, and was told that by this stage they reckoned on having a lead of a full lap based on the 1929 schedule – how WO must have smiled inwardly when recalling the comments

of his frustrated drivers when confronted with the "slower" signs! Neubauer, the head of Mercedes' racing team, told the press that the battery had completely discharged owing to a short circuit in the dynamo. The Speed Sixes were then slowed down, still followed by the Supercharged cars in third and fourth places.

The Blown cars had a completely separate pit organisation run by Bertie Kensington Moir. Ramponi became ill during the night, suffering from fever and delusions, and was only persuaded with difficulty to drive for a lap after working on the car to prevent disqualification under the regulations. Birkin later retired with a broken valve and a rod through the crankcase. He was soon followed by Benjafield, whose reward for driving single-handed for ten hours was a collapsed piston, a performance which did little for the Supercharged cars' reputation for reliability.

Barnato was delighted with his third win, the Barnato/Kidston pairing finishing first on outright speed and second on Index of Performance. Clement and Watney finished second on speed and fifth on Index of Performance, and the British triumph was completed by the Talbots finishing third and fourth on speed and first and third on Index of Performance. The two Bentleys finished first and

second in the Rudge-Whitworth Biennial Cup, having qualified in 1929. It was, of course, noted by some of the Press that the British cars had met with little real opposition.

Within days of Le Mans, the press was carrying a new story – "A Motoring Surprise – Will Bentleys Stop Racing?" ran the headline in the *Daily News* of 1st July, anticipating an official announcement from the Company that they were to retire from racing. The following day a statement was issued: "The motoring world will learn with some surprise that Bentley Motors have decided, for the present at any rate, to discontinue racing. Since the first Bentley car was produced the Company has had brilliant successes on both road and track. Their most recent triumph at Le Mans represented the fourth consecutive win and the fifth in all the eight years since this international reliability trial was inaugurated. This racing experience has been invaluable to Bentley designers, a fact which is undoubtedly proved by the popularity and prestige which the Bentley car enjoys in the market. It is now felt, however, that sufficient data have been acquired. Their consistent success would suggest that there is little more to learn either in speed or in respect to reliability at the present moment, and it is the desire of the Company to concentrate the

LEFT **Closing the circuit after the race. Kidston drives the Speed Six, with Barnato in the passenger seat, a tired looking Dudley Benjafield, almost ghoulish following his ten-hour solo effort, is seated in the back with the two gendarmes. Note that the brake lever for the front brakes is right back, indicating that the linings are completely worn out.**

ABOVE **The Speed Sixes on the way back from the race, somewhere in France: the repaired No.3 car GF8511, with the No.2 car GF8507 with damage to the rear wing from a thrown tread still in evidence. On the right is EXP 5, with Gurney Nutting touring body with the mis-matched rear wings and standard 4½ litre tank in place of the 'D' tank used with the earlier body.**

whole of their attention to their regular production. This decision means that the entries for the Irish Grand Prix at Dublin next month are withdrawn, and also that Bentleys will not race at Le Mans next year. With regard to the latter race the Company feel that an early announcement of this decision is called for out of courtesy to their good friends the Automobile Club de l'Ouest. Should conditions change within the next year or two Bentleys will reconsider their decision. Bentley Motors' decision does not cover the team of Supercharged 4½ Litres under the control of Captain H.R.S. Birkin, this being privately owned and not connected with the Company in any way." Bentleys compiled their last Le Mans victory booklet, *Plus Four*, and as an introduction included a résumé of policy in which the Company's reasons for retiring from motor racing were set out (see Appendix II).

Basically, the Works Team had become so successful that that success had become a liability in itself. Another win was just another win, while failure or retirement became newsworthy – the same reason for Jaguar's retirement from racing roughly a quarter of a century later. Shortly afterwards Barnato issued a statement to the effect that he too was retiring from racing – as Chairman of Bentley Motors he could scarcely compete in another firm's car. Carruth gave Barnato's reasons for retiring – "Captain Barnato believes the time has come for younger men to uphold the racing prestige which he and others have helped to gain for Britain in the eyes of the world. I understand he is continuing to race motor boats on the Welsh Harp." Barnato piloted *Ardenrun V*, built by Saunders Roe and fitted with a 3 Litre Bentley engine, in the Duke of York International Gold Trophy held on the Welsh Harp on 26th June. It sank in practice, was fished out after seven hours in the water and cleaned up, but seems to have dropped a valve through a piston in the race itself. Barnato later told the *Daily Telegraph* that he was not going to take motor boat racing very seriously. According to Rivers Fletcher, "Bentley Motors Ltd. was never really a truly commercial proposition, being always dependent on Barnato's beneficent finance. With his last Le Mans victory he had achieved all his motor racing ambitions and now had other fish to fry. He was absolutely fair about the situation, giving the Company plenty of warning. He had said that after his racing retirement he wouldn't be putting any more money into the Company and that they would have to become self-sufficient."

The financial aspects of the Works teams have also come in for a degree of scrutiny. The figures

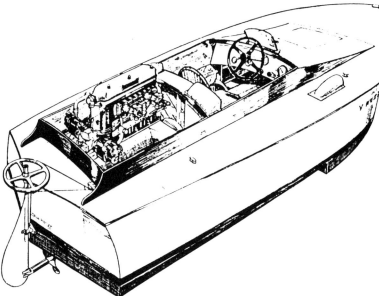

ABOVE **Barnato in his 3 Litre Bentley-engined speed boat 'Ardenrun V' on the Welsh Harp for the Duke of York's International Trophy held over the weekend of 26th/27th June 1930. 'Ardenrun V' capsized and sank on the Saturday, but was fished out and repaired at the Cricklewood works overnight. It then dropped a valve on the Sunday.**

quoted by WO in his autobiography of less than £3,000/year over the years 1927–30 do not seem to fit. This is not to question WO, but the method by which the figures were derived. Allowing for three new cars in 1928 with a chassis price of £1050, to one in 1929 of £1700 and two in 1930 of £1800 each, plus preparation of cars from previous years and the cost of bodywork, WO's quoted costs have gone already. (Admittedly, it is difficult to be sure who paid for the chassis, as they were all privately owned soon after they were built.) If the cost of running the Racing Department and paying the mechanics, the costs of transporting and maintaining the teams abroad, and the costs of tyres, parts and other expenses, are considered, substantial figures would have been spent. Deduction of starting money reduces these figures somewhat, but they are still higher than those quoted by WO. It is possible that the Team was substantially supported from Barnato's own pocket, and this is hinted at in the 1929 Annual Accounts in which the £2,487 spent on racing expenses was refunded to Barnato as it was felt he need not bear the full sum out of his own pocket. Reading between the lines of the 1928

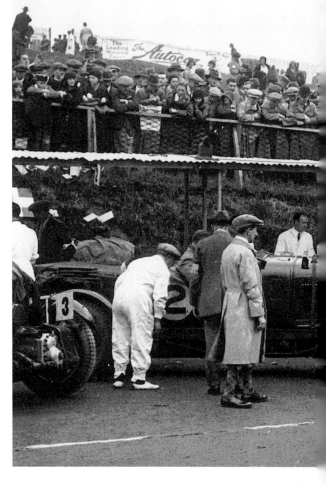

report it is clear that WO's figures were indeed the above-the-line figures, after deducting contributions from the drivers and suppliers. After each major race the component manufacturers tended to advertise their involvement with the victorious team, and at the end of race reports the magazines typically listed the suppliers of tyres, electrical equipment and sometimes other components used by the first-placed cars.

It is conceivable that the cost of maintaining a racing team was too much of a financial burden for the Company to bear in the economic climate of the day, balanced against the dubious publicity

The Birkin team at the pits before the start of the 1930 Tourist Trophy race at Ards, Belfast. No.1 UU5872 Birkin, No.2 YU3250 Moir, and No.3 UR6571 Benjafield. Chassagne can be seen in the flat cap and dark jacket directly below the 'Control' sign.

value of future wins and almost certain damaging publicity accruing from potential failures. In view of the downturn in sales that had been noticeable in the summer of 1929 (which was becoming worse) and the dire financial situation, it is likely that Barnato's financial advisers were set against any more racing. It was certainly the obvious aspect of the Company's activities to dispense with, and as Barnato himself became increasingly unenthusiastic about the sport, retirement became an obvious option. Barnato was well aware of the risks, and never raced again after Dunfee's death in the 1932 500 Miles Race. It is also clear that by mid-1930, Company policy was changing from the manufacture of sports cars to the manufacture of town carriages.

The Racing Team was dissolved, the mechanics going to the Service Department, which was their traditional winter habitat anyway. Nobby Clarke had less and less to do with racing, owing to the

pressures of work in the Service Department, which were too much for Hubert Pike on his own. The Racing Shop building at Kingsbury continued to be used by the Service Department.

Although the Works team was completely disbanded, the Birkin team continued to operate from their Welwyn Garden City base. All three road cars were entered for the Irish Grand Prix, Birkin driving his favourite of the short chassis cars UU 5872, Chassagne UR 6571 and Harcourt-Wood YU 3250. The Alfa-Romeos as usual posed a threat on handicap, as did the 7 litre Supercharged Mercedes of Caracciola, Howe and Campbell. Chassagne and Harcourt-Wood had lubrication problems even before the start, at which Caracciola got away to an early lead followed by Birkin. Campbell had clutch trouble, even using his fire extinguisher on it, but Birkin was going extremely well. The Mercedes gave the Bentleys two laps on handicap, and with the race run over 70 laps Caracciola did not pass Birkin until lap 34. Birkin's pit stop was faster than Caracciola's at 65 seconds to 69, but smoke and a jet of oil heralded a broken oil pipe. As usual

ABOVE **Possibly the great Blower success – Birkin in the No.4 car UR6571 after his epic drive to finish second in the French GP at Pau. Note the extra strapping to the front shock-absorbers.**

ABOVE RIGHT **The last appearance of the Birkin team was at the BARC 500 Miles Race at Brooklands in October 1930, in which the single seater and the two short cars were entered. Birkin and Duller drove the blue single seater, seen here high on the banking, with one of the road cars in pursuit.**

all the oil lines (and fuel lines) were doubled with the spare taped on and fitted with a set of end fittings, but the 2 minutes 48 second delay was enough to let Caracciola into the lead. The weather turned to torrential rain, conditions under which Caracciola was almost unbeatable, and he went on to win from Campari and Howe, setting the fastest lap in the process. Birkin had to stop twice for minor problems, dropping back to 4th place. Both the other Supercharged cars retired.

Birkin's situation by the Ulster Tourist Trophy

s at all Events PRATTS

on 23rd August was becoming rather difficult. Dorothy Paget had already indicated that she would be withdrawing her sponsorship at the end of the season, and there was no way that Birkin's Works at Welwyn Garden City could support the racing team in the manner to which they had become accustomed. Whether Birkin's Works ever made a profit in their own right is not clear, or whether despite glowing articles in the weekly press they ever undertook any experimental or development work for other firms. Birkin probably had little interest in the business, and was very possibly no more of a businessman than WO himself. Birkin took on Mike Couper as his partner and many years later Couper said that he was "done" by Birkin but that he did get some nice cars to drive! Some good results were needed before the end of the year to convince potential sponsors of the viability of the team, and the only major road race remaining was the Ulster Tourist Trophy at 'Ards near Belfast. The three road cars were entered, Birkin driving UU 5872, Harcourt-Wood UR 6571 and Benjafield YU 3250. However Harcourt-Wood had to

withdraw because of septic tonsils, so Benjy took over the second short chassis car and Bertie Kensington Moir was persuaded to drive the long car YU 3250. Bertie had not raced since his drive with Benjafield in the 3 Litre at Le Mans in 1925, because he had married shortly afterwards and been banned from racing. The three 1750 cc supercharged Alfa-Romeos driven by Nuvolari, Campari and Varzi were almost unbeatable on handicap. Three 7 litre Supercharged Mercedes were also entered, but Caracciola was disqualified before the start of the race because the overall dimensions of the supercharger on his car were greater than those on the other two. Eddie Hall entered his own 4½ Litre registered UV 3108, and drove a very steady race to finish 12th overall. Bertie Kensington Moir finished one place in front, but Birkin crashed into a low wall at Ballystockart Bridge – Whitlock, his riding mechanic, looked down at the floorboards and Birkin looked over for a second too long. The Alfa-Romeos ran away with a 1-2-3 win, Benjafield still running at the end of the race but having only covered 27 of the 30 laps.

Dudley Froy in the No.2 Birkin Blower UU5872, still wearing number 38 from the 500 Miles race, at Brooklands on 31 October 1930. The Blower was tackling the six hours record. Note the reinforcing plate along the outside of the chassis rail.

It was most certainly not the result that Birkin needed at that time.

The circumstances surrounding Birkin's entry of his Supercharged 4½ Litre into the 1930 French Grand Prix would be interesting to know. Grand Prix racing in 1929/30 was run on a fuel consumption formula and had fallen somewhat into disrepute. With virtually no entries for the Grand Prix the joint organizers, the Automobile Club de France and the A.C. de Rosco-Beauvais ran the event as Formula Libre. As usual hordes of Bugattis entered, 16 in all, with an Aries, a la Perle, a Delage, two Montiers and the Bentley making up the field. Birkin drove the No. 4 car, UR 6571, presumably because his favourite of the two short chassis road cars UU 5872 had not been fully recommissioned after the accident in the Tourist Trophy race. Even in stripped form the Supercharged car weighed 2 tons and stood head and shoulders over the 18/19 cwt Bugattis. Top speeds

of both were more or less equal at 135 mph, and the Bentley was faster on the straight but slower on the winding section. The French crowd were dismissive of the Bentley until the race got going, Birkin moving into 3rd place on lap 8 with seventeen to go. Chiron caused Birkin some anxiety by looking down at his oil pressure gauge while wandering down the centre of the track. Using the horn failed to alert the Bugatti driver, who was finally brought up by Birkin's yelling. Birkin's was possibly the only Grand Prix car ever fitted with a horn!

One of the privately-entered Bugattis driven by "Sabipa" came to grief, leaving the driver prostrate across the road. There was practically no room between his body and the edge of the track, but not knowing whether he was alive or dead Birkin elected not to run him over but to try and squeeze through. The crowd were absolutely silent as Birkin managed to get through the gap by such a close margin that blood stains were later found on his

After the racing season had finished at Brooklands, the BARC again held their final race of the year, the 500 Miles Race on the outer circuit on the 4th October. Three of the Supercharged cars were entered, the single-seater UU 5871 driven by Birkin and Duller and the two short chassis cars, UU 5872 driven by Benjafield and Eddie Hall and UR 6571 by Eyston and Harcourt-Wood. The wet weather conditions were more of a handicap for the large cars than the small, and the diminutive 747 cc Austin Sevens were soon lapping at around the 80 mph mark, Sammy Davis leading at 200 miles lapping at 83 mph. The single-seater was distinctly off colour, sounding more like a motorcycle on occasions. The short chassis cars lapped at about 118 mph until Harcourt-Wood and Eyston retired due to a sheared magneto drive. The lead was held by Sammy Davis, followed by Bird's Riley, then Dunfee's 2 litre Sunbeam when Bird dropped out. Dunfee suffered a broken half-shaft which sent the wheel and hub assembly across the track, narrowly missing the Davis Austin, allowing the Benjafield/Hall Bentley into second place. Although the Bentley put in a fastest lap of 122.97 mph to the Austin's 87 mph the handicap proved unassailable and the Bentley finished second at an average of 112.12 mph, flinging the tread off the nearside rear tyre right on the finish line with a bang that could be heard all the way across the track. Birkin and Duller finished ninth.

It is rather ironic that the two best results achieved by the Supercharged cars, second places in the French Grand Prix and the 500 Miles Race, both occurred after the team was under threat of closure due to a lack of sponsorship. Dorothy Paget had had little return for her money, and by May of 1931 all four cars were up for sale. Birkin advertised them in *MotorSport* in that month for sale on behalf of the Hon. D.W. Paget, each car guaranteed to attain 125 mph in racing trim, with complete sets of racing spare parts available. The latter were probably used to create the No. 5 Birkin car UR 9155 with the 1929 body from UU 5871. No prices were mentioned in the advert – just "offers invited". However, Dorothy Paget retained the single-seater until 1939, continuing to allow Birkin to race it in 1931 and 1932 at Brooklands, until finally persuaded to sell by Peter Robertson-Roger as a source of engine replacement for the No. 4 car UR 6571. The single-seater was then rebuilt after the war with a 2-seat body and was not returned to its original form until the 1960s. The four Birkin-Paget Blowers were finally reunited in 1969 after the rebuilding of UR 6571.

tyres. "Sabipa's" wife later thanked Birkin in his hotel room because "Sabipa" had been conscious throughout and confirmed the closeness of the Bentley. Zanelli went into the pits to refuel, allowing Birkin into second place behind Etancelin. With the Bentley closing on him Etancelin missed his last pit stop and crossed the line with a practically dry tank and five of the six clutch-retaining bolts broken – the sixth breaking on the way home! Had the race continued for another lap Etancelin would have run out of fuel allowing Birkin into the lead, but it is quite likely he would have been passed by Zanelli who was only seconds behind. After the race the smiles were wiped off the French crowd's faces, and the veteran correspondent Charles Faroux said, "I have seen the Bentleys at Le Mans and I know. I am Faroux. I am not a bloody fool." Birkin had driven the race of his life, and the repercussions if he had won the race would have been fascinating to observe.

15 The Works

W ELCOME TO THIS unearthly tour of Bentley Motors, the date being 18th February 1930. Obviously we cannot see all of the factory, but most of it is here – and imagination will have to suffice for the rest. The Supercharged 4½ Litre, 6½ Litre and heavy crank 4½ Litre are currently in production. The 8 Litre and the SV (4 Litre) are still very much in the experimental stages.

Author's note.

These photographs comprise a series taken by Chas Bowers on 18th February 1930, after the completion of the extensions to the original buildings and the new office block. Some were in fact taken slightly later, extracted from an album of WO's marked ★ in the following text.

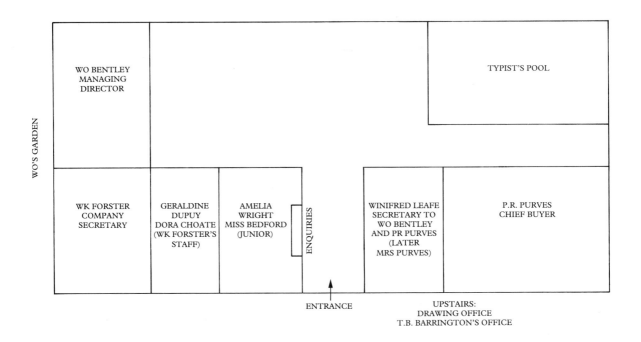

ABOVE **Bentley Motors, new office block** *c.* **1930.**   RIGHT **Works layout, February 1930.**

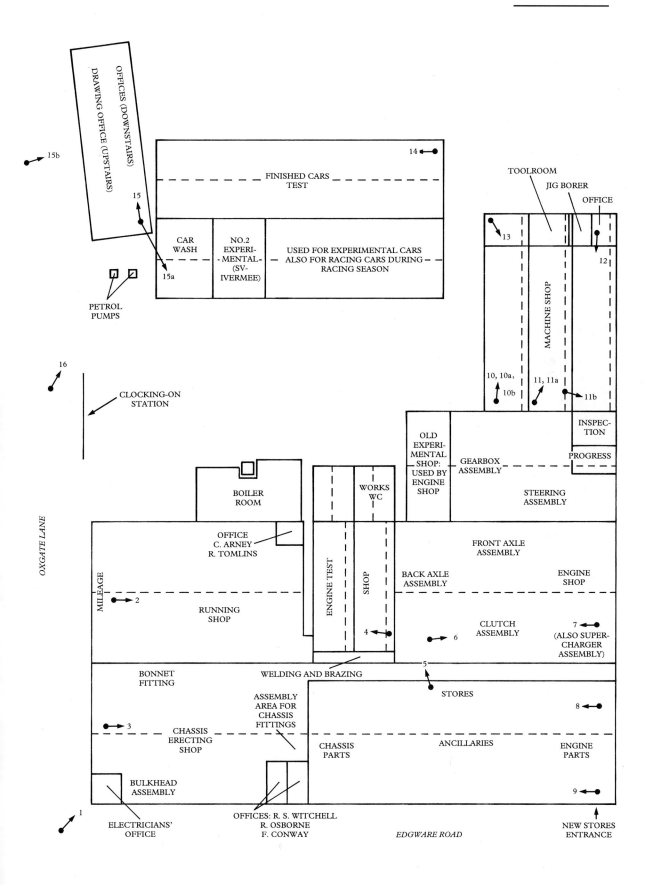

DRAWING OFFICE (UPSTAIRS)
OFFICES (DOWNSTAIRS)

15b

15

15a

PETROL
PUMPS

CAR
WASH

NO.2
EXPERI-
-MENTAL-
(SV-
IVERMEE)

USED FOR EXPERIMENTAL CARS
ALSO FOR RACING CARS DURING
RACING SEASON

FINISHED CARS
TEST

14

TOOLROOM

JIG BORER

OFFICE

13

12

MACHINE SHOP

16

CLOCKING-ON
STATION

10, 10a,
10b

11, 11a

11b

INSPEC-
TION

PROGRESS

OXGATE LANE

BOILER
ROOM

OFFICE
C. ARNEY
R. TOMLINS

MILEAGE

2

RUNNING
SHOP

ENGINE TEST

SHOP

WORKS
WC

4

OLD
EXPERI-
MENTAL
SHOP:
USED BY
ENGINE
SHOP

GEARBOX
ASSEMBLY

STEERING
ASSEMBLY

FRONT AXLE
ASSEMBLY

BACK AXLE
ASSEMBLY

ENGINE
SHOP

6

CLUTCH
ASSEMBLY

7

(ALSO SUPER-
CHARGER
ASSEMBLY)

BONNET
FITTING

WELDING AND BRAZING

5

STORES

8

3

CHASSIS
ERECTING
SHOP

ASSEMBLY
AREA FOR
CHASSIS
FITTINGS

CHASSIS
PARTS

ANCILLARIES

ENGINE
PARTS

9

BULKHEAD
ASSEMBLY

1

ELECTRICIANS'
OFFICE

OFFICES: R. S. WITCHELL
R. OSBORNE
F. CONWAY

EDGWARE ROAD

NEW STORES
ENTRANCE

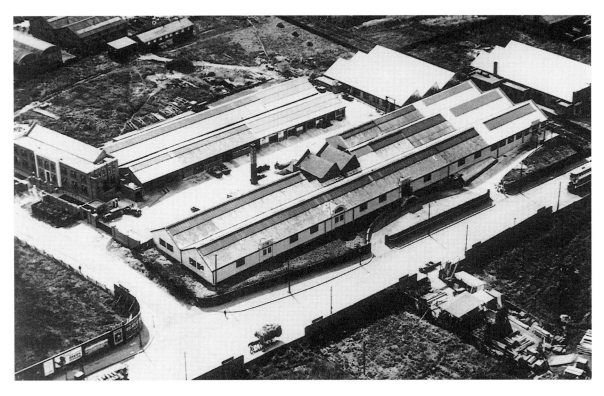

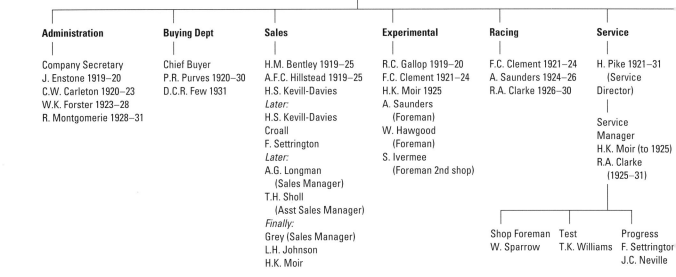

MANAGING DIRECTOR – W.O. Bentley
W.O. Bentley and Marquis de Casa Maury (1927/29)
J.K. Carruth (1930)

General Manager – G.A. Peck

Administration	**Buying Dept**	**Sales**	**Experimental**	**Racing**	**Service**
Company Secretary	Chief Buyer	H.M. Bentley 1919–25	R.C. Gallop 1919–20	F.C. Clement 1921–24	H. Pike 1921–31
J. Enstone 1919–20	P.R. Purves 1920–30	A.F.C. Hillstead 1919–25	F.C. Clement 1921–24	A. Saunders 1924–26	(Service
C.W. Carleton 1920–23	D.C.R. Few 1931	H.S. Kevill-Davies	H.K. Moir 1925	R.A. Clarke 1926–30	Director)
W.K. Forster 1923–28		*Later:*	A. Saunders		
R. Montgomerie 1928–31		H.S. Kevill-Davies	(Foreman)		Service
		Croall	W. Hawgood		Manager
		F. Settrington	(Foreman)		H.K. Moir (to 1925)
		Later:	S. Ivermee		R.A. Clarke
		A.G. Longman	(Foreman 2nd shop)		(1925–31)
		(Sales Manager)			
		T.H. Sholl			
		(Asst Sales Manager)			
		Finally:			
		Grey (Sales Manager)			
		L.H. Johnson			
		H.K. Moir			

Shop Foreman	Test	Progress
W. Sparrow	T.K. Williams	F. Settringtor
		J.C. Neville

Aerial views of the Works at the time. It should be possible to relate these to the floorplan on p.277 (which is not, of course, to scale).

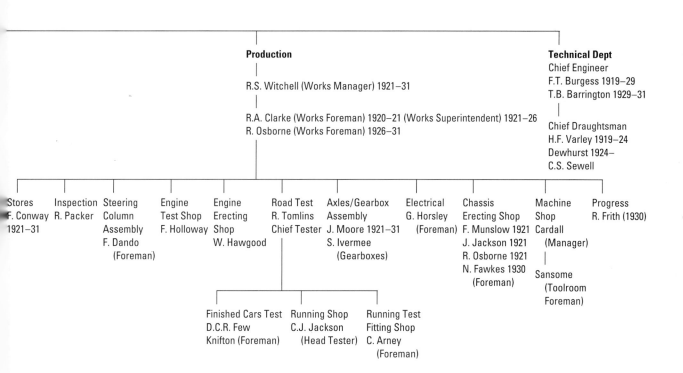

Production | Technical Dept
Chief Engineer
F.T. Burgess 1919–29
T.B. Barrington 1929–31

R.S. Witchell (Works Manager) 1921–31

R.A. Clarke (Works Foreman) 1920–21 (Works Superintendent) 1921–26
R. Osborne (Works Foreman) 1926–31

Chief Draughtsman
H.F. Varley 1919–24
Dewhurst 1924–
C.S. Sewell

Stores
F. Conway 1921–31

Inspection
R. Packer

Steering Column Assembly
F. Dando (Foreman)

Engine Test Shop
F. Holloway

Engine Erecting Shop
W. Hawgood

Road Test
R. Tomlins
Chief Tester

Axles/Gearbox Assembly
J. Moore 1921–31
S. Ivermee (Gearboxes)

Electrical
G. Horsley (Foreman)

Chassis Erecting Shop
F. Munslow 1921
J. Jackson 1921
R. Osborne 1921
N. Fawkes 1930 (Foreman)

Machine Shop
Cardall (Manager)

Sansome (Toolroom Foreman)

Progress
R. Frith (1930)

Finished Cars Test
D.C.R. Few
Knifton (Foreman)

Running Shop
C.J. Jackson (Head Tester)

Running Test Fitting Shop
C. Arney (Foreman)

1

1. As you approach, from a vantage point on the opposite side of the Edgware Road, you will see on the left the new Office Block. The buildings immediately in front of you are the 1920/21 shops, built straight after the first Engine Shop. The W.H. Wagstaff sign is still here, as they have only just finished the new extensions. From here, if you will follow me, we go over the Edgware Road, up Oxgate Lane and right through the gate into the yard. Now we go across the yard, past the clocking station on the right and through the double doors into the Running Shop.

2. Standing inside the Running Shop, we are looking north towards the Engine Test Shop. These are 1930 Speed Sixes, NH/LR series, with bolted strut gear, stainless steel handbrake, C type box, internal battery trays and B&D shock absorbers – the hydraulic rear pattern can be seen on the chassis in the foreground and that to the left. This is more or less the last batch of short chassis (11′8½″ wheelbase) to be built, before we switch over to the 12′8½″ chassis at chassis LR 2790. The chassis are received here from the Erecting Shop (the double doors through which the chassis are wheeled, just out of sight, to the right) and are handed over to the mileage men, Willy Lytton and Tiny Earl, behind us. They drive the chassis for about 100 miles, handing them over to the testers, under their chief, Bob Tomlins. Tomlins' office, shared with Charlie Arney, the foreman of the shop fitters, is in the far left corner. In the far right corner are racks for bonnets.

After mileage, the chassis have the brakes, tappets and magnetos adjusted and are then handed over to the testers. At the moment, the testers are Bill Bowie, Jimmy Jackson, the Head Tester, Bert Hedges and George Hawkins. Each has two bays with two fitters, who in turn have two mates each. Bowie is nearest us, then Jackson, Hedges and Hawkins last, nearest the Engine Shop end. The chassis are run on test and then worked over by the

fitters until satisfactory, when Jimmy Jackson does the final test before the test card for each chassis is filled up and signed off. For testing, each chassis is fitted with a scuttle, two seats, and, on the 6½ (and later 8) Litre, a concrete weighted rig over the rear wheels. Once passed off test the chassis will be driven to the coachbuilder by the delivery driver, Charrington. Leaning over the engine of the nearest Speed Six is Bill Durnford, one of Bowie's fitters. In the far right distance, George Hawkins is standing in front of the left-hand bonnet in the rack. Just to the right of the bonnets in beret is Eric Sewell, one of Charlie Palmer's mates – Charlie Palmer is one of Hawkin's fitters.

3. Going through the door on our right, we are now standing in the Erecting Shop. Here we have 1930 specification 4½ Litres, with the heavy crank engine. The engine on the stand is a heavy crank unit, probably intended for the chassis on the left. This chassis is probably at the stage of having its bonnet fitted. All the bonnets are hand-fitted by one fitter, sub-contracted from Ewarts, the sheet metal company who make all the bonnets, petrol tanks, etc. The Ewarts' fitter is the bloke on the far left. The bonnets are then numbered off to ensure that they stay with their chassis. Note the bonnet side leaning on the chassis, and more bonnets in front of it. Note also on the bench the dumb-iron cowl fitted to the latest 1930 specification Speed Six chassis. Further back to the left are more four cylinder chassis, on the right two Supercharged chassis, another bare chassis, and a Speed Six. The Supercharged chassis are readily distinguishable by the front tie-bar and the cut-away radiators: note the heavy pattern B&D shock absorbers.

The powers that be are still working on the chassis specification – you will see that these chassis have been fitted with Autovacs. It will be ages before the drawings for the bulkhead blanking plates (to cover the hole for the Autovac) and the brackets for the Autopulse pumps will be delivered to the shops (it was May 1930 before the drawings were issued for production). In the background to the right are the offices of Mr. Witchell, the Works Manager, Bert Osborne, the Works Foreman, and Fred Conway, the Head Storeman, in front of the Stores. Looking to the left, the corridor leads along the back of the Engine Test Shop into the Detail shops. The chassis are wired up in this shop and delivered complete through the doors just out of sight to the left into the Running Shop, to the mileage men.

Leaning over the nearest $4\frac{1}{2}$ Litre chassis on the left is Jack Baker, an apprentice fitter. If you look back to the next $4\frac{1}{2}$ Litre chassis on the left, standing in the centre of the shop are Fred Wilmot in cap and dark coat with the charge-hand, Norman Fawkes, immediately next to and behind him in lighter coat with no cap. Looking to the right are Fred Kirby the electrician, sideways on to us, then John Morrison, chassis fitter, on the other side of the bench, and George Wills, chassis fitter, just behind him and slightly to the left.

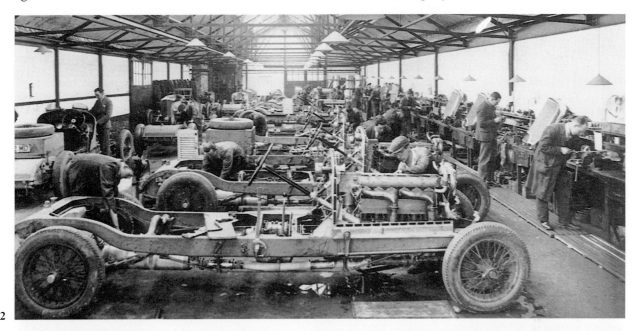

2

3

4. From the Chassis Shop we pass into the Engine Test Shop itself. This is, of course, the very first building to have been put up here. The engines are four cylinder, and most of them are for supercharged chassis. Slave carburettors are fitted for test purposes. You will notice the early, unribbed supercharger visible to the right of the foreman, Mr. Holloway, making notes on his pad. This engine is connected up to the second of the two Heenan & Froude dynamometers, and is on full power test. Note the oil pressure gauge, next to the magneto. The engine to it's right is another Blower engine, currently being run in with slave carbs. The next engine along is another Blower unit, passed off two days ago on full power test at 181 bhp at 3900 rpm. Watson, the tester, is adjusting the top bevels. This engine also has slave carbs fitted.

Working in here are, on the left hand row of engines, from the left (back) "Dae the Mine", Walter, either Bill Zamit or Harry Jackson, and Jock Ferguson. Then Fred Holloway, the foreman, with pad, not sure about the chap behind his head, possibly Cheeseman in the light coat, I think George Mason in cap, Dennis Rumbold, not sure about this one either, Arthur Watson, then one of the labourers with his back to the wall, and lastly Taffy Evans on the far right. Note the two old 6½ Litre pattern petrol tanks on the wall, for fuel consumption tests. The double door on the left leads back into the open, and then back into the Running Shop. Visitors are asked to avert their eyes from the Experimental 8 Litre engine on the far right, which Taffy "Sparks" Evans is testing electrical mods on, using the special switchboard in front of it. Note the massive main bearing caps and strongbacks used with the 80 mm crankshaft of these engines.

The reason why there are so many supercharged engines in the shop at the moment is because of the problems caused by the lubrication system. Note on the sump in the foreground, the extra plate on the front of the sump, for the Blower oil pump with the scavenger pump. The lubrication system caused problems with the engine, and there is now a backlog of Blower engines for the production

chassis. You will recall from the Erecting and Running Shops, that while all the six cylinder chassis have their engines fitted, there is a significant shortage of four cylinder engines at present. (This shop was re-arranged later in 1930, and a second Heenan & Froude dynamometer installed.)

5. If we go through the double doors behind and on into the Detail Shops, we are now looking back over the Back Axle Shop. The first two racks nearest are standard 4½ Litre back axles – note the liberally applied sealant. On the farthest rack is (from the left) a 6½ Litre/Blower back axle, the Supercharged cars using the 6½ Litre diff, of course, thus making the axles indistinguishable, three empty banjos and an experimental worm drive back axle. This didn't work, because some boffin forgot that larger = faster = hotter, and the ball bearings were nearly a foot across. On about the second trial run the car came to an 'orrible grinding halt, the races having blued and the cages and balls skidded into a real mess. The solution? Use a hypoid diff and on

the far left you will see an early hypoid diff being assembled. But that, of course, you haven't seen either.

If you look to the left of the fitter nearest us you will see chalked on his bench AD3656/AD3659/AD3663, the numbers of the axles he is building up at present – these are for chassis that will be built up in the Erecting Shop in March (next month). Before we leave the Axle Shop, note the two 6½ Litre/Blower diffs being assembled on the second bench from the right, and the queue of engines outside the Test Shop. This shop used to finish on a level with the upright behind the cupboard in the background, there being an open space between the end of this shop and the double doors into the Engine Test shop (this open space was built over sometime between 1927 and 1930). The Shop Supervisor, John Moore, is immediately to the right of the crownwheel hanging on the far wall. To his left is C.J. Collins. This shop used to be the Engine Shop (see p.88) and on the other side of the wall is the first Experimental Shop.

5

6. Moving just a little to the right, we can now look down the length of the Detail Shop. On the far left, just visible, is the Gearbox Shop. You will note the gaskets hanging on the bench, the square for C box lids and the rounded ones for D boxes. To the right in the distance is Steering Column assembly, the two visible on the benches being for four cylinder chassis. Behind that is Inspection – you can just see the drawing files in the middle distance, and then the Progress Department in the far corner. Progress Department manages the workload of the new Machine Shop, on the other side of the wall to us towards the far left corner.

In the middle foreground, we have Front Axle Assembly. Of the axles on the floor, the nearest looks to be a 6½ Litre unit, with standard brakes and the very heavy beam with integral jacking pads. Behind this are three four cylinder axles with self-wrapping brakes, one assembled, one in pieces, and one complete with drums. This last is for a Blower, as it has the finned drums. The other two axles are 6½ Litre, with standard brakes and with-out jacking lugs. The front axle being worked on the bench in front of the row of four cylinder headlamp stand castings is a self-wrapping unit, for a Blower chassis, again with ribbed drums.

On the right is Clutch Assembly. Three cardan shaft assemblies, two with their spiders, can be seen on the nearest bench. The fitter behind them is holding the friction disc for a plate clutch. On the shelf of the bench are pedal shaft parts, with a line of clutch stop horse-shoes with their friction pads rivetted on. At the bench behind, the fitter is assembling steering column controls. On the bench behind him are hub/drum assemblies and a line of assembled plate clutch shafts and friction discs.

7. Walking along the corridor in front of the Stores, we are now looking back over the Detail Shop, from the Engine Assembly end. In the centre distance you will see the double doors into the Engine Test Shop, at the back of the Back Axle Shop. On the far left is a line of front axles in front of the Stores.

6

Immediately in front of you is a line of engine top halves and blocks. From the left, these are AD 3674, destined for The Honourable R.L. Salmon's Gurney Nutting coupé bodied 4½ Litre in four month's time. Next to it are three Supercharger units, SM 3920, SM 3919, and SM 3918. Next are bare four cylinder and six cylinder block castings.

The two castings on the bench are for early unribbed superchargers, and the parts on the shelf are for the blow-off valves. The superchargers are built up complete here, and then delivered to the Engine Test Shop for coupling-up to their respective engine, for test purposes.

On the other side of the bench are four cylinder engine top halves, with a complete engine on the right. To the left is a complete small-sump 3 Litre engine being overhauled. To the right is a six cylinder bottom end, with the five gallon barrel-sided sump and it's crankshaft in the vice. Further back are more four cylinder engines in various states of assembly, and the row of 4½ Litre headlamp brackets visible in no. 6. The fitter on the far right, on the other side of the bench nearest us, is George Armstrong. George has a thriving sideline in cigarettes and sweets, all blessed by the management. The fitter on the left is Fred Lusted.

The engine build has recently been changed, in that it has been broken down into stages performed separately rather than the fitter and his mate doing the whole job. In the past, the fitter and his mate would build up a complete engine from parts issued from the stores, with the exception of the vertical drive turret on the four-cylinder cars. This needs machining of the bushes, and has always been done as a sub-assembly.

Under the new system, a fitter dresses the crank-case casting, and then hands it to another fitter, who fits up all the main bearings to a 0.002″ oversize mandrel. The connecting rods are similarly mandrel fitted. Another fitter assembles up all the cylinder block water jacket plates and the drain tubes, and then the block is pressure-tested for leaks. The engine is then assembled up as a unit.

7

8

9

8. Stepping smartly to the left, we can now look back along the length of the Stores. You will appreciate that at Bentleys we are predominantly an assembly and test concern and a large store is essential. The front axles are for four and six cylinder cars, with and without jacking lugs, both standard (six cylinder) and self-wrapping (four cylinder). Note on top of the second row of shelves a pile of redundant main bearing oil galleries for 3 and light crank 4½ engines, and just behind Con-

way, the storeman's head, three Blower inlet manifolds with blow-off valves. Conway is handing a plate clutch cover to George Armstrong.

9. Moving to the left again, we are now standing on the Edgware Road side of the Stores, looking back along the front, in the new part of the building that has only recently been finished. We now go back through the Engine Erecting Shop, past Inspection and out through the double doors into the yard, and immediately left into the new Machine Shop.

10. Standing just inside the door looking over the shop, you can see how spacious and well set out the new machining facility is. Visitors are asked to ignore the fact that the 4½ Litre engines dotted around are merely there for effect – and the three 4 Litre cylinder blocks on the floor straight in front, behind the 4½ Litre sump! Messrs. Wagstaffs have only recently finished this shop. A supercharger inlet manifold can be seen on the machine in the bottom left, with a heavy crank bottom end on the machine in front, clearly identifiable by the spacing of the main bearing studs. Immediately above this machine can just be seen the Toolroom Shop's jig borer, in it's own room by the offices, in the distance.

*10a. This is a slightly later view, showing 4 Litre cylinder blocks and crankcases, and a re-arrangement of the machines down the right-hand aisle.

10

10a,
10b

11

11. Milling machines, and more four cylinder bottom ends. Note the Blower manifold on the mill to the left of the con rods, and another on the floor by the machine. I'm sorry? Yes, indeed, it does look like a 4 Litre cylinder head on the next machine along. In the far corner is the office.

*11a. The same view as (11), but later, showing 4 Litre crankcases and sumps being machined.

*11b. This is the jig borer, seen on the left of 12, being used to machine a 4 Litre crankcase. Note

the three mountings for the tube supporting the crankcase, cast in. Note also the pile of 4 Litre cylinder heads to the left of the machine.

12. Looking back towards the Detail Shops (Progress is on the far side of the wall immediately in front), we have the larger machines. To the left, two capstan lathes for repetition turning, three more lathes to the right, then horizontal milling machines in various sizes. Connecting rods seem to be in considerable demand, as well as some drums to skim.

11a

11b

12

13

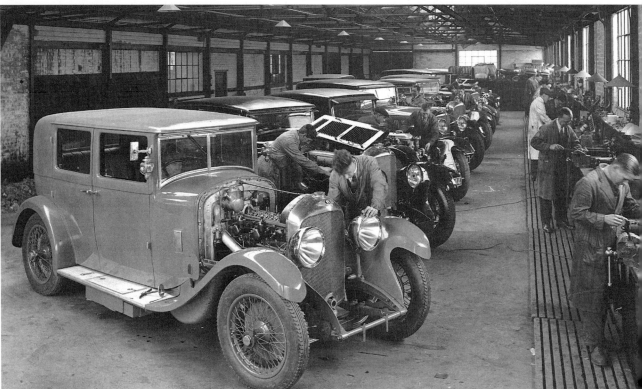

14

13. We are now standing in the far corner behind the wire mesh partition, in the Toolroom section. Here we have grinders, lathes and small milling machines. The fitting benches are along the wall to the left. If you look towards the left are the offices, containing the jig borer, then an office with two jig and tool draughtsmen, then Mr. Cardall's office in the far corner. Mr. Cardall is the Shop Manager, Mr. Sansome is the Toolroom foreman.

14. Leaving the Machine Shop we can now traverse the short distance to the Finished Cars Shop. Here, the cars have returned from the coachbuilders to be inspected before issuing the Five Year Guarantee. The cars will be weighed, and appropriate road springs fitted. Visitors will appreciate that cars can vary in weight by several hundredweights on the same model of chassis, so we have to fit springs to suit. Of the nine cars in at present, all but one is fitted with closed coachwork, on 6½ Litre and 4½ Litre chassis. The 4½ Litre nearest is having attention to the dynamo drive, and the dynamo can be seen being worked on by the fitter on the far right. This car is a 1930 specification chassis with the short arm Bentley & Draper shock absorbers, with the large Smith's headlamps introduced on the 1929 chassis. We are now using Lucas P100s on the six cylinder chassis in place of Smiths.

The Speed Six saloon next along has a specially lengthened bonnet, with a third set of louvres. The 4½ saloon next to it is probably Evelyn Baring's, his third. If it is it's GC 6344, a 1929 HB series chassis with an Arthur Mulliner saloon which got it's guarantee three days ago on the 15th. The shop running parallel to us, on the other side of the wall to the left, contains the second Experimental Department – strictly out of bounds! (See p.304.) This shop is also used for housing experimental and other cars, and is used when the racing cars are here for engine rebuilds.

15. Passing from the Finished Cars Shop, we are now on the second floor of the new Office Block, in

15

the Drawing Office. From front to back, Messrs Snow and Snell, who worked on the Golden Arrow project before joining Bentleys, then Messrs Harvey, Cyril Trim, Frank Ayto, Ken Wise, Eric Easter, and Gilbert Whitlock, Harold Whitlock's brother. Not here at present are the Senior Draughtsman C.S. Sewell, the Chief Draughtsman Fred Dewhurst, nor the Chief Designer, Colonel Barrington.

15a

16

*15a. The view from the Drawing Office window, looking back over the Works. Beneath us from the front the car wash, second Experimental Shop and the racing/experimental cars building, then the new Machine Shop with the crane over the door. Then the old buildings containing the old Experimental Shop, Engine Shop, Detail Shops, Stores, Engine Test Shop in the old brick building, and then to the right the boiler house, Running Shop nearest us and the Erecting Shop towards the Edgware Road. On the far right is an 8 Litre chassis. The saloon is WO's experimental 8 Litre, MH 1030.

*15b. If you look up, you can just see the special Bentley weathervane on top of the office block.

*16. Looking back at the Office Block, from the other side of Oxgate Lane, this concludes the trip.

15b

16 The Writing on the Wall

AFTER MAKING a profit of £28,467 in the year up to March 1929, for the year to March 1930 the figure was a profit of just £1,023 – which should have been a substantial loss if adequate provisions had been made. However, it had been decided to show a small profit "so that we may not have further embarrassment from the creditors" (Carruth). Unfortunately, no copy of the Director's Report for the financial year ending 31st March 1930 can be found, but other papers show all too clearly the serious nature of the position. In outline, the situation was a re-run of 1925. Sales were insufficient, leading to stocks of unsold cars. Sales in the financial year to March 1930 were 116 6½ Litres and 157 4½ Litres, compared to 134 6½ Litres and 244 4½ Litres in the year to March 1929. However, while in 1925/26 the 6½ Litre was due to be launched, in 1930 there was no such potential saviour. The 4½ Litre Supercharged model was in production, but this was merely a stop-gap. The Board were well aware that the four cylinder cars had reached the end of their life, and had indicated this clearly in the approach to Ricardo for a more modern, six cylinder engine design, in August 1929 (see p.242).

In the first half of 1930 after a Board meeting Barnato instructed Carruth to spend some time in the Bentley Works, to evaluate the position. While Carruth was a director and shareholder, he does not appear to have held an executive position in the Company until later in the year. Carruth's report to Barnato was completed on 24th June, and was marked "Confidential Memorandum to Captain Barnato from Mr. Carruth". In view of the contents, this is hardly surprising. The contents are significant in understanding the position of Bentley Motors in the period March 1930 – July 1931.

Not all of the report is relevant, but parts are here reproduced, interspersed with comments:

"The general unfinished stock at the 31st March was £74,000. Of this sum, £45,000 is what I term "quick stock", that is, capable of being turned into money in immediate production.

"Going on, a preliminary test of the quarter to June 30th shows that if the whole output for this period is sold at ordinary trade discounts, the result for the quarter must be a loss of approximately £12,000. At the close of the last financial year [to 31st March 1930] we had a stock of 29 6½ Litre chassis, 7 standard 4½ Litre chassis and 3 Supercharged 4½ Litre chassis.

"At the end of June [1930] the arranged programme is for a further 14 Supercharged 4½ Litres, and thereafter to complete the present sanction of 50 of this model, a further six must be balanced and produced as soon as they conveniently can. Apart from this we are entirely dependant on the production of new models to pay our way and as it is not possible for these to contribute any revenue in the quarter to September, we must look forward to a further loss of say £20,000 in that period. Faced with this difficult question of production, or one should rather say lack of it, I have considered the question of producing a modified 6½ Litre chassis for 1931, or alternatively to give a further run to the standard 4½ Litre, but I have dropped both these propositions as impracticable."

There are several important points here. First of all, the Company was carrying a substantial financial burden in unsold stock, and would continue to do so right to the end. This factor negated the possibility of continuing to manufacture existing models. Secondly, there is no question of sanctioning a further batch of Supercharged chassis – the possibility is not even mentioned. In the case of the 8 and 4 Litres it is easy to predict the next chassis series because there are YH 8 Litre engines (next series YH 5126-YH 5150) and VP 4 Litre engines (next series VP 4026-VP 4050) but the second series of 4½ Litre Supercharged engines go from MS 3926-MS 3953, the latter three tacked on to an existing series to finish off the Supercharged cars without needing to start a new series of chassis numbers.

The possibility of a modified 6½ Litre for 1931 is interesting, as there are hints that this was a possibility. The Barnato "Blue Train" Speed Six, chassis HM2855 (from the last series of chassis, HM2851 – HM2870) has the comment "Steering column extended to enable steady to be fitted as 1931" (implying a 1931 model) in its Service record. As Carruth was so concerned about transmission noise he would doubtless have wanted to see the F box and a hypoid back axle in a modified 1931 6½ Litre, and this is supported by the project codes on drawings BM7181 and BM7211 (see pp.73 and 87).

It is also abundantly clear that the Board had accepted that the 4½ Litre was obsolete and that a modern, six cylinder chassis was certainly needed to replace it.

"We now come to the consideration of the Company's preparations for producing the 1931 8½ Litre [sic – the author has altered subsequent references to "8½ Litre" to "8 Litre"] six cylinder model and if we can here agree two points which are to my mind vital to a proper appreciation of our problems, it will be very helpful.

"1. Broadly speaking, our main troubles in the past can be traced to the delivery of new models to the public before adequate tests of production have been made. Let me at once emphasize that I do not for a moment suggest that this course has been deliberately adopted from choice. Too often the till has been empty and, in such circumstances, I readily admit how easy it is for a harassed management, beset by many difficulties, to fall into such an error of policy. But it is an error of policy, and may even be a fatal error of policy.

"2. Our whole production scheme should be drawn on the basis that faults must be eliminated before our cars reach the public."

It is perhaps significant that Carruth came into Bentleys with Barnato early in 1926, as the 6½ Litre was being produced. In the year to 31st March 1928, £6,378 was written off against up-dating 6½ Litres at Bentley's expense – and that figure only covered ¾ of the 196 chassis affected. This reduced profits to £1,049.10.4, and must have struck Carruth forcibly. The teething troubles with the 6½ Litre have already been discussed. No such problems are known with the production 4½ Litre, although such modifications as strengthened frames were production changes forced onto the Company. The costs of such modifications had to be borne by the Company, and would obviously displease an accountant. The "quality" philosophy espoused by Carruth is a remarkably modern one.

"Whatever the management of the Company may do, it seems to my mind that we are inevitably faced with a loss by December of, say, £40,000. This can be faced in two ways.

"We can adopt rush tactics and get these cars [8 Litre and 4 Litre] into the hands of the public at the earliest possible moment, repeating the exact conditions which have caused our troubles in the past and thereby destroying our one remaining chance, or realising that such a loss is in any case inevitable, we can face it and spend our money wisely, getting in return a definite result at a given date.

"The one great necessity is time, and as time in this instance means money at the rate of £2,000 per week it is imperative that safety be balanced with speed as far as possible. I therefore suggest that the following policy and procedure is the best expedient we can find, bearing in mind that our difficulties are considerably influenced by the Show date.

"a. One demonstrator must be got ready for the Show even if it is the only car of its kind in existence. We know that if this model is to make any reasonable impression outside our present market it must be pretty good. I put it very strongly that it must be a vehicle in which the present transmission noise of the 6½ has been eliminated to that degree which will produce sales in the "town carriage" market to the extent of 100 a year. It is therefore essential that Barrington be given further time for the gearbox. [Refers to the new F box.]

"b. Two experimental cars must be prepared as soon as possible, bodies being ordered now to ensure delivery at the proper date. These two cars ought to be available about the middle of August.

"c. Seven further chassis are necessary for the Show. So far as I am concerned these can, if necessary, be merely skeleton chassis, the bodies being demounted after the Show, and the chassis completed later. ALL THESE SHOW CHASSIS MUST BE BOUGHT IN BY US, and retained until the general release of this model takes place.

The cost of doing this will be fairly considerable, but I think Longman can help me by making fairly good arrangements on this point.

"d. The technical staff should immediately examine the production position in detail, and decide on a programme of progression which will keep the Works going at the best possible rate on this model...so that at least FORTY COMPLETE CHASSIS CAN BE READY BY THE 5TH DECEMBER. The two experimental models will, by that time, have been in our hands for nearly four months, and Witchell will have had considerable experience of production chassis.

"e. Following on the above, and to give production a chance, our present sanction of 8 Litre must be increased from 50 to 100 as follows:

40 produced by December 5th

4 per week until February 13th

2 per week thereafter.

"From my latter remarks it will be seen that this sanction of 8 Litre, and a sanction of 100 SV [at seven per week from 13th February] will both run out in the month of May [1931].

"f. This plan is far from ideal but it seems to me to be the most hopeful procedure in all the circumstances.

"g. Looking further ahead, no attempt should be made to show the SV chassis at Olympia. Instead we should show the standard 4½ Litre remembering that there are still 50 of these to be disposed of, including present agents' stock.

"h. It should be privately agreed with Barrington that we will be satisfied if he delivers last production drawings of the SV by 15th October [1930]. These plans should be carefully scrutinised by the Managing Director [WO] and Purves from an economy standpoint, as the price of this chassis is very important.

"i. Purves should then progress the SV model carefully on the basis of 100 sanction at the rate of seven per week, production to start at this rate without undue haste in the week ending February 13th [1931].

"j. I think I am right in saying that present commitments should provide at least two experimental SV chassis by the end of October or early November [1930], and proper bodies should be ordered for these experimental chassis in good time.

"k. No general release of the SV must take place until at least 20 chassis have been passed off test by Witchell, and as this model has not been shown at Olympia, there need be no pressure for delivery, so long as finance is provided to carry the position."

Again, there are several important points raised by this. First of all, there is a very real break with previous policy, of a radical nature. In backing the 8 Litre and SV models, the Company was clearly moving from the manufacture of sports cars into the manufacture of luxury closed cars. The manufacture of sports cars had ruined the Company in 1925 and had effectively done so again, and it seems that the Board was set on a head-on fight with Rolls-Royce for the "town carriage" market. That such a market existed is supported by Rolls-Royce's sales figures of 2,940 20HP chassis in the period 1922–29 and 2,212 Phantom I chassis between 1925 and 1929. They went on to sell 1,672 Phantom II chassis between 1929 and 1935, and 3,827 20/25 chassis in the years 1929 to 1936. That this was indeed the aim is made clear by WO: "We had the new 8 Litre on the stocks, a car intended to compete directly with the Continental Phantom Rolls-Royce...What we did instead was to design and market a new car, the 4 Litre, of unhappy memory. 'We must have something to compete with the small Rolls', was the Board's verdict." This policy of building town carriages is indicated by the main thrust of production in the early part of 1930. Of the last 29 Speed Sixes, 24 were built on the 12′8½″ wheelbase. These were not intended to be sports cars. Of the other five, two were Team cars, one was built for Barnato, and one to Team specification for Mrs Scott.

Taking the above points in turn:

a. By September 1930, the first 8 Litre chassis had been built (chassis YF5002, used by WO). YF5002 was running by October. In November, WO drove Northey in YF5002, WO telling him that at that time only two cars had been made.

b. YF5001 and YF5002, both of which were completed and bodied by September/October.

c. Six 8 Litres appeared at the Show (YF5004 – YF5009). All were retained by Bentleys, and none were passed off Final Test until December 1930.

d. By 3rd January 1931, the 52nd 8 Litre chassis YM5027 was in Road Test; the 40 chassis were probably pretty much ready in time.

e. There exist chassis specifications for "Sanction I" and "Sanction II" 8 Litres. These presumably refer to the two batches of 50 8 Litres that were ordered and built. The 4 Litre (SV) sanction was evidently reduced in 1931 – this will be referred to later.

f. Carruth's plan is astute, if overly ambitious in terms of targets. A production programme based on selling 350 4 Litre cars a year is very optimistic when peak 4½ Litre sales were 244 in the year to

March 1929. Against that, they were aiming at a new market.

g. The 8 Litre was the star attraction at Olympia. It would have overshadowed the 4 Litre.

h. Note the emphasis on economy (see p.297).

i. The first SV chassis were built up in the Erecting Shop late February 1931. VF4002 was delivered to Road Test on 4th March 1931. (VF4001 was the Works' experimental car.)

j. Unfortunately, no information can be found on this point.

k. The 4 Litre launch was held on the 14th May 1931. VF4014, the 14th chassis, was in Road Test by 26th March 1931, so it is likely this target was reached. Carruth's report continues:

"Finance.
It will be obvious from the foregoing that the Company's present resources are inadequate. Our commitments for the six months to 1st December [1930] will, on the above programme, be £207,000. This can only be relieved by the sale of existing models and the Service revenue, and the limit of this relief is £130,000. The prima facie deficiency is therefore £77,000, but for practical purposes, if this scheme is thought to be commercially worthwhile, credits of £100,000 should be available.

"If you are satisfied that the technical side of these two models will be alright, I on my side am satisfied that a market exists for these cars in the numbers which we require for minimum production.

"If, bearing these points in mind, you [Barnato] are prepared to provide this credit, I think it should be done in the following manner:-

"Baromans Ltd. should formally enter into the contract and should undertake to provide such credits, up to a limit of £100,000."

This is followed by details of share manipulations. What is clear is that Baromans were being asked to provide funds to the Company of up to £100,000, as an unsecured loan. Baromans Ltd. was a finance house, incorporated in 1922, with offices at Pollen House, Cork St, W1. It is therefore no surprise to find, in a later letter from Carruth: "I may say that Baromans Ltd. is a company consisting solely of Captain Barnato and myself, we are the sole shareholders and the sole directors." Baromans continued to operate from Pollen House until the lease ran out in 1941, but unfortunately for our purposes the records in the Public Records Office relating to this company were destroyed in 1957.

"Sales.
I have had a detailed report from Longman dealing with the position of his department as a whole. I think this report is good and without unnecessarily elaborating this memorandum I can say that I look forward to a speedy improvement in this department. I mean by that we shall either get better value for the money we now spend, or the costs of running the department will be reduced [no copy of this report has survived].

"Buying.
I am not at all certain that in the past we have appreciated how difficult it has been for Purves...owing to the way production has been chopped and changed about. This will be automatically cured when the re-organisation of Production comes into being.

"Service.
All we can hope for is perhaps a little better detailed organisation when the new building is ready, and that some use is made of the valuable data which automatically comes to the Service Department, at the Works end.

"I should like to point out that in my opinion there is an undisclosed liability in the Balance Sheet in respect of the future Service. This figure I put at £28,000 and I suggest it represents an unfair dead weight on future production. It should, to my mind, have been provided in the Trading years in which the cars which are having the benefit of it were produced."

Sales was extensively reorganised in 1931, when Longman and Kevill-Davies were replaced by Grey and Johnson – both from Rolls-Royce. The latter were possibly hired for their experience in the market that Bentleys were aiming at. The Service liability relates to the costs incurred under the Five-Year Guarantee on cars made in earlier years. Carruth obviously felt that a reserve against guarantee work should be maintained on the balance sheet each year against cars produced that year.

"Production.
At the moment the production programme can only be said to be in a state of chaos, but I have laid before everyone concerned – including WO – what I consider to be the proper procedure for the future and none of them have raised any practical difficulties. On the contrary they have all expressed approval. The system is based on the fact that at the moment the Motor Car Trade, owing to the fact that we have the Olympia Show once a year, is a

seasonable one. Actually the selling season consists of eight months from October to May and in the remaining four months demand is very much curtailed. This to my mind suggests that the 31st May or thereby must be the last day in each year on which any current models are produced. Consequently, as the Works must have a non-interrupted run, the 1st June must be the first day in any year on which new models are produced. The proper scheme of things would therefore seem to resolve itself as follows:

"Design must understand that all desirable modifications have to be produced, experimentally passed and production drawings ready for the Buying Department by the 1st February in each year. Purves then has from the 1st February until early June in which to progress his buying, a period which is ample for this operation to be done economically. From early June until the Show we must accumulate our new models so that by the time the Show comes, we should have roughly 100 chassis available, some of which will have reached the coachbuilder's hands in August. It is therefore clear that if experimental work has been passed in time for the production drawings to be prepared by February and if the experimental cars have been further tested during the whole period that the Buying Department is progressing the production, and further if Witchell is actually producing cars without any undue pressure four months before they reach the public's hands, it is almost impossible for any serious fault to reach the retail buyer. Now that, to my opinion, is one of the important aspects of the whole manufacturing game.

"If, for instance, by the end of July, we have discovered that something in these suggested models is not right, the cost of putting it right can be clearly ascertained. It is the cost of the labour, new materials and scrap materials involved, not a penny more and not a penny less, but if, as has been our habit in the past, we have discovered these faults after the cars are on the road, there is no-one alive who can tell you what the cost of such an error is to the Company in goodwill and otherwise. It is my firm belief that only by adhering to such a scheme as this and seeing it is definitely administered can one achieve success in such a difficult market."

Again, Carruth's comments have a remarkably modern tone. Any management consultant would recognise the concepts, even if they would use jargon such as "Total Quality" or "Right First Time". It is difficult to reconcile WO's scathing reference to Carruth as "Barnato's crony" with the foregoing. Again it is necessary to look at circum-

stances, but Carruth's plan only really failed because of time, aggravated by the slump, and perhaps the sales performance of the 4 Litre.

The perception of the Bentley as a racing car was not an image desired for the forthcoming 8 Litre, a car unlikely ever to have been made into a racing car, or for the 4 Litre. It is abundantly clear that it was designed with a view to reducing costs (ref: Hives's memo of 3rd February 1931 "We have heard that Bentley will in a few months announce a new 6 cylinder, 4 Litre car, which will be a direct competitor with our 25HP. The design is being carefully gone into to reduce costs. We understand that for this reason the engine is to have side by side valves") with the Ricardo-inspired overhead inlet/side exhaust valve F head engine.

On this basis, it is easy to see Bentleys' evolution in the 1930s with the 8 Litre competing directly with the Phantom II Continental, and gradually evolving into the proposed $6\frac{1}{4}$ Litre (Napier) Bentley with, eventually, independent front suspension. The 4 Litre would have competed with the 20/25 HP Rolls-Royce, and should have evolved into a car very similar to the $3\frac{1}{2}$ Litre Derby Bentley. The Company in this form would have retired from racing for some years (with no suitable product to race with) and built up an image as a solid, respectable manufacturer of high-class carriages with sporting performance, again not dissimilar to the Derby Bentleys whose slogan "The Silent Sports Car" would have been equally applicable.

George Oliver's remarks in the Bentley $3\frac{1}{2}$ and $4\frac{1}{4}$ Litre "Profile" are perhaps pertinent: "During the twenties the discriminating motorist of means, who sought a combination of high performance with easy running, and comfortable springing with good handling and roadworthiness, was obliged to choose from the automobile elite – from cars of such size and splendour as the 40/50 Rolls-Royce, the 37.2 or 45 hp Hispano-Suizas, the straight-eight Isotta-Fraschini, the Lanchester 40, or the $6\frac{1}{2}$ Litre Bentley, for example.

"But these cars were at once costly to buy, and to a greater or lesser degree ownership implied a certain formality of living that was beginning to become irksome to a growing number of people. By the end of the decade [ie by 1930] road conditions had greatly improved in many parts of Europe and attitudes towards long-distance travel by motor car were quickly changing; better roads encouraged greater use of them and a demand slowly grew for a medium-sized car that would combine high performance, comfort and ease of running." It was at this market that the 4 Litre was aimed.

17   8 Litres

IT WAS AT THE Olympia Motor Show, opening on Thursday 16th October 1930, that Bentleys unveiled the "Big Daddy" of them all – the 8 Litre. Achieved by extending the stroke of the 6½ Litre to give dimensions of 7983 cc from 110 mm stroke by 140 mm bore, the 8 Litre was at the time the largest car on the British market. No less than six 8 Litres appeared at the Show, with coachwork by Thrupp and Maberly, H.J. Mulliner, Gurney Nutting, Park Ward and Freestone & Webb. With a chassis price of £1850, the same as the short wheelbase Phantom II, the 8 Litre was obviously pitched directly at the Rolls-Royce market. (The long wheelbase Phantom II had a chassis price of £1900.) WO's comment that "to the dismay of our rival company [Rolls-Royce] we priced it a shade above their car" must have been a failure of memory. It is arguable that the 8 Litre was the closest any competitor ever got to producing a car better in all-round terms than the then current Rolls-Royce. Rolls-Royce had their own problems with the Phantom II, the engine of which was much less susceptible to development than that of the 8 Litre. The engine of the Derby car was referred to by Rolls-Royce's own testers as being "gutty rough" because of its comparative roughness compared to the Ghost, and numerous changes were made throughout the life of the car to improve matters.

It is no wonder that Royce was concerned about the performance advantage of the 8 Litre over the Phantom II Continental when any further attempts to increase performance would probably lead to unacceptable levels of engine noise and harshness. The Phantom II also suffered from interior noise or body "booms" when fitted with large closed coachwork, a problem which the 8 Litre was far less susceptible to. WO was quoted as saying that "I have wanted to produce a dead silent 100 mph car, and now I think we have done it." It has been said that WO wanted to produce a road-going locomotive, and WO's description of a train ride on p.12 could easily be re-cast as a night drive in an 8 Litre.

Indeed, WO later referred to the 8 Litre as the best car Bentleys ever made: "I personally hated the noise of the four cylinder Bentley and when the 6½ Litre six cylinder came along we sold one to an old Bentley owner [Foden] who was always very nice to me about his 4½. The next time I met him I eagerly asked him how he liked his 6½. I was looking forward to hearing what he had to say because he was very pleased with that car. He looked slightly embarrassed and said: "Oh, I like it, but I do miss the bloody thump!" Later he changed back to a new 4½. We called the hotted-up edition of the 6½ Litre the Silent Speed Model and both this and the 8 Litre, which in my opinion was the best car we ever made, were silent for their day." (WO in *The Autocar*, 28th April 1943.)

The first production 8 Litre on the road was the second chassis No. YF 5002 on the 12′ wheelbase, fitted with an H.J. Mulliner saloon body registered GK 706. This became WO's personal car. GK 706 was driven by both *The Autocar* and *The Motor*, who were full of praise. For the first time the coachbuilders were given full rein with a chassis capable of carrying a full 7-seat limousine body at over 100 mph, and some of the most elegant formal coachwork of the vintage era was constructed on 8 Litre chassis. As so often in the past the dual nature of the Bentley was emphasized: the docile town carriage on one hand capable of accelerating from walking pace to over 100 mph in top gear, the

ABOVE **Barclay's showroom at 12a George Street, with 8 Litre chassis, Supercharged 4½ Litre with Vanden Plas sports coachwork, and a Phantom II saloon. This illustrates the enormously high radiator of the 8 Litre, and reinforces the Bentley position in late 1930/early 1931. The only cars they had to sell were the expensive 8 Litre, and the Blower, the price of which was slashed in January 1931 in an attempt to sell some cars. The Rolls-Royce is an interesting comparison, in terms of such features as bonnet height and wheel sizes.**

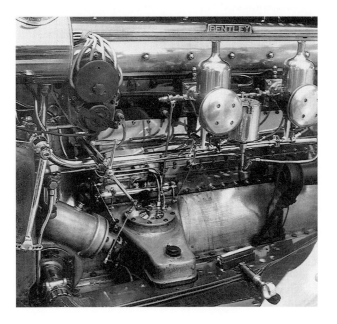

RIGHT **Inlet side of the 8 Litre engine, showing the twin SU HO8 carburettors with square throats and four bolt mountings. This also shows the Bosch magneto, adjustable accelerator pedal, and the slotted adjustment on the flange of the steering box bush. This bush is machined eccentrically, so that the gear meshing can be adjusted. It is then locked with a plate and locking tab.**

LEFT **Blower 4½ at Jack Barclay's showroom. This chassis would appear to be fitted with the high-sided Vanden Plas touring coachwork, fitted to two 8 Litres, two 4 Litres and several Supercharged chassis.**

BELOW **Typical 8 Litre coachwork, in this case a six-light saloon by Park Ward on the 13′ chassis, No. YF5022 registered GN4110. This shows the evolution of metal-panelled coachwork with waist-line and shaped wings, to conceal the radiator height. The body comes right down to hide the chassis. Lucas P100 lamps are fitted.**

RIGHT **WO's saloon at the Montlhéry track, with Jean Chassagne (centre). Chassagne was in charge of Bentley's French service facilities, and it is possible that this was a reconnoitre, as there were some plans to use Montlhéry for record runs with a Speed Six. These plans came to nothing in the end.**

sports car on the other, capable of very high point-to-point averages. *The Autocar* described it as "Motoring in its very highest form" and there is little doubt that the competition, particularly Rolls-Royce, must have been somewhat concerned when *The Autocar*'s road-test of GK 706, published in their issue of 5th December 1930, with an all-up weight of 48 cwt, recorded a maximum speed of 101.12 mph over the ½ mile. Interestingly, according to Graham Robson an exhaustive survey of all *The Autocar*'s road tests of the 1930–1939 era showed that only one car was faster – an 8C 2300 Alfa-Romeo which recorded 106.8 mph. It is, of course, likely that *The Autocar* did not test every performance car available but the figures are still

very revealing. By the standards of the era, the Bentley must have been staggeringly quick. The 8 Litre remained the fastest production Bentley until the advent of the R-Type Standard Steel saloon which managed 101.7 mph in October 1953. The first chassis YF 5001, with H.J. Mulliner six-light saloon coachwork, was delivered to matinee idol Jack Buchanan, registered GK 672, in October 1930.

It has often been observed that the 8 Litre was not the most suitable car to be selling in a depression, a fact that Bentleys themselves must have been well aware of. In order to sell more cars, a plan was formed to increase export sales. This new approach was first made public by the *Sunday Chronicle* on the 5th October, announcing the

formation of a new Continental sales organization with large capital resources (these resources can only really be regarded as a figure of speech as the Company had no money – the resources were Barnato's own). Further articles the following day gave the name of the new company as the Société Européenne Bentley Automobiles SA, and announced that a depot was to be opened in Paris in a few weeks' time. This would be staffed by mechanics trained at the London Works, and would carry a full complement of all necessary spares. The intention was to follow the Paris depot with one in New Zealand, "arrangements [for which] were concluded a few days ago." Barnato, as Chairman of the new Company, sailed to America to make arrangements for similar sales and servicing depots to be opened in the United States and Canada.

At the Bentley Luncheon at Olympia on the 17th October, Hubert Pike in a speech said that: "We are making a bid to expand our market and we feel that America may be able to help. That is why our Chairman, Captain Woolf Barnato, has gone to the United States. It is idle to maintain that times are not bad, but we believe that there is always a market for the best and we are trying to cater for it. Nine years ago we produced our first car. Today the name 'Bentley' is a household word. Enthusiasm permeates the whole of our organisation".

Despite the difficult sales position, the Board were obviously still prepared to take some initiatives in finding solutions. However, one has to question the wisdom of opening service facilities when a

quick glance at the figures shows that the Kingsbury facility lost between £8,557 and £14,642 every year between 1926 and 1931.

The Supercharged 4½ Litre continued to be available, not because more were being made but because there were still unsold chassis. Much the same could be said of the 4½ Litre. Production of the 4½ Litre had ceased by June of 1930, and many of the later chassis were bodied and sent to the Pollen House showrooms as stock. That the 4½ Litre had died a death is shown by the fact that only 11 cars were passed off Final Test in 1931, and five of these were passed off after the Company had gone under. Final test date was the date that the complete car was passed off with coachwork fitted, meaning that the chassis themselves had been completed earlier and were simply being bodied to be sold. The demise of the Supercharged car is heavily underlined by a reduction in chassis price from £1475 in October, 1930, to £1150 by January, 1931. The latter figure was just £100 higher than the standard 4½ Litre chassis which was still catalogued at £1050.

The 1931 model 4½ Litre was basically unchanged from the 1930 model; indeed, the "1931 model" was largely a figure of speech. The chassis was fitted with the heavy crank engine, and more and more of the aluminium castings were replaced by Elektron. The crankcase, sump, magneto turret, differential casing, gearbox casing and steering box were made of this material, and many parts were made from stainless steel, then referred to as

"Staybrite". Chromium plating introduced from America in 1929 was specified on more and more parts, a finish easier to clean and maintain than nickel, but in appearance much brighter and harsher. On the very late cars the bright work was a mixture of nickel (generally parts supplied by other manufacturers) and chrome (parts made by Bentleys). On the subject of original finishes it is interesting to note that although most of the coachbuilders did not start spraying with cellulose paint until about 1930, Vanden Plas had spraying facilities since 1925 and used cellulose on almost all their cars after that date.

Coachwork was at an evolutionary phase, the Weymann patent method dropping almost out of sight – by the 1931 Show fabric bodies had become a thing of the past. The early examples were suffering from the fabric cracking and were looking rather tatty, and the angular shapes dictated by the limited curvature available were going out of fashion. The fabric was not liked by chauffeurs because it was impossible to polish, and although smooth-grained fabrics had been tried these had a short life, and were an unsatisfactory compromise. Mass production cars were already moving towards the American-inspired principle of pressed steel bodies welded together, which were cheap, light, and free from rattles. In short, a pressed steel body offered advantages over even the finest coachbuilt bodies at a fraction of the price if adequate numbers were constructed. The 1930s were the zenith of the coachbuilders' craft, after which they fought a rear guard action which has not completely finished, but is nevertheless on a vastly diminished scale to past years. At the 1930 Show all the Bentleys shown were panelled, and very few fabric bodies were built on either 8 or 4 Litre chassis. The 1930 Olympia Show and the cars presented there marked the end of the qualities that define the vintage car. Even the Vintage Sports Car Club has accepted the better pre-war 1930s cars but it was many years after the war before connoisseurs began to accept that modern cars possessed qualities the equal of the better vintage cars. Now, of course, the modern car is so good (albeit boring) that one looks back through rose-tinted spectacles.

There were surprises at Pollen House, where an extensive shake-up of the sales staff was conducted. Longman and Kevill-Davies left, to be replaced by K.C. Grey and L.H. Johnson. Grey was the new Sales Manager; both were from Rolls-Royce. Johnson was the son of Claude Johnson, Rolls-Royce's Managing Director until 1926. Kensington Moir disbanded Kensington Moir and

Bob Montgomerie, appointed company secretary in October 1930, on holiday at Scarborough. Montgomerie was later the liquidator of Bentley Motors.

Straker to re-join the Company. Grey held daily sales meetings in his office and it soon became clear that all was far from well. These meetings became more serious as the sales figures dropped each month. The travelling sales team was dispersed, and Frank Clement was seconded to Dex at Newcastle in a sales capacity.

There were surprises, too, in the management of the Company. WO said in his Autobiography that finally "the Bull was taken by the horns" and himself and Maury were displaced from their positions as Joint Managing Directors, Maury resigning from the Board in October 1930. Ramsey Manners died in January 1930, at the early age of 37, indirectly from wounds received during heavy fighting in France in 1915. Carruth became Managing Director in December 1930. WO continued as a director, and as Chief Engineer. Barnato further tightened his hold on the Company, by appointing Robert Montgomerie as the Company Secretary, in place of W.K. Forster. Barnato thus directly controlled the two most important executive posts in the Company.

W.K. Forster, who had been the secretary since 1923, was eased out of his position in a rather underhand way. Forster signed the cheques for the bills, and when he went on his annual holiday in Cornwall, these were sent down by his secretary Geraldine Dupuy to be signed. In 1930, this was used as an excuse to push him out, and the job of dealing with the cheques passed to Peter Purves, the Chief Buyer. However, Purves, who had been the Chief Buyer since 1920, left to be succeeded soon after by his deputy, Darren Few. Purves and WO shared the same secretary, Winifred Leafe, who spent much of her time in WO's office, behind two firmly-shut doors, the inner covered in green baize. However, Winifred Leafe became Mrs Purves, evidently to WO's displeasure. WO is recalled as having an eye for the ladies. From a letter written by Freddie Settrington: "W.K. Forster related a splendid tale about WO concerning Peter Purves – whom apparently he disliked intensely – it seems the latter had just struck a lamp post in a 3 Litre and, according to some member of the luncheon table – had been badly cut by a piece of glass that "came off the screen and struck him". "That's a strange thing" said the teller "I would think that anything falling off a moving car would fall forward, not back." So they asked WO (the oracle) who, after the inevitable lengthy pause replied "I don't know, but Purves is a bloody fool!" He was quite unique."

Frank Ayto suggests that Purves' services were dispensed with because he was no longer needed and Few stepped in merely in order to keep things going in the interim as the Company ran down.

R.S. Witchell and Sir Walrond Sinclair joined the Board in January 1931. Witchell had been Works Manager for Bentleys for many years, and Sinclair was the Managing Director of the British Goodrich Rubber Co. Ltd., whose Goodrich tyres Bentleys were using exclusively.

This removal of WO from the head of his own Company must have been a shattering vote of no confidence. It seems that WO was in some sort of state of shock from about the middle of 1930. Mute testimony to this comes from his own personal notebook, in which he had meticulously recorded sales/production figures every week for every model from April 1926 on. The notebook does not suddenly end, as if he had started a new one – the headings are there up to December 27th 1930, but the figures peter out after the end of June. The figures for July to September are unheaded, but consist mainly of a string of zeros. After that, it is not filled in at all. Knowing perfectly well the way the sales were going, WO must have seen the writing on the wall. He had achieved the ultimate with the 8 Litre, after which the design work foundered with the 4 Litre that WO washed his hands of.

It is easy to understand the sort of frame of mind that WO might well have been in. His Company was going under, which spelt personal ruin as well

as imminent unemployment (when interviewed many years later, WO said "of course, one lost all one's capital"). The disbanding of the "Bentley Boys" coincidental with the end of the racing deprived WO not only of the racing itself, but also the social life attached to it. His marriage was breaking up, the divorce coming several months after Bentley Motors was sold. It is perhaps in itself significant that WO does not even mention the name of his second wife, Audrey Gore, in any of his writings. One of the Bentley family commented to the author that "She was more of a society lady: not the sort to want to sit on an oil drum in order to talk to her husband." WO said "We weren't temperamentally very well matched." He had lost HM and Burgess, the two bastions of the Company in the earlier days. WO had been working continuously, under great stress, since 1912 – some 18 years. Under such intolerable pressures, any human being would be in a state of retreat. However, the situation in which the Company found itself needed strong direction from the top, and in the absence of such direction from WO, it had to come from somewhere else. Hence the removal of WO and Maury and the appointment of Carruth.

It is clear that Carruth was appointed by Barnato to try and salvage what he could. Before this, although he had been a director since 1926, Carruth had not held an executive position inside the Company. His appointment probably stemmed from the June, 1930, report, and it is little wonder but that WO resented Carruth's appointment.

One of the first actions of the Board, under Carruth's direction, was to secure the means of borrowing even more money. To this end, a notice was sent to shareholders. Article 47 of the Company's Articles of Association read as follows: "Borrowing Powers. The Directors may raise or borrow money for the purpose of the Company's business, and may secure the repayment of the same, together with any interest or premium thereon by mortgage or charge upon the whole or any part of the assets of the Company (present or future) including its uncalled or unissued capital, and may issue bonds, debentures or debenture stocks, either charged upon the whole, or any part of the assets and property of the Company or not so charged, but so that the whole amount borrowed or raised and outstanding at any one time shall not, without the consent of the Company in General Meeting, exceed the share capital of the Company for the time being issued or agreed to be issued." In other words, they had already borrowed so much money that it would have been *ultra vires* for the Directors to have borrowed more without the consent of the Company to change the relevant article. Borrowing in such circumstances would be void, and the Directors themselves could be sued by the lender (*Topham's Company Law*, 1919). The Board requested a change in the wording to include "from time to time at their discretion" and removing everything from "not so charged" onward. The amendment, of course, effectively gave the Board unlimited borrowing facilities. The Company was borrowing heavily from Baromans (they had a £100,000 line of credit with them) and Barclays Bank; these unsecured loans were personally guaranteed by Barnato. One of Carruth's first actions was to pay off all the Company's trade creditors, and change from buying on credit to buying for cash. Commercially, this put them in a much stronger negotiating position with suppliers, in order to push costs down.

Ever since the June 1923 mortgage for £40,000, the Company had managed to sustain the interest payments to prevent the mortgages being called in, but they never paid off any of the capital sums. The October 1927 and 1929 mortgages with The London Life Association required the interest payments to be made within 15 days of the 15th of each month, or The London Life would call in the capital sum of the loan. In effect, they absolutely had to pay The London Life, so they were now in the position of borrowing from Peter to pay Paul.

In the second Experimental Department work was well under way on the new 6 cylinder 4 Litre engine. This second Experimental Shop was set up in part of the Finished Cars Test Shop, which had been extended greatly in 1929/30 after the old wooden office block had been demolished and replaced by the new office block fronting Oxgate Lane. This new shop worked principally on the SV project, under Stan Ivermee. A number of single cylinder prototypes were made, giving up to 17 bhp, or 102 bhp plus from the final engine. These prototypes were made up to the ideas of the various consultants who were called in to work on the project. The earliest that these can be dated to is January 1930, when Arthur Watson recalled seeing a single cylinder of "F" head configuration – overhead inlet valve, side exhaust; this apparently was the configuration favoured by Whatmough. A decision had been taken to use this head configuration to reduce production costs. Apparently, this first single cylinder displaced roughly 665 cc, or 3990 cc for six cylinders, with a compression ratio of the order of 6 to 1. When first run in the Test Shop, the completed engine gave 117 bhp at 3800 rpm, but

apparently WO was not very happy with it.

WO claims that he refused to have anything to do with this design, stating that Bentley customers would have nothing to do with a side valve engine and Weslake, Whatmough and Ricardo were all called in to work on it. As Michael Frostick put it in his book *Bentley – Cricklewood to Crewe*, it is unlikely that people like Tallulah Bankhead would fully appreciate the subtleties of the overhead camshaft four valve per cylinder power unit that they drove behind. WO was against push-rods and felt that side-valves had been a limiting factor in extracting power from the DFP engine. It is possible that WO was more piqued at being over-ruled over a top end design that he had good reason to be proud of.

Of those mentioned above, Ricardo was chosen to work as a consultant on this engine. In many respects the 4 Litre engine was a very good one, albeit handicapped by a lack of development work. Frostick also tells us that Ricardo was called in largely after Weslake had failed to induce any great power output from the original side valve engine. Ricardos said that the head design was the lightest, most powerful and cheapest they had worked on. The engine produced 120 bhp from 3915 cc which compared very favourably with the 4½ Litre's 110 bhp in standard tune and 124 bhp in semi-Le Mans tune. However, the 4 Litre engine could not easily be tuned, partly because the gap between the piston crown and the inlet valve was so small that there was no easy way of increasing the compres-

Sketch of the Ricardo cylinder head design for the 4 Litre.

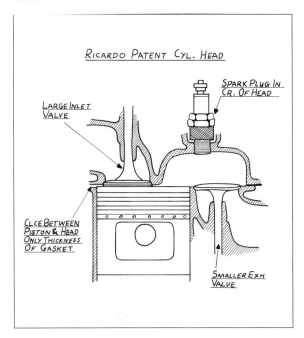

RICARDO PATENT CYL. HEAD

SPARK PLUG IN CR. OF HEAD

LARGE INLET VALVE

CL.CE BETWEEN PISTON & HEAD ONLY THICKNESS OF GASKET

SMALLER EXH VALVE

sion ratio. Ricardo records show that a 652cc single-cylinder engine on test at Cricklewood early in 1930 gave a torque of 120lbs/sq in BMEP and a maximum power of about 25 bhp at 4200rpm – corresponding to about 150 bhp from a six cylinder engine! However, the two six cylinder engines run on test from June 1930 failed to live up to this promise, suffering from problems with valve behaviour and crankshaft bearings. One was sent to Shoreham, but little further development work was undertaken either by Bentleys or Ricardos.

The level of WO's involvement in the 4 Litre engine design is a matter for speculation. It would be very interesting to know whether WO produced the initial design schemes and then called in outside experts because he was unhappy with the results, or whether he refused from the start to countenance a side-valve or an "F" head engine. The author is inclined to speculate that WO did indeed produce the first designs and that they were not succesful. This is not pure hypothesis, but relates to implications from the writings of those present at the time. It is also important to realise that WO had lost his Chief Engineer, Burgess, who was obviously very much a guiding light in the Design Department. When deprived of HM, the Company finances fell apart – it is equally conceivable that when deprived of Burgess, the design side fell apart as well. Confirmation of this is given by the 4 Litre, which was Bentley's only design failure. Nobby Clarke said of Barrington that "he was no Burgess." Ricardos only supplied the scheme for the cylinder head. The detail design and the rest of the engine was pure Bentley Motors, directed either by WO himself or by Barrington.

Although Barrington was the Chief Designer for the last 21 months of the Company, there are virtually no references to him in any of the writings on the Bentley Marque. WO does not even mention Barrington in any of his books, implying that he had very little to do with him. We do know that he worked for Napiers and Rolls-Royce before Bentleys, and there is correspondence between Colonel Barrington and Hives at Rolls-Royce in January 1931 over an arrangement for Bentleys to try the latest experimental Phantom II Rolls-Royce and Hives and Elliott the latest Bentley (presumably an 8 Litre). He is also mentioned in the list of creditors of the Company in liquidation, in which Lt. Colonel T.B. Barrington was owed no less than £1536.9.6. This would suggest that as his contract was for three years at £1500 pa from 1st November 1929, this figure represented a claim for the balance of his contract. Barrington later went back to

Rolls-Royce, and was Chief Designer – Aero Engines in 1934. He was later sent to the States to deal with the production work on Merlin engines, and died there in 1941.

The prime consideration with the new SV 4 Litre model was the cost price of the chassis, and hence the profit margin. WO was well aware that the manufacture of a smaller car to the same design and to the same standards yielded virtually no cost savings (from WO's *My Life and My Cars*: "When the clouds of the depression had been seen on the horizon from Derby [Rolls-Royce], work had begun on a small 2¾ litre bread-and-butter economy car...It turned out eventually to be as complicated and refined a piece of machinery as its bigger brother, and as the only saving was in the weight of metal involved, the term *economy* was a misnomer."). Hence the new car had to be simpler to be cheaper. Accepting that there was little scope for cost savings on the chassis, the obvious area for attention was the engine – and hence the cylinder head design.

Weslake's biography implies that Bentleys were working on an unsuccessful single-cylinder side-valve engine in 1927/1928. This engine produced only 11 bhp from 600 cc. When the final 6 cylinder engine was run up on the Heenan and Froude dynamometer, Arthur Watson tells us that "WO attended several times, but was not altogether satisfied, and modifications were made. Jock (Jock Ferguson, who was operating the dynamometer) said 117 bhp at 3800 rpm was obtained several times." None of the participants makes any mention of Barrington. Although the author is sceptical of some of Weslake's comments, some credence is lent to his story by the fact that he was working with Whatmough, who also worked on the 4 Litre, at the Automotive Engineering Company, whose managing director, Hewitt, has already been referred to. It is perhaps significant that the SV design was under consideration for so long – since before August 1929.

It would not be unthinkable that WO should have had a failure in his career. Indeed, not to would render him unhuman. Donald Bastow stated in his lecture on WO in 1988 that there were indeed failures, mentioning certain aspects of the 2.6 litre Lagonda engine, but this will be looked into further as the events of late 1930/1931 unfold.

On the subject of the inadequate development of the 4 Litre engine, Conway, the Head Storeman, remembered well many years later the problem of trying to find a valve spring for the engine that would not break up. Bastow briefly discussed the 4

Litre engine in *W.O. Bentley – Engineer* and mentioned problems finding a suitable inlet valve spring capable of going over 4100 rpm, 100 less than the engine's maximum recommended speed of 4200 rpm. At 4100 rpm the bhp of the engine was still rising, the 4 Litre engine also giving practically all its torque output at 1000 rpm. A 4 Litre engine that had been on the test-bed for more than six months was sent to Rolls-Royce's Derby factory for tests in January 1932. This engine was fitted with a standard camshaft BM 8743, and with the engine was a modified shaft BM 9290, fitted to all the chassis disposed of to Jack Barclay and Jack Olding. According to Witchell, "So far we have no valve springs for use with this latter camshaft which give satisfactory results."

The bottom end, however, was immensely rigid with mains and journals of such a large diameter that they overlapped, producing a massively rigid crankshaft. This was picked up by Bastow as a pointer to indicate that WO had little to do with the 4 Litre engine. The increase in main bearing diameter from 55 mm to 80 mm from the light to heavy crank 4½ Litre engine reduced the power output by some 5-10% and generally made the car more sluggish. The implication is that WO would never have used seven such massive bearings on a smaller, higher revving engine, even though it was then possible to dispense with the crankshaft damper (the 4 Litre had mains and journals of 75mm and 65mm respectively). According to Rivers Fletcher the new, light chassis design was on the way as well, but it never seems to have seen the light of day. Among WO's papers at the BDC there is a manuscript of unknown date, part of which reads: "The fault of the 4 Litre was its weight and this was caused by our using the 8 Litre frame, gearbox, axles etc., with a 4 Litre engine. Given the time we could have used up all the 8 Litre chassis parts on the 8 Litre and put the 4 Litre into a modified 4½ Litre frame using those axles, etc., which would have produced a good car" (this is reproduced in full in Appendix IV).

There is, though, a specification for a proposed Napier-Bentley that was relayed many years ago by C.W. Sewell to Harold Hinchliffe, then the editor of the Bentley Driver's Club *Review* (Sewell worked in the Design Department at Cricklewood from 1921 to 1932, and was responsible for the design of the "self-wrapping" front brakes). Quite what this car represents or what became of it is not known, but as far as is known, no drawings survive and in view of Snagge's cessation of all motor car work at Napiers from May 1932 onwards, it be-

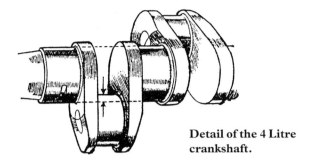

Detail of the 4 Litre crankshaft.

comes even more mysterious. However, Sewell was "Chief Designer, Car Division" at Napiers from 1932 to 1935, and the 4 Litre Napier specification dates from 1932–33, so at some point Napiers must have lost interest and ceased design work.

It was the 8 Litre, though, that was worrying the competition and the memos in the Rolls-Royce files make very interesting reading. The Speed Six was already faster than the Phantom II Continental, and Royce himself was obviously concerned about this, commenting that they should only be 5 mph slower on top speed rather than 15. Royce's proposal was to use twin carburettors and "double top", ie top gear above 1:1 and an additional gear just below 1:1. Rolls-Royce were experiencing problems with booming of bodies, a problem which Bentleys seem to have solved by rubber-mounting the engine and insulating the body from the bulkhead as well. Rolls-Royce also questioned the Bentley's ability to do the "magic ton" with saloon coachwork, and invited Jack Barclay to bring a car down to Brooklands to do some timed runs. In the event the 8 Litre was timed at 99 mph, and Rolls-Royce had to admit that it was indeed a true 100 mph car. This is not to say that the 8 Litre did not have its faults, and Rolls-Royce criticized the springing, commenting that the pitching was the worst they had encountered in a big car, and the low speed running was somewhat rough, a typical occurrence with SU carburettors.

There was a far more serious problem, though, that did not get out at the time, and that was a tendency to severe axle tramp at very high speeds. First to draw WO's attention to this was Nigel Holder, who owned chassis YF 5007 fitted with a Gurney Nutting coupé body, who pointed out that at high speed the car lurched severely to the left. With Holder driving, WO in the passenger seat and Kensington Moir in the back, at 97 mph the car lurched just as Holder said, WO saying, "For God's sake, Nigel, don't play the fool with the car at this speed," at which Moir pointed out that Holder's knuckles were white and it was all he could do to hold the car on the road. With the 8 Litre already in

	4 Litre Bentley	Napier Bentley	3½ litre Derby Bentley
ENGINE.			
Numbers of Cylinders	6	6	6
Bore & Stroke	85 × 115	85 × 115	82.6 × 114.3
Capacity	3915cc	3915cc	3669cc
Max bhp	120 at 4000rpm	120/130 at 4000rpm	115 at 3800rpm
Max torque	175lb/ft at 2500rpm	—	—
Compression ratio	5.1:1	5.8:1	6.5:1
Valve gear	Overhead inlet, side exhaust	SOHC, gear and chain drive	OHV, by push rods
Cylinder block		wet liners	
CHASSIS.			
Wheelbase	11'2" or 11'8"	10'6" or 11'6"	10'6"
Track	4'8"	4'10"	4'8"
Ground clearance	7"	7"	6"
Frame type	ladder	—	ladder
Tyre size	650 × 20"	650 × 18"	550 × 18"
Back axle	4.58:1, 4.75:1 or 5.18:1	4.0:1 hypoid	4.1:1 or 3.9:1 hypoid
Suspension	semi-elliptic leaf springs, friction/hydraulic shock-absorbers	Leaf springs	semi-elliptic leaf springs, hydraulic shock-absorbers
Gearbox	Dog clutches 3rd & 4th	Dog clutches 3rd & 4th	
No. of speeds	4 forward	4 forward	4 forward
Ratios	4th 4.58:1 3rd 6.15:1 2nd 7.81:1 1st 14.85:1 (4.58 axle)	4th 4.0:1 3rd 5.2:1 2nd 7.38:1 1st 12.84:1	4th 4.1:1 3rd 5.1:1 2nd 7.08:1 1st 11.3:1 (4.1 axle)
Chassis weight	32½ cwt	27/27 cwt (estimated)	20 cwt
Brakes	Bentley mechanical, self-wrapping to front	Lockheed hydraulic	mechanical with servo

production this was a symptom of a very serious and potentially damaging problem. The initial design of the 8 Litre chassis had the shackles for the front springs at the front, and the early cars were built in that manner. Very quickly the shackles were moved to the rear of the springs, in which position they were used on the 4 Litre as well. It would seem that this was one move in attempts to eradicate the problem. WO told Holder that if they could not solve the problem on his car they would take it back and break it up and Bentleys would give him a replacement chassis and money towards another body. As it was the car survived, but Holder did indeed buy another 8 Litre chassis YX 5112, the chassis of which was delivered on a fate-ful day which loomed on the horizon. WO himself had a very nasty experience when tramp set in on an 8 Litre that he was driving at about 80 mph. This occurred just before the first cars were delivered and was cured by various modifications including the use of rubber bushes on the bolts securing the front of the body to the bulkhead. In view of the serious nature of this problem, WO personally tested all the first cars delivered until he was sure that the problem had been solved. Deliveries of some of the early chassis were held up for some weeks while the problem was cured, which for some reason was more pronounced on chassis fitted with very heavy coachwork.

Although a considerable amount of correspondence was generated internally at Rolls-Royce, we have no real comment from Royce on the car. A.F. Sidgreaves, MD of Rolls-Royce wrote to Royce on the 22nd January 1931:

"I am sorry to say that this car appears to be becoming a formidable competitor of the Phantom II.

"We are hearing rather more about it than we like from people who have tried it, and as an indication of the sort of thing, I attach a memo relating to an interview which one of our people recently had with the Maharajah of Rewa who is an owner of one or more P.II cars.

"There have been other instances of a similar nature.

"It will be remembered that, when we were making tests of the early P.II cars, I arranged to borrow an 8 Litre Bentley car for trial and Hs [Hives] and Rm [Robotham] agreed that the engine was extraordinarily good and free from the troubles in regard to vibrations and torque reactions which, as indicated in my Sg12/E22.1.31 are doing us considerable harm with our big car business.

"I was keen at that time and put forward a suggestion that we should buy a Bentley so that we could ascertain why they were free from these troubles in their engine, but you will recollect that you did not favour this proposal.

"Admittedly there were many things about the Bentley car other than the engine, which were nothing like as good as on the P.II, but the engine was the point about which we were really concerned.

"I do not think that any of our technical people have really tried the 8 Litre Bentley. We could doubtless arrange this through one of our coach-building agents in the same way as we arranged for the trial of the 6 Litre [sic], and I should like to have your views on this subject. It seems to be very important not to under-estimate one's competitors.''

Doubtless Bentleys themselves maintained such files on Rolls-Royce and comparable marques, but unfortunately no such material has survived. The memo relating to the interview with the Maharajah of Rewa reads as follows:

"re: H.H. The Maharajah of Rewa.

"The above called in here on Saturday morning in regard to his Phantom II car which he has now decided to take back to India with him. He informed me that he was only doing this as he could not get what he called a good price for his car in part exchange for an 8 Litre Bentley.

"He has apparently tried all the fast cars including the Mercedes and the 8 Litre Bentley and he considers the latter to be the best car he has ever driven. He thinks it is far superior to the Phantom II in every respect. He particularly pointed out that the steering pleased him very much; that he did not get any movement on the steering wheel at all and also he did not get any road shocks transmitted through to the steering wheel, and he suggested that we should have an alternative type of steering to give to customers who did not like our present form of steering.

"He told me that the gearbox on the Bentley was quiet; the springing was very comfortable and the whole car was silent and flexible. He himself drove the car at 100 mph according to the speedometer and he said it was quite easy to control at this speed.

"He ended up by saying that unless we had a very good answer to the 8 Litre Bentley we should find ourselves not in the premier position, as he felt sure that a number of the Indian Princes once they had experienced the Bentley car, would buy it in preference to anything else."

Would that more people had felt the same in 1931 – and had had the money to act on the impulse! After comparing an 8 Litre and 6½ Litre lent by Jack Barclay with two Phantom IIs, 26 EX and 27 EX – the former the original Continental, in April 1931, Hives concluded his report by saying: "The Bentley is the best competitive car we have ever tried. It impressed us far more then the 8, 12 or 16 cyl cars we have tried. The torque reaction was manifest. There was also a vibration at 2800 rpm which was objectionable, this we could not properly locate. At a genuine 75 mph the absence of any roar or fuss was very remarkable."

With its short production run the 8 Litre was subject to very few modifications, the principal changes being a redesign of the pinion assembly in the back axle after the first 25 cars and a stiffening up of the crankcase after the first 50 cars. This was perhaps a benefit accrued from Carruth's new policy.

Despite the financial problems, far from reducing the Company's activities there were even moves for expansion afoot. In March 1931 *The Motor Trader* carried the following announcement: "Bentley Depot in Paris – Opening this Month. At the end of the present month Bentley Motors are opening a service depot and Continental sales organization at 21, Rue des Graviers, Neuilly, near Paris. A complete stock of spare parts will be maintained, shops will be equipped for rapid and complete service to Bentley cars, this work being entrusted to M Jean Chassagne, who for several years has been associated with Bentleys in races and competitions on the Continent. It is understood that an active commercial programme will be developed from Paris for the various Continental countries, this being under the control of a former export manager from the Voisin company." The Company already had a depot at 3bis, Rue Charcot, also in Neuilly. The new facility must have been the first result of the plans announced back in October, 1930, mentioned earlier. As far as is known, this project never got off the ground and with such notices appearing in the press so soon before the crash, it is no wonder the latter were so stunned when Bentleys went under!

Although the racing was over, the cars were still competitive and most of the later Team cars were still in the hands of their initial drivers. Birkin went over to Alfa-Romeo and Maserati because of the lack of competitive home-grown machinery (something which obviously upset him deeply), but not until he had entered a Speed Six in the 1931 Double Twelve, held on the 8th/9th May. Quite why Birkin entered a Speed Six when he could probably have got one of the Supercharged Team cars is not clear, but perhaps he realised which was the better bet for endurance racing. Whatever the reason, Humphrey Cook's car, GF 8511, the No. 3 car from the 1930 Double Twelve and Le Mans race, was entered by Jack Barclay to be driven by Birkin partnered by B.O. Davies. Birkin was first round on the first lap, but the Ramponi/Eyston 2½ litre Maserati which was putting in occasional laps at over 100 mph was soon seen to be gaining on the Bentley. The duel did not develop, however, as the Bentley ran a bearing after 2 hours, denying Davies a drive. The Maserati retired on the first day with a broken axle, leaving the MG Midgets to take the first five places on handicap in a rather uninteresting race.

There was still one further model to be presented to the public – the 4 Litre. It is easy to speculate that the 4 Litre should have been introduced late in 1931, perhaps at Olympia, with a properly developed engine. Acording to Bastow, WO said that Barrington was very much responsible for the 4 Litre design scheme (*W.O. Bentley – Engineer* p. 82). Bastow further tells us that Barrington did all the liaison with Ricardos and was responsible for the work in fitting the 4 Litre engine to the 8 Litre chassis. Presumably this referred to the design work involved in the subframes that the 4 Litre engine was mounted on, and their attachments to the chassis side-rails.

It is quite conceivable that the 4 Litre was not the makeshift that has been suggested in the past at all. Carruth's plan was to have 20 chassis completed by the launch date, and this target had probably been met by the May launch. The engine and chassis design work was completed well in time and production progressed to plan before the model was

Final fling – the new 4 Litre chassis in May, 1931. It is interesting to see the revival of the dashboard support brackets, first seen on the 3 Litre in 1921! The steering column used on the 4 Litre chassis was originally designed for the Supercharged 4½ Litre, but was not fitted to that model. Note the Bosch lamps and starter (behind the steering box), outside front brake rods as 4½ Litre, F box, 6½ Litre pattern differential and the massive underslung tubular cross-member between the rear spring front mountings.

launched. A full chassis specification was drawn up, and a large number of parts that are unique to the 4 Litre model designed and manufactured. There is nothing about the 4 Litre chassis to indicate that it was a lash-up; quite the reverse. All the evidence suggests that the 4 Litre was produced exactly as it was designed and intended, and that its commercial failure was partly a matter of circumstances and partly that the concept was ill-judged.

It is perhaps ironic that the car Bentley Motors should have made in 1931 was the 3½ Litre Derby Bentley, which in 3½ and 4¼ Litre form sold 2,411 chassis between 1933 and 1939 at a rate of about 400 per annum: a rate that, coupled with maybe 85 8 Litres per annum, would have made the Company a viable concern on the basis of Carruth's calculations.

The 4 Litre has long been regarded as a scapegoat by Bentley enthusiasts, an unwanted makeshift forced on WO by a faceless Board, ignorant of motor cars. This litany is repeated everywhere: "A car that will undersell [the small Rolls]. It doesn't matter much about the performance, so we'll use the 8 Litre chassis – we've got a lot of them hanging about the works. And we don't want any of WO's expensive cylinder heads with four valves per cylinder. Push rods will do." "When people ask me why Bentleys went bust, I usually give three reasons: the slump, the 4 Litre car and the Blower 4½s in proportions of about 70, 20 and 10% respectively." (*WO – An Autobiography*.)

"That 4 Litre was a...bastard of a car. WO wouldn't have anything to do with it. I don't blame him. The Board wouldn't listen to him, and look what they got." Frank Clement in *The Other Bentley Boys*. "We didn't think of it as a Bentley – a shocking thing, it was. The engine wouldn't go into the 8 Litre frame to start off with; we had to cut a whole length right down the side of the frame to get it in." Wally Saunders, in *The Other Bentley Boys*.

The Board consisted of WO, Pike, Barnato, Carruth, Sinclair and Witchell early in 1931. Pike and Witchell had been friends of WO since before the Great War. WO says of his relationship with Barnato that "we were good friends". While WO certainly had little good to say about Carruth, it was hardly a Board either ignorant of motor cars or likely to steamroller their Chief Engineer.

Indeed, WO's comments are so far from true that they have to be largely disregarded. The Company was effectively bust by June 1930 without Baromans' loan (out of Barnato's pocket). Carruth's task was already to salvage what he could. Experimental expenditure on the 4 Litre was, literally, peanuts at that date. The 4 Litre could have saved Bentley Motors. To attribute 20% of the blame for the failure of the Company to it is manifestly unfair. To attribute 10% to the Supercharged 4½ Litre is tenable.

It is important to remember the bias contained in WO's accounts, in his books. It is perhaps relevant to consider that WO's position was less secure than might be thought. That WO's position was not inviolable had already been shown by the appointments of Maury in 1927 as Joint Managing Director and then WO's and Maury's replacement by Carruth. That he might even have been dispensable was darkly hinted at by Barnato in 1946, in legal papers put together by Rolls-Royce for the Lagonda/Bentley case: "Mr. W.O. Bentley became Managing Director and afterwards in 1928 [in fact 1927], when things were not going happily Joint Managing Director with the Marquis de Casa Maury...If Mr. W.O. Bentley had ceased to be associated with the Company in any way the only difference that would have been made as far as the outside World was concerned would have been that his name would have been deleted from the name of Directors on the letter paper...It was some consideration which led to his [WO] services being retained by the Company in 1928 when there were differences of policy. It was never considered that his continued employment was essential to the production of the car or that the use of the name [Bentley] was in any way dependent on his continued association with the Company."

Frank Ayto's comments on WO are revealing. "WO might have appeared to lose his direction, but he had probably lost confidence in his fellow directors. He gave the impression of a feeling of – "what is happening?" WO had previously had the final say on anything to do with design, and was then out on a limb, in a position of no control." Diana Barnato-Walker is specific that Woolf Barnato and WO did not get on, although whether this was the case before 1931 or this came later is not clear.

WO's comment on 8 Litre chassis hanging around the Works gives an equally false impression. 100 8 Litre chassis were sanctioned, and 100 were built. The 4 Litre was designed and built in its own right, and all the evidence indicates that it was built to a mature design. It was no lash-up. It is simply that the aim of the 4 Litre was new – it was a town carriage. It was not a sports car. The error lay in public perceptions, grossly exaggerated by the sports car enthusiasts of the time, certainly inside the Company, and by the Bentley fraternity ever since. There was no viable market for a 4½ Litre type car in a progressively more sophisticated market. The 4½ and 6½ Litre cars were obsolete. And the sports car market was insufficient to support the Company as a viable manufacturing concern.

So how did the Board meet this challenge?

In late 1929, they conceived a six cylinder chassis with a modern engine design with quiet valve gear.

This would power a refined, town carriage type of chassis.

Subsequently, they aimed to develop the 6½ Litre into a luxury town carriage at the very top of the market.

The SV project, in fact, came first, and was neglected later. The latter project was easier, as the engine technology was readily available.

The considerations for a town carriage were:

1. Capable of carrying closed coachwork with a high level of comfort.

2. Silence.

3. Reasonable performance.

These considerations were tackled by producing an enormously rigid chassis with much more sophisticated suspension than the 4½ Litre model. All the considerations listed for the 8 Litre on p.315, in fact. Hence it is hardly surprising that the frame and suspension on the two models were so similar. Both the 8 and the 4 Litre frames were based on the same side-rails, to drawings BM7191/7192 and BM 7193/7194 (see p.315) varied in length by changing the length of the parallel portion of the frame. Indeed, although it has a numerically later drawing number (BM7701/7702), the 11'2" 4 Litre frame sidemembers were drawn six weeks before the 8 Litre sidemember drawings. Whether or not it was standard practice is not known, but certainly the 4 Litre frame was tested in the Erecting Shop under Norman Fawkes' (shop foreman) direction, by loading the frame and noting deflections. WO and Witchell were both in attendance at the time.

The engine had several advantages over the 4½ Litre unit. It was quieter, and had a pre-engaged starter, which is again quieter in operation. Driving the dynamo off the nearside of the engine eliminated a unit that, being positioned under the scuttle in the 4½ Litre, is a source of noise and heat in a closed car. The engine was mounted on rubber insulated sub-frames bolted into the chassis side-members, which were redesigned to give optional wheelbases of 11'2" or 11'8". The F type gearbox was used, or the G type (known as the modified F type), the latter basically an F type box with the casing made in Elektron. The F box was quieter than the earlier boxes, and easier to operate with dog clutches on top and third. The back axle used a 6½ Litre differential assembly in place of the 8 Litre hypoid unit. Doubtless the hypoid differential would have been fitted, but perhaps wasn't for economy reasons. The front brakes were operated by rods outside the chassis as on the 3 and 4½ Litres, as opposed to the internally positioned tubes used on the 6½ and 8 Litres. Tecalemit one-shot lubrication was used as on the 8 Litre, and Bosch electrical equipment with a combined starter motor/dynamo unit which also drove the water pump. The new steering was lower geared and required less effort.

All in all, the chassis was well-conceived as a town carriage. The fault was its weight, but although the weight of the chassis was widely published in descriptions at the time, not one of the motoring journals made any comment on it. They did, though, praise the strength and rigidity of the

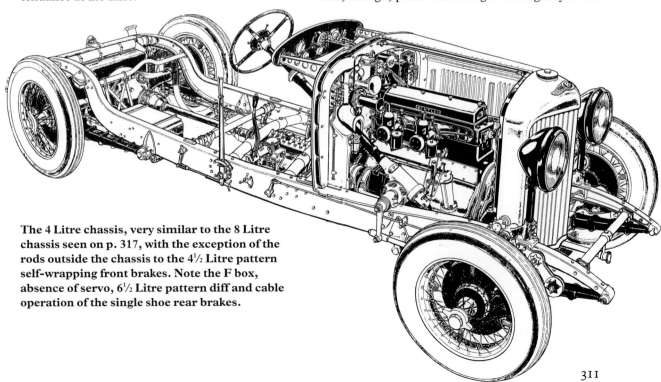

The 4 Litre chassis, very similar to the 8 Litre chassis seen on p. 317, with the exception of the rods outside the chassis to the 4½ Litre pattern self-wrapping front brakes. Note the F box, absence of servo, 6½ Litre pattern diff and cable operation of the single shoe rear brakes.

frame and the "supple springing, designed to combine comfort with roadholding" (*The Motor*, 19th May 1931).

In short a basically sound engine was fitted to a grossly overweight chassis intended for a car with twice the engine capacity and 80% more power. As the frontal area and rolling resistance of the cars were similar, and the only real weight saving the reduced length of chassis and body, it is no wonder that the performance in outright terms was somewhat uninspired for a Bentley. It was still as fast as a 20/25 HP Rolls-Royce and offered a good standard of refinement, but has never really been accepted as a Bentley, almost entirely because of prejudice surrounding its conception rather than an objective and balanced view of the car itself. Those who have driven a 4 Litre have often been known to express with some surprise the view that it is a very pleasant motor car, but that it doesn't really feel like a vintage Bentley.

The difficulty here is historical perspective. George Oliver expressed the problem succinctly in *Cars and Coachbuilding*: "H.I.F. Evernden, who served Rolls-Royce throughout the twenties as an exceptionally able and well-informed technical ambassador to their numerous accredited coachbuilders, has told of the difficulties then encountered. These were not of a mechanical nature, needless to say, but all to do with bodywork that simply shook itself apart in extreme examples, or gradually broke up under the stress of high-speed running for hours over the indifferent or poor road surfaces of favoured routes through France – to the Riviera, for example. Problems of this nature were not exclusive to Rolls-Royce; it just happens that their one-time body engineer took the trouble to note them down.

"Like the wheel tramp difficulties that all makers of large and speedy motor cars encountered once they began to fit low-pressure balloon tyres, these troubles were not talked about publicly at the time. Those intimately involved, however, could all tell tales of popping panel pins, doors flying open (or off!) at speed, and body frames that came apart at the joints. To some extent the bodybuilders were to blame...But the designers and makers of car chassis must also take their share of responsibility. Sound structures depend upon sound foundations, and these were not always provided for the bodybuilder."

Bentley Motors succeeded in producing a rigid chassis, but sacrificed weight in the pursuit of one virtue. "It was thought (incorrectly, as it turned out) that our customers would want a large spacious body similar to that supplied on the 8 Litre" (Rivers Fletcher, *Bentley – Past and Present*).

There exists a weight comparison made at the time, between the 4 Litre and the 4½ Litre. It is

Cutaway of the 4 Litre engine. Note the large driving gear on the front of the camshaft, mounted low down in the crankcase. The single distributor fires the six plugs.

Detail of the inlet valve.

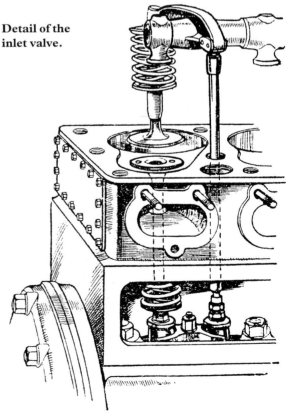

significant that the parts on the 4 Litre that are substantially heavier are those in which improvements (in terms of the design brief) had been made: engine, differential/back axle, gearbox, frame and suspension. The final result is that the 4 Litre was approximately 20% heavier than the 4½ Litre, with perhaps 10% more power. By comparison, the 20/25 HP Rolls-Royce chassis weighed approximately 2,700lbs to 3,600 lbs for the 4 Litre. The 4 Litre was still the faster car of the two, with comparable coachwork.

On the subject of weight, the power/weight figures for the Bentley chassis are interesting:

Model	Wheelbase	Weight cwt	BHP	BHP/ ton
3 Litre Speed Model	9'9½"	23	80	70
6½ Litre Standard	12'7¼"	33¼	143	86
6½ Litre Speed Six	11'8½"	32¾	180	110
4½ Litre standard	10'10"	27	110	81
4½ Litre semi Le Mans	10'10"	27	124	92
4½ Litre S/C	10'10"	28½	181	127
8 Litre	13'	36¼	220	121
4 Litre	11'2"	32	120	75

However, allowing for coachwork, the figures for the 4 Litre are even less favourable. For the same type of saloon body on an 11'2" 4 Litre chassis, or a 12' 8 Litre chassis, H.J. Mulliner quoted figures of 9½ cwt and 9½–10 cwts respectively in February

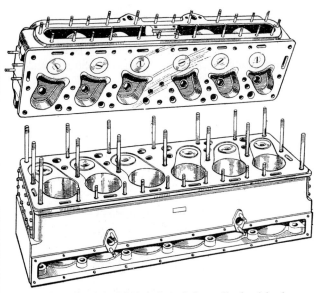

4 Litre detachable head and the cylinder block, showing the "F" head arrangement and the large exhaust valve.

1931. Coachwork on the 4 and 8 Litre chassis added 9 to 13 cwts to the chassis weight, depending on the type of body chosen, with effectively no difference between the smaller and the larger chassis. Hence 4 Litres weighed 41 to 45 cwts, to a maximum weight under guarantee of 36 cwts for a 4½ Litre with closed coachwork, widening the gap from 5 cwts between the chassis to 9 cwts for a saloon car.

The following parts (at least) were unique to the 4 Litre chassis:

Engine and most accessories (fan, water pump, dynamo and drive, inlet manifold, exhaust manifold, all engine mountings, distributor drive and HT conduits).
Radiator.
Headlamp bracket/Perrot bearing/radiator trunnion castings.
Bulkhead casting and toe-boards. Accelerator pedal and countershaft.
Steering column and mountings.
Chassis side-rails, engine front cross-member and gearbox front cross-member.
Front brake reversers.
Petrol tank.
All three crownwheel and pinion sets offered for the model.

It would be impossible to compile such a list if the 4 Litre had really been created in the way that is implied by WO and others. As Tom Threlfall put it in *Classic & Sportscar*, November 1989, after driving two 4 Litres: "...that is about as logical as blaming your Labrador for not being a Greyhound." It should perhaps be noted, though, that the 4 Litre steering box had already been designed and was intended for the Supercharged 4½ Litre chassis. This pattern of steering box is listed in the build sheets for the Supercharged 4½ litre, but all the production Blower chassis were fitted with the 4½ Litre steering column.

The first 4 Litre chassis were assembled in the Erecting Shop in February 1931. On the 4th March, the first production chassis VF4002 was delivered to the Running Shop, followed by VF4003 on the following day. VF4001 was built up as the Works Experimental car, but the first chassis to be finished and bodied was VF4004, which was passed off Final Test on the 14th May. Fitted with a lightweight H.J. Mulliner saloon body (despite which the car still weighed 2 tons 2 cwts) this car was used as a demonstrator, and would have been present at the Pollen House launch. For some reason, and supposedly on WO's instructions, the

The disappointment – the Bentley that wouldn't 'pull the skin off a rice pudding'. This is the showroom demonstrator, chassis No.VF4019 registered GP5193, fitted with the high-sided two door touring body by Vanden Plas and delivered to Bentleys in May, 1931. This car was lent to *The Autocar* for a road test, but was returned because of gearbox problems caused by a bearing breaking up. By the time it was repaired, the road test had become superfluous.

bonnets of all the 4 Litre chassis were sealed when they were sent to the coachbuilders, so that no-one should see the engine. On the 19th March, George Hawkins of Road Test was sent to H.J. Mulliner to seal up bonnets, which was done by wiring up the catches with locking wire and then fixing the lead seals used on engines and axles by the Works.

The chassis price was £1225, and Royce was quick to observe that there would be a fine splash and then disappointment, and that their customers would come back to them. In general the tone of the Rolls-Royce memos reads that their initial concerns over the introduction of the 4 Litre were unfounded, and there was a certain degree of relief on seeing the end product.

The rush feature of the launch of the 4 Litre, at a cocktail party at Pollen House on the 14th May, was the catalogues. The chassis and engine photographs were taken at the beginning of April, but no cars had been completed so coachwork drawings were included rather than photos. The catalogues were barely ready, and were delivered on the day of the launch, having to be sneaked in the back door halfway through. The first descriptions appeared in the press in mid-May, but no road tests were published. *The Autocar* borrowed a car fitted with high-sided four seat coachwork by Vanden

Plas, but one of the bearings in the gearbox failed. By the time it had been repaired, it was too late. However, the published descriptions were positively glowing. "Following upon the production of the Eight-litre it [Bentley Motors] has brought forth the Four-litre. Simply described, it is just a marvel. It has the characteristic Bentley performance – more than which no one in reason could desire – but over and above all that, it is sweetly smooth and quite incredibly noiseless. There are features of design in this vehicle which will, I expect, be the basis of much conjecture and a good deal of controversy. But of those, for the moment, I am not permitted to speak. This, however, I can assert, that it is the most logical motor car I have ever seen." Such was the 4 Litre heralded by *The Graphic* on the 9th May, 1931. One other paper's comments are interesting: "This occasion [the unveiling at Pollen House] was the new Bentley's first appearance in its public capacity. Up to that moment it had been one of the 'secrets' of the motor trade, as everybody had turned a blind eye to its testing runs on the road for the past year or so." Evidently motor manufacturers were not plagued by the "scoops" that so often feature in today's motoring press.

Inevitably, development work was still underway on the 4 Litre that never saw fruition. Sewell was looking at a proposed ENV axle and also at redesigning some of the gearbox internals, principally the constant mesh to reduce inertia. Teething problems with the 4 Litre seem to have been limited to serious steering wobble at 75 mph, which was cured by Nobby Clarke and Stan Ivermee by increasing the front end of the second leaf in the front road springs by $\frac{3}{8}$". More serious faults in later life encountered by private owners have been premature big-end failures at high speeds, believed to be due to undamped crankshaft vibrations, and cracking of the cylinder bores through to the exhaust ports due to thermal stress. It is possible that Bentleys were aware of problems in this area, as the plain exhaust manifold fitted to the first forty engines was changed to a finned manifold to improve heat dissipation from the cylinder block.

The drawings tell of the problems with the engine that were not properly ironed out – the cylinder block, drawn October 1929, was at issue 9 by January 1931. The crankcase, BM7265, was drawn in January 1930, modified June 1930, and was at issue 11 by December 1930. This crankcase was only used on the first 40 engines and it was redrawn in October 1930 as BM7860 for subsequent engines. The new October 1930 drawing was up to issue 11 by May 1931. It is evident that there was substantial design work in hand on the engine well after it should have been finished and in series production. With this sort of design commitment, it is highly likely that the car had a lousy reputation within the Company well before it was announced to the public. That reputation has been perpetuated by WO himself and some of the Bentley staff. This sort of design work would, of course, have caused endless headaches to Buying, the Machine Shop, Stores, etc, etc.

It is interesting to speculate whether the 4 Litre would ever have gained public approval. It's introduction within weeks of the demise of the Company virtually ensured that nobody would have time to decide whether or not they wanted one before the old firm went under. The press reactions tend to suggest that compared with its contemporaries the 4 Litre was probably a good car, providing one does not make unfair comparisons with the vastly more expensive 8 Litre or the overtly sporting character of the 4½ Litre.

There was no doubt that the situation was desperate, WO commenting "by early summer [1931] the writing was on the wall as clear as a Guinness poster, and despair in all the departments." The Drawing Office had already run down, and even on the shop floor there was some level of awareness of the problems. At Service, the workforce was kept busy on such jobs as cleaning up the fabricated 8 and 4 Litre chassis cross-members, brazed up at the Works. The surprise was the suddenness with which it happened. On Friday 5th June, the news finally broke and the Works effectively closed down. Most of the employees were given their cards, with no notice. George Hawkins noted in his diary "Bentleys close down – out of work again. Rotten wet afternoon makes a black Friday."

Developing the 8 Litre from the 6½ Litre.

As has been discussed in Chapter 16, Bentleys had a very definite model policy for 1931. It is clear that the 1931 chassis was to be substantially updated from the 1930 model Speed Six, with the F type gearbox and hypoid back axle. The frame was designed very early in 1930, the siderails being BM7191/7192 for offside and nearside members respectively for an 11'9" wheelbase chassis and BM7193/7194 o/s and n/s respectively for a 12'9" chassis. These were drawn by Sewell on the 2nd January 1930, but were not issued. The production wheelbases of 12' and 13' were arrived at later, on drawings BM7251/7252 (13') and BM7253/7254 (12'). These were for "1931 S" (6½ Litre code) and were drawn on 19th March 1930. The 1931 "S" model would have had the 6½ Litre engine if the 8 Litre engine had not been ready.

It is important to realize that the 8 Litre is a development of the 6½ Litre, representing both Bentley's increased knowledge and experience gained from the 6½ Litre and the technical and marketing pressures of the time. The changes to the chassis frame came about because of pressure from coachbuilders and customers, and higher expectations of comfort and silence in bigger cars. The changes in the transmission reflect pressures to produce a quieter car that required less effort and skill to drive. The chassis was considerably updated from the 6½ Litre: the frame was basically similar, but the pressed gearbox cross-members were replaced by tubular ones and a new tubular member fitted between the dumb-irons bolted into flanges in the channel to stiffen up the front end. The frame was double-dropped at the front and back to reduce the floor height and hence the overall height of the car, but the body was still built on top of the chassis frame – deleting the upward cross-member fitted to the 6½ Litre chassis assisted in dropping the floor height. Springing was by long semi-elliptics front and rear, the rear springs outrigged from the frame which allowed a steeper kick-up of the frame over the back axle, with the kick-up positioned much farther back. Bentley and Draper shock absorbers were fitted all round, friction to the front and hydraulic to the back.

The idea was to ease the coachbuilder's task by giving him a large, flat area of chassis and making it possible to seat the rear passengers in front of the back axle. However, the coachbuilder's task was made much more difficult by the enormously high radiator. Many of the luxury cars of the early Thirties were distinguished by their monstrously tall

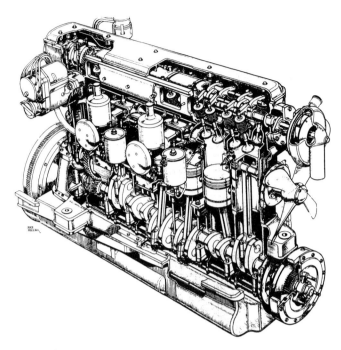

Cut-away of the 8 Litre engine, showing how similar it was to the 6½ Litre. The sump capacity had been increased from 3½ to 5 gallons by barrelling the sides.

radiators and pillar-box slit windscreens to reduce the roof height. It also gave the coachbuilders enormous problems in trying to disguise the vast areas of flat door panels caused by the depth of the body sides. Study of photographs of 8 Litre cars show that virtually all the closed cars had waist-line mouldings in order to reduce this apparent depth of door. Bosch electrics were fitted, the radiator incorporating thermostatically controlled shutters that were balanced to resist the effects of air pressure at high speeds.

Steering by worm and wheel was basically as the 6½ Litre, but with a slightly lower ratio and the column stiffened up to reduce whip. The steering column was redesigned, reflecting problems with the 6½ Litre design. On driving chassis KD 2112, a 1928 Standard Six fitted with Hooper limousine coachwork, Rolls-Royce severely criticised the amount of flex in the steering column that could be felt by the driver. The 6½ Litre steering column does indeed have a very long unsupported length above the bulkhead mounting (see p.280). The new steering box used the same casting as the 6½ Litre, BM5519, seen at p.142. The 8 Litre box was made by putting a 3 mm wide cutter through the parallel portion of the casting, machining the ends of the two pieces, and then sliding a length of steel tube over

each bit. The three parts were then held by a pair of split ring clamps at each end, with a bolt dowelling through each clamp, the tube and the Elektron pieces. This considerably stiffened the steering column, further helped by fitting the top mounting to the bulkhead casting nearer the steering wheel.

The engine differed only in detail from the 6½ Litre, with ignition by one Bosch magneto and a Delco-Remy coil. A new camshaft was fitted, BM7130, with performance characteristics midway between the standard 6½ Litre and Le Mans Speed Six shafts. The crankcase and other parts were cast in Elektron, a magnesium alloy, to reduce weight. The lightness of the new material was emphasized by publicity photographs showing a man holding an Elektron camchest on the end of a spring balance – a feat he would have been pushed to achieve with an aluminium one! The crankcase casting was stiffened up after the first 50 engines to prevent creeping cracks. Twin SU HO8 carburettors were fitted, now back on offside where they should have been on the 6½ Litre. Apart from simplifying the controls this arrangement prevented heat from the exhaust system coming back through the pedal slots and roasting the driver's feet.

The biggest changes from earlier chassis were in the transmission area, with a completely redesigned gearbox and back axle unit. The gearbox was the F box, which was radically different from any previous Bentley box in that the casing was made in two halves split vertically down the centre-line of the car. This enabled the fitting of a bearing between virtually every gear, stiffening up the whole assembly and reducing noise. Top and third gears were engaged by dog clutches, and the whole was mounted at three points from a pair of tubular cross-members. The reverse catch was tidied up to a button on top of the lever rather than a rod down the side. For some reason there is no record of an E type box, the series going A, B, BS, C, D, F, G – only a hint in Clutton and Stanford's *The Vintage Motor Car* to the effect that the E box was a still-born experimental affair. The F box was designed under Barrington's direction, and was a massively heavy affair intended to be as silent as possible with a distortion-free casing. The F and G boxes were approximately 40% heavier than the earlier designs.

The input gear of the hypoid bevel pair was offset 2″ below the centre line of the back axle, making it possible to reduce the floor height and hence the overall height of the car. As far as is known, the first hypoid back axle was fitted to 6½ Litre chassis MD 2469 in mid 1928: "Hypoid rear axle fitted with special Spicer shaft ratio 15/47 [4.07:1]." This back axle was very quickly removed and fitted to chassis FA 2520, another demonatrator. A second hypoid back axle was fitted to chassis BR 2371, a 1928 saloon belonging to Barnato, on 3rd November 1928. This axle was again a 15/47 ratio. It was not run for very long, as the Works refitted a standard back axle to this chassis on the 26th January 1929. These hypoid gears caused problems initially, with the first three sets scoring after less than 2000 miles. Experts were called in from ENV to investigate. This problem was aggravated by inadequate quantities of lubricant in the casing. The ENV representative put in more than double, without overflow onto the rear shoes, which Witchell was very much afraid of. With only the acme throwers on the halfshafts to prevent oil from the differential getting into the drums, Witchell's concern was more than understandable. This problem is less prevalent with the baffled banjos used with the later cars than with the unbaffled banjos of the early 3 litres. The problem of the gears scoring was reduced by improved hypoid oils. It was found that after setting up the gears on the bench they had to be checked again in the chassis to ensure correct meshing. Of the first three casings, two were malleable, one aluminium.

On the production cars, malleable casings were used on the first 25 chassis and then aluminium alloy (RR 50, or Hiduminium, newly evolved by Rolls-Royce metallurgists at Derby for the Schneider Trophy engines) on the remainder. ENV's representative said that "they are very anxious to make a success of the hypoid job." A lot of effort was expended on the hypoid back axle, the design of which had been worked on since early in 1928. It is evident from the "S" and "1931 Sports" codes on the drawings (see drawing E3266/ BM7181, p.73 and 81) that had 6½ Litre production continued in 1931 it would have been fitted with the hypoid unit. The testing procedure for

the 8 Litre chassis was varied from the earlier chassis, in that a closed saloon body, with four seats but no trim, was fitted to the chassis in the Running Shop to check for gearbox and axle noise. The body was an unpainted, aluminium shell fitted with a flat vertical screen and two doors. Hawkins took out 8 Litre chassis YM 5027 on 3rd March 1931 with this body, observing that "axle sounds bad with body." Obviously, testing a chassis with no coachwork fails to show up transmission noise at the speeds attainable on the 8 Litre chassis.

The rear brakes were operated by cables, from the compensator to an idler lever so that the hand and footbrakes operated the same, single pair of shoes. These shoes were wider than the previous shoes, because of the need for four instead of eight with separate foot/handbrake shoes. For the first time the Tecalemit "one-shot" lubrication system was fitted, consisting of a reservoir fitted to the bulkhead with pipework to shackle pins, king pins and other lubrication points. This was operated by a pedal high up under the dashboard, one push lubricating all the chassis fittings and eliminating the need to go round with an oil gun every couple of months – the pedal was supposed to be operated every day that the car was driven, and providing all the oilways were kept clean proved satisfactory. Rolls-Royce used the same system on their cars.

The massive 8 Litre chassis. It seems rather strange to go to a hypoid differential to lower the transmission line and then put a tubular crossmember over the Spicer shaft – it would seem more sensible to have put it underneath. The double drop of the frame can be seen, as well as the F box and the Dewandre servo to the right of the gear level gate. Note the outrigged rear springs, and the front shackling of the front springs. This was very soon changed to shackling at the rear.

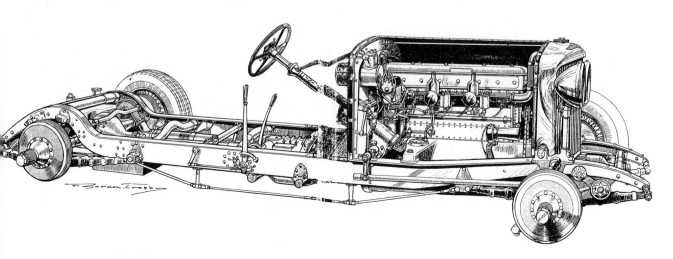

18 Out of the Frying Pan and into the Fire

WITH WO REMOVED from the senior position, there was to be a twist in the tail. On Tuesday 9th June 1931 Carruth telephoned Sidgreaves, also a financial man and the Managing Director of Rolls-Royce, with a view to an amalgamation of the two companies. Carruth wrote to Sidgreaves on the 10th, setting out the position in detail, marking his letter "Strictly Private and Confidential".

"Dear Mr. Sidgreaves,

"Since my telephone conversation with you on Tuesday morning, I have felt the value of our meeting would be considerably enhanced if you had some indication of the nature of the business which I wish to discuss with you. It is so much easier to put a long story in logical sequence in black and white, and I feel it would be only right and fair for you to have at least a little time to consider the position beforehand, and to know what prompted me to make an appointment with you.

"As Captain Barnato's Personal Business Manager I was nominated by him last December as Managing Director of Bentley Motors for the purpose of improving the commercial organisation there, and managing the financial side of the business.

"We had produced the 8 Litre which has made its mark as a particularly fine motor car, and one of the problems was to keep the Works going satisfactorily while the new 4 Litre, which has been recently announced, was being prepared and progressed. This was done, and our buying was put on a cash basis at a substantial discount.

"Frankly, the sales of the 4 Litre have not been up to expectations, although the car is probably the finest piece of mechanism on the road. I would go so far as to say that it is about 98% perfect but it must be admitted that the high "revving" engine with which it is fitted is sluggish at low revolutions and it would undoubtedly require a strong sales effort in these depressed times to overcome this at its present price. In view of this, and entirely on my own responsibility, production at the Works was materially reduced last Friday, and I have since been giving the most earnest consideration to the future before calling my Board together for a decision.

"The finished chassis available at the Works amount to approximately 100 in the proportion of two 8 Litres to one 4 Litre.

"Captain Barnato is a very determined man on any matter which affects his personal prestige, and he is well provided with means to give effect to any of his desires. He may say that the large quantity of material purchased but not yet erected has to be completed and put on the market at reduced prices, so that under cover of the realisation of stock this slightly unfavourable feature of the 4 Litre can be put right. That, I think, is the view likely to be taken up by him.

"My own experience and temperament would probably lead me to an entirely different decision.

"When I am faced with a position of this nature I like to cut my loss, get rid of any consequential liabilities which may be about, and if expedient, make a fresh start. This would involve in the present circumstances of Bentley Motors the appointment of a Receiver, who might adopt the same tactics outlined above, but who would in any case be compelled to realise his finished stock.

"In considering this position it would appear that either contingency must have a very great effect on Rolls-Royce, and as the total trade creditors of Bentley Motors are only about £15,000 with, possibly, contingent liabilities for unexpired contracts, which could be settled, in my opinion, for not more than £20,000, it seemed to me that ordinary business prudence demanded that details of this position should be disclosed to you before anything further is done.

"One can imagine the immense benefit which would accrue from a working arrangement between the two companies. Say, for the sake of argument, among others:

"Firstly, a common sales policy

"Secondly, a common service policy

"Thirdly, Derby could probably manufacture at least two-thirds of the components which Bentley Motors at present buy outside, say £200,000 per annum, which must, of course, be a very useful addition to turnover there.

"Fourthly, Rolls-Royce having a large say in the future of the Company, it need only be a question of time until the models of each company cease to compete, and are definitely designed to cater for different aspects of the luxury market.

"Fifthly, Mr Bentley is a designer of the first rank. He is a young man who originally became eminent as a designer of aeroplane engines, and I have no doubt at all that Rolls-Royce would find his co-operation extremely useful at the present time.

"The advantage to Captain Barnato, with whose interests I am primarily concerned, is that the Company on these lines should go onto success, and any question of his own personal prestige would be obviated.

"Apart from the small trade liabilities mentioned above, there is a secured mortgage on the Company of £105,000 which is amply covered. The rest of the loan and share capital is controlled by Captain Barnato, who is prepared to go a very long way to maintain the name of Bentley, and I personally feel it would be little short of a national tragedy if such a name were to come under a cloud.

"If the foregoing is of interest I would suggest that each Company should nominate two Directors as a small committee to carry on preliminary negotiations, with a view to submitting their recommendations to both Boards for their consideration and decision.

Yours very truly,
For and on behalf of BENTLEY MOTORS LIMITED,

J.K. Carruth, Managing Director."

That Carruth was proposing the continued existence of Bentley Motors as a separate identity could hardly have appealed to the Rolls-Royce Board, but nevertheless they discussed the proposal on the 16th June.

On the 15th June, the monthly interest payment to The London Life Association became due. This had to be paid within 15 days. The Company had no money to pay them, and Barnato declined to do so himself. WO said: "It was late in June, 1931, when I heard that Barnato wasn't prepared to put up any more money, and I realised what must inevitably follow."

On the 17th June Sidgreaves informed Carruth that they were interested, but needed more information. This was given as follows, in a letter on the 18th:

"First mortgage (London Life) £65,000
Second mortgage (Capt. Barnato) £35,000
Owed on supplies delivered £19,700
Orders placed £92,000
Overdraft £91,500

"Sales of 8 Litre........6/month since introduction (10/30)
"Sales of 4 Litre.......19/month since introduction (5/31)

"*Stocks.*
"53 8 Litres, including 4 Company cars and 36 not allocated
"32 4 Litres, including 4 Company cars and 21 not allocated

"*Costs.*
"8 Litre chassis – £935
"4 Litre chassis – £761

"*Suppliers' accounts.*
"Goods invoiced – £17.629.18. Bills payable – £2,128.3.10.
"Orders placed for £92,000...I think....all these contracts could be disposed of for a sum not exceeding £20,000. This is taking a superficial view that one would simply cancel all these contracts.

"*Staff Contracts.*
"Mr W.O. Bentley is under contract as Chief Engineer of the Company for life so long as royalties amounting to not less than £1,000 per annum are paid to him.
"Col. Barrington has a three year contract from 1st November 1929 at £1,500 per annum.

"In the case of Mr Grey and Mr Johnson who joined us from your staff a short time ago, it was agreed that they should be given a minimum of three months' notice."

The overdraft was guaranteed by Barnato and Baromans. Carruth's estimate that the £92,000 could be reduced to £20,000 fairly easily would have been achieved simply by cancelling all the orders on the last 50 of the sanction of 100 4 Litre chassis. Sales of 6/month for the 8 Litre are realistic – WO said they sold 63 before the bankruptcy, so assuming we can work from the 1930 Olympia Show in October, 6/month equates to 55/60 cars by mid-June. Sales of 19/month for the 4 Litre were well below Carruth's production target of 7 per week.

Captain Turner of Rolls-Royce examined the balance sheet, which presented a sorry spectacle;

Liabilities	£477,939

Consisting off:

Issued capital	£171,320
Loans	£282,186
Sundry liabilities	£24,433
Assets	£316,501

Consisting of:

Fixed assets	£98,827
Current assets (stocks and work in progress)	£217,674
Balance	£158,919

Consisting of:

Losses	£92,000
Goodwill and experimental expenses	£47,000
Expenses/underwriting charges	£19,500

Hence there was good reason for believing that sale of the assets would not even pay off the outstanding loans. It was considered that the current assets were practically valueless (which proved more or less correct – those notional assets of £217,674 translated into not much more than £40,000 in hard cash terms, assuming the £98,827 of fixed assets were realistically valued).

On the 23rd June the Rolls-Royce board decided against the proposal, and on the 24th Sidgreaves wrote to Carruth to convey this information. The gist was that it was considered that £40,000 was inadequate, and that Bentley Motors would be a drain on Rolls-Royce's resources. As Sidgreaves said,

"The main point however is that assuming it did [ie that Carruth's estimate of liability worked out correctly] the most favourable aspect is that £42,000 is required forthwith to meet these two liabilities. This point is important because, you will remember that in our conversation you mentioned that if we were prepared to put up a sum of between £30,000 and £40,000 you thought we could be given control. In our view however such a sum would only be a temporary palliative, as apart from the liquidation of stocks, there would be no other cash available for current expenses."

The two liabilities referred to were £22,000 to pay invoices and interest owed to Barnato on his £35,000 mortgage, and £20,000 to terminate existing contracts. Rolls-Royce's overwhelming concern was that either of the two mortgage holders could call in their mortgages, in which case, even if Rolls-Royce's loan was secured third in line, the money would be lost. Their estimate would have been realistic: had they proceeded and events carried on, they would have got back less than £12,000.

Again we come back to the old problem – more cash flow had to be generated, which had to come from selling cars – and there was no imminent danger of a great rush of such activity. Sidgreaves' letter of the 24th ended the correspondence between Rolls-Royce and Carruth. In fact, they were in the same position they had been in late 1925 – but this time there was no fairy godmother to bail them out. Rolls-Royce were indeed worried about the 8 Litre as a direct competitor (and were to be even more concerned about the proposed 6¼ Litre Napier-Bentley backed by all the engineering and financial backing of an organisation the size of Napiers), and must have been pleased to find that a close rival was obviously no longer in a position to compete. This aspect of the whole liquidation of Bentley Motors and subsequent take-over by Rolls-Royce seems to have remained secret for many years. Carruth was presumably acting off his own bat, probably without consulting the Board. It is highly likely that only Barnato himself would have known what was going on, and highly likely that Barnato supported the move as a painless way of discharging himself with minimum loss, as the Company, as a going concern, would have been worth considerably more than in liquidation. That Barnato wanted to divest himself of Bentleys is clearly shown by the subsequent events, the failure to sell off the Company as a going concern, forcing the liquidation and increasing Barnato's (and everybody else's) losses.

The Le Mans 24-Hour Race had come round again on the 13th/14th June and for the first year there were no Works or semi-works Bentley entries. The foreign teams turned out in force, Mercedes and Alfa-Romeo among them. Barnato was there in an 8 Litre to close the road, a privilege awarded to him because of his win in 1930. Birkin won in a 2.4 litre Alfa-Romeo partnered by Lord Howe, and was less than amused to receive a telegram from Mussolini congratulating them on their win "for Italy". Birkin's partner Couper entered Birkin's old 4½ Litre YV 7263 driving with Bevan, but the Bentley broke its crank after 29 laps and had to retire. A Supercharged 4½ Litre was entered in 1932 and 1933, the No 4 Team car UR 6571 driven by Trevoux and Mary in 1932 and Trevoux and Gas in 1933, but it was not until 1948 that a Bentley was to finish at Le Mans again and that was a Rolls-Royce built car, the 4¼ Litre Derby Bentley owned and driven by Soltau Hay.

The financial position of the Company had deteriorated to such a point that something had to give. With no imminent change in the Company's fortunes and Rolls-Royce having rejected Carruth's proposals, the only real option open to Barnato was to go into voluntary liquidation and minimise his losses. The Company could then be sold, with Barnato disengaged from his liabilities. The only alternative open was to continue to put money into the firm in the hope that the Depression would end in the near future, and sales of both the 8 and 4 Litres pick up. It would have needed an optimist to back the latter option, with the 4 Litre a bit of a dead duck, and sales were very slow on the 8 Litre, albeit at a large profit on each one that found a customer.

The 8 Litre chassis cost £935 to build, and sold at £1850. The 4 Litre cost £761 or £830 to build, selling at £1225 (the figures of £761 and £830 come from different, equally well-informed sources). Both of these were subject to trade discounts ranging from 15% to 22½%. Perhaps adequate cash flow could have been generated by ruthless cutting down in all areas until the depression passed, but it is arguable that Barnato no longer had any reason to sustain the Company. He had retired from racing after the 1930 Le Mans and never raced again. The Company had not been and did not look likely to be a commercial success in the foreseeable future, so there was no reason for him to maintain his support.

In short, accepting WO's opinion of Barnato as a shrewd businessman quick on the uptake and quite ruthless in action, continued association with Bentley Motors could only do Barnato's reputation harm. It has been said that Barnato was beginning to wish that he had been less publicly associated with Bentleys, because their failure was a slur on his reputation as a shrewd investor. Despite assertions to the contrary, Barnato was in fact in England at the time that Bentleys went into receivership, and issued statements to the press explaining his position. Barnato appointed Carruth as his Receiver, with effect from 12am on Friday 10th July. In the morning of that day WO rang up Nigel Holder and told him to come over right away and collect his 8 Litre chassis, saying that events were imminent that he could not discuss over the telephone. Longman informed the sales staff that the company was going into receivership, and would stop trading. Moir leaned towards Rivers Fletcher. "Stop trading? They never started, Rivers, never started!" J.C. Neville of Sales drove a 4½ Litre laden with goods and papers out of the basement at Cork Street to Oxgate Lane that day, just as the receiver entered the showrooms above.

Faced with Bentley's inability to meet the interest payment by 30th June, The London Life Association had no option but to call in the mortgages for £65,000. Under the original terms, the interest and the principal had to be called in. The Company notified The London Life Association that they could not pay the wages, due on the 10th July and amounting to roughly £1,000. Taken in conjunction with the default on the interest payment, an application was made in the Chancery Division by The London Life that a receiver be appointed.

WO says that "on the 11th July a news item on the City pages of *The Times* announced that The London Life Association had applied for the receiver, the sums of £40,000 and £25,000 being due under two mortgages. Captain Woolf Barnato, it stated, was unable to meet these debts." WO's account is misleading – it was stated in court and reported by *The Times* that "the Company, which had been financed by Captain Barnato, no longer had that source of income." Patrick Roper Frere of Frere Brown & Co., 12 Old Square, Lincoln's Inn, WC2, was appointed Receiver for The London Life Association on the 10th July and we enter the last phase of the original Company's existence during which the Company was wound up and the assets disposed of. Frere was given leave by the Court to borrow a further £1,500 from The London Life to pay the wages, to keep the Company going.

Barnato explained his decision to *The Star* on the 10th July. "People can no longer afford to buy expensive motor cars. Motoring is either a necessity or a luxury – there is now a limited public who can afford luxuries. If it is a matter of getting from one place to another, third class gets you there as well as first class.

Patrick Frere, appointed as receiver and manager by the High Court for The London Life Association. The man that WO wished they had had as managing director in place of Carruth. Frere effectively directed Bentley Motors from 11th July until 17th November 1931, when the sale to Rolls-Royce was finalised.

"The other joy of motoring is the joy of ownership – but the present depression is no good to an industry producing luxury cars of this type.

"I understand that the application to appoint a receiver will be made in the courts today. In my opinion, this was the fairest thing to do in view of the gravity of the situation. Rumours have been rife ever since we stopped production, and it is just as well that the position should be made clear. There is no doubt but that since these rumours started we have lost business which we would otherwise have had, and people have been getting anxious about the service part of the business. I can say definitely that the service department, which does the repairs for owners of Bentley cars, is being fully maintained. It is obviously one of the most valuable assets of the Company. I imagine that all the employees in this department will be kept on.

"I feel very keenly disappointed, as you may imagine. I have put a tremendous lot of work into Bentley Motors – in fact, it has been my 'baby'! And it seems a shame that this situation should have arisen just at a time when we were producing the finest car ever made – the 8 Litre.

"It remains to be seen what action the receivers will take. It is felt, of course, that in the present circumstances it will be far better for the business to be carried on by a receiver, who will have the fullest opportunities of investigating the situation and finding the exact position.

"If the position is entirely uneconomical, I suppose the business will have to close down. But I cannot imagine that such a contingency will arise, and I sincerely hope they will find it possible to carry on. But no mistake, the service department will be carried on full blast in any case."

Carruth looked after Barnato's interests as the holder of the second mortgage on the Company's assets. It is important to understand from the above that the finances were such that Barnato had only to refuse to put up more money at any time, and that by the 30th of the month The London Life would be forced to put the Company into receivership. That, of course, is precisely what happened. It is also important to realise that the oft-quoted assertion that Barnato was in America at the time of the appointment of the receiver is not true. "Mr. Gossip" in his column in the *Daily Sketch* of 29th July reported attending a dinner thrown by Barnato at Ardenrun, concluding his report with "The host told me he is going to America this week for some months." (See also p.332.)

WO later wished that they could have had Frere as Managing Director rather than Barnato's crony Carruth! It is understandable that WO should have resented Carruth when he had supplanted WO as Managing Director of his own company and then been appointed to wind it up, but Carruth was employed by Barnato to increase his wealth and look after his interests, not sustain a failing business without good reason. As a final and generous gesture of goodwill, Barnato paid all the staff a month's salary. Frere presumably liked Bentleys, as he bought himself a secondhand 3 Litre, chassis 755, fitted with a Harrington touring body, in 1932. Frere had, in fact, been friends with HM and Hardy Bentley for some years before, so it is hardly suprising that they should all have got on so well.

It has been suggested that Barnato wanted to

maintain the Bentley name. Carruth stated quite strongly in his correspondence with Sidgreaves that that was the case. It has been suggested that Barnato intended to let the Company go into liquidation and then buy the assets himself. This would enable him to wipe out all the Company's debts and liabilities (which amounted to more than the assets) and carry on as before. He would still have lost about £200,000, but for maybe £105,000 (Napier's final bid was for £104,775) he could have bought a Company that proved to be worth £148,000, and carried on as if nothing had happened. Taking this line for a moment, had Barnato paid £105,000 for the Company, adding his losses of about £200,000 he would have paid £305,000 for a company worth £148,000. His notional loss would then have been £157,000. This would not, however, have alleviated any of the Company's financial difficulties, and the author doubts very much that such a course of action was ever seriously considered.

Frere continued to run the business on a much tighter basis. Kingsbury was kept open, making enough money to more or less run the greatly reduced operation. Frere managed to sell three 8 Litres and one 4 Litre between July and November. Witchell and Conway retained enough staff to finish off a number of incomplete chassis. The Design

staff were also kept on, although as WO later admitted, it would have been better if they had laid off virtually all the staff and turned out the odd 8 Litre to order (before the events of June/July 1931). WO's feelings concerning his workforce and his obvious enjoyment of wandering around the Works during difficult times, where he could be surrounded by people who understood and respected him are well known, but were not compatible with the harsh economic reality. It is difficult to find a better summary of the economic climate of the times than that of C.H. Wilson and W.J. Reader in *Men and Machines – D. Napier and Son 1808-1958*, as applicable to Bentleys as to Napiers:

"In circumstances where there appeared, as yet, no prospect of halting or increasing the steady decline of markets, the only answer for public and private authorities seemed to be retrenchment. In practice it meant that, if a business was to survive at all, attention must be concentrated on making ends meet, and if the income end could not be pulled nearer to costs, the costs must be made to meet diminished income. For Napiers, as for other firms similarly situated, financial management was paramount and had to be given priority over purely technical considerations."

It is abundantly clear that Bentley Motors did nothing of the sort.

19 An Arranged Marriage

BY THE 17TH JULY, Frere had assessed the position of the Company and reported to the court the following:

"Book debts, estimated to realise £12,000
Freehold land and buildings at
Oxgate Lane, estimated. £45,000
Plant, machinery, equipment, patterns, dies, jigs and tools; book value per Balance Sheet to 31st May 1931 ....£42,300
Cars, completed chassis, work in progress and raw materials; as valued by the Managing Director [Carruth] as a going concern £167,000
Two Endowment Policies (London Life) on the life of W.O. Bentley, and one Endowment Policy (London Life)on the life of H.M. Bentley – Surrender Values £16,000
Leasehold premises, Pollen House (part), Cork Street.
Rent £1,700 per annum expires 1941.
This appears to be a liability 0
Service Dept., Kingsbury La., N.W.9
Rent £750 per annum, expires 1944;
This appears to be of no value 0
Service Dept., Glasgow, Leasehold
This appears to be of no value 0
Total ....£282,300"

Frere needed to find a company that would take over the Bentley concern and a suitable suitor soon appeared in the form of D. Napier and Son Ltd. of Acton. During the years 1908-11 Napiers had produced probably "The Best Car in the World" before that slogan passed to their rivals Rolls-Royce, in no small measure due to the latter's aggressive marketing principles and their one-model policy enshrined in the famous Silver Ghost.

In the early post-war years Napiers had re-entered the car industry with the T75, but Montague Napier had already determined that the firm would specialise in aero-engines with the "Lion" engine and that car manufacturing would be short-term. It certainly proved to be that, largely helped by a series of errors concerning the production of the T75 which delayed launch until the post-war boom had died down, and the first slump set in. WO's opinion of the T75 was that it was "an overhead camshaft six-cylinder machine of excellent design, great weight and indifferent performance." Montague Napier died in 1930, leaving the firm to be run by a board of directors who had a lot of ready cash but nothing to manufacture with it. The "Lion" engine had reached more or less the end of its life and their next generation of aero-engines would not be ready for some years, so alternatives were examined by the board to find short-term projects to occupy the works and generate money. One such suggestion was to go back into the motor industry once more, and the availability of Bentley Motors (and more importantly the services of WO himself) must have seemed an ideal solution.

An intriguing, but unfortunately unsubstantiated, aspect of the Napier deal was suggested by Harold Evernden of Rolls-Royce, then on Royce's personal design staff at West Wittering and Le Canadel. Apparently "the Government was urging Napier to buy Bentley and go back into car production, because Napier's aero engines were by no means obsolescent and it was costing the Government a lot of money to keep Napier going for defence reasons." (*Motor*, 26th May 1979.)

The Financial Times of 28th July announced that

Napiers were interested in buying Bentley, and then in *The Autocar* of 14th August 1931, a news item appeared to the effect that although negotiations between Bentley Motors and Napier had been proceeding for some weeks and were at an advanced stage, nothing definite had been agreed. The London Life were keen to sell, but things were not that simple. Arthur William Evans, The London Life's assistant manager, had applied for Bentleys to be sold on the 24th July, but no action was forthcoming. Napier's approach to Bentleys resulted in an offer of £84,000 for the whole concern. It cannot be a coincidence that H.M. Bentley was a personal friend of H.T. Vane, the chairman and managing director of Napiers. Several meetings were held at 3, Hanover Court, during this period between HM, WO, Vane and Frere, at which the details of the takeover must have been hammered out.

With the Napier negotiations at an advanced stage, Frere gave WO leave to work on a new 6¼ Litre Napier-Bentley. Most of the design staff were kept on at Cricklewood for the purpose. Frere continued to pay WO's royalty of £1,000 per annum, thus keeping him on as Chief Engineer and maintaining the terms of the March 1926 agreement binding WO to Bentley Motors for life. This agreement included "provisions for the transfer of the employment to the Successors in business of Bentley Motors Limited who should for the time being be entitled to the designs and patents and any future designs and patents as referred to in the said Agreement." In other words, although WO referred to it as a "Napier-Bentley", the designs would belong to whoever bought Bentley Motors, as long as Frere maintained the agreement by continuing to pay WO.

This proposed 6¼ Litre car was basically an improved and updated 8 Litre, with a bore and stroke of 110 x 110 mm, and a new and stiffer crankshaft to eliminate engine periods that could be felt on the 8 Litre. WO must have been well aware that the shaft of the 4 Litre was so rigid that it did not require a damper. The camshaft drive was also to be altered to take up less space (one could speculate that the three-throw drive could have been replaced by a single drive with one connecting rod), the whole fitted to a lower and lighter frame. Plans also included the possibility of a new aero engine.

With the negotiations well under way, an Extraordinary General Meeting was held at Oxgate Lane on the 9th September, at which the resolution was passed "That the Company by reasons of its liability cannot conduct its business and that it is advisable to wind up the same, and that the Company be wound up voluntarily." Robert Montgomerie, the Company Secretary, was appointed as the liquidator. The arrangements were published in a statement in *Motor Sport* November 1931, which recounted that negotiations between Bentleys and Napiers had been concluded following Barnato's return from America. Barnato was to be connected with the Company, as was WO, with production of the 8 Litre to be continued and "a new 3½ Litre [sic] car is said to form part of the new programme." The Cricklewood works were completely closed down except for WO's office, from which he was still working – the Cork Street showrooms were shut as well, but the Service Department remained operational.

On the 20th October Mr. Justice Maugham approved the sale of the Company. A Conditional Contract of Sale was drawn up and approved by Napiers, Frere and The London Life, dated 21st October 1931, valuing the Company at £84,000. All that remained was for the court's approval to be given. By this date the negotiations had gone on for so long that on the 28th October, the court extended Frere's management of the Company up to the 1st February 1932. Indeed, one has to ask why these negotiations took as long as they did. Perhaps Napiers were indeed being pushed reluctantly into the deal.

Although The London Life were keen to see the sale go through, as there would be enough to pay them off, there was evidently some feeling in the Bentley camp that Napier's offer was distinctly ungenerous, and in particular that the offer should be higher, to reduce Barnato's liabilities somewhat. On the 4th November, Bentleys submitted their own valuation, conducted by H.G. Alaway of Alaway & Partners, Auctioneers and Estate Agents of 20 Bloomsbury Square, London:

"Land and buildings at Oxgate Lane £44,190
37 8 Litre chassis @ £600 .... £22,200
25 4 Litre chassis @ £250 ....£6,250
Less to complete to specification £4,450
Total £24,000
Machinery, tools, fittings, etc at Kingsbury ..£300
Machinery, plant, equipment etc at Cricklewood
....£12,500
Plus; Spare parts, book debts, cash at bank and in hand, assets at Contractors, complete cars."

The conclusion: "In my judgement and opinion the sum of £84,000 in the contract mentioned is far below the value of the property and assets proposed to be purchased and I am further of the opinion that if the property is sold separately the same will realize a sum in excess of £84,000." On the same basis, Montgomerie calculated the total value of the Company as £140,697.11.3. The offer for the chassis was definite, from Jack Barclay. Montgomerie pointed out that the Service Department had made a profit of £1,000 per month under Frere's management, and that Napier's offer was far too low, leaving nothing for the unsecured creditors. He suggested that liquidating some of the assets would raise enough money to pay off The London Life and keep the Company going until a better offer was forthcoming.

On the 6th November, Napiers raised their bid to £90,000, plus an additional £3,500 to cover certain debts, provided the sale contract was endorsed by 18th November. Frere was keen to discharge his responsibilities, pointing out on the 9th November that Montgomerie's figures valued items that were really only of scrap value at their book value, and that the profit made at Kingsbury was "owing to a more precise working to the terms of the guarantee." Napier's offer brought the total receipts from sale to £109,825, consisting of £93,500 from Napiers and £16,325 from cashing the insurance policies. In the absence of offers from other concerns, Frere urged the acceptance of this offer. It is evident that Napiers had Bentleys over a barrel, and were trying to push the sale through as cheaply as possible.

On the 11th November, Carruth summarised Barnato's position in all this. "The Company is indebted to him [Barnato] in respect of unsecured loans amounting together to £176,224.3.6 guaranteed by him since the date of the said debenture [June 1927], as to £88,028.5.2 in respect of loans by Barclays Bank Limited and as to £88,195.18.4 in respect of loans by Baromans Limited. The Defendant Woolf Barnato will in view of the deficiency above be legally liable to pay the full amount of the sums so guaranteed. Notwithstanding the heavy loss which will be suffered by the Defendant Woolf Barnato if the offer made by Messrs. D. Napier & Son Ltd. as stated in the Receiver's last affidavit is accepted, he has authorised me to say that he approves of a sale on the terms of the said offer and is of the opinion that no better offer is likely to be obtained. I am of the same opinion."

However, in his Centenary Lecture to The Institution of Mechanical Engineers on 16th September 1988, Donald Bastow made the following observation: "From then on [9th September 1931]

two sets of negotiations proceeded in parallel. Frere did not know of the non-Napier talks, from which an agreement late in October allowed Rolls-Royce to make detailed plans for setting up Bentley Motors (1931) Ltd. after buying the assets from the Receiver." Bastow also confirms that Montgomerie was Barnato's nominee (see p.302). Carruth continued to act as Receiver until 26th March 1932 (along with Frere) so Carruth, possibly in conjunction with Montgomerie, could have continued negotiations with Rolls-Royce or any other party throughout this period on Barnato's behalf without Frere knowing. Barnato would have been in touch with the proceedings, because according to *The Financial Times* for 14th September 1931, when Montgomerie was appointed as the Liquidator, a committee consisting of representatives of four of the principal creditors and Woolf Barnato was set up as well to oversee the liquidation.

It is perhaps significant that the initial approach to Rolls-Royce in June 1931 was made by Carruth, Barnato's Business Manager as well as MD of Bentley Motors. It has been suggested to the author that Barnato intended Bentleys to go to Rolls-Royce. One can only speculate on the precise motivations contained in WO's comment that "By 1934, Barnato, who had bought a substantial number of Rolls-Royce shares shortly before the liquidation, was on the new Bentley board." Unfortunately, the Rolls-Royce share records from this period seem to have been destroyed during the Second World War. It is a very strong possibility that Barnato ensured that Bentley Motors went to Rolls-Royce, and the author is inclined to subscribe to this hypothesis.

That there were indeed two sets of negotiations going on is confirmed by Lloyd's *Rolls-Royce – The Years of Endeavour*. The order to sell Bentley Motors was made on the 20th October 1931. "News of this reached Sidgreaves, who immediately decided that Rolls-Royce should endeavour to prevent this from happening. The reasons for his decision are obvious and need no elaboration. Had Napier acquired the Bentley assets and re-entered the motor car field in this way, this would have presented a much more serious competitive threat than that which would have arisen as the result of the disposal of the Bentley stock of chassis. [Rolls-Royce's] solicitors, Claremont Haynes, were instructed to outbid the Napier company in court, but it was realised that this would be a much more expensive business if it was known that they represented Rolls-Royce. It was thereupon arranged to appoint agents who would not be associated with

Rolls-Royce in any way and the British Equitable Trust were briefed to act in this capacity." F.C. Champneys, a partner in Claremont Haynes, said that it was decided to keep Rolls-Royce's identity secret, as "if disclosed it was feared that the price might be exaggerated." Champneys also reveals that Rolls-Royce made a search for the name "Bentley" on Sidgreaves' instructions before they bought the Company. Sidgreaves instructed Claremont Haynes to register the name Bentley and the B in wings device as soon as they were in a position to do so.

Although an agreement had been drawn up by Rolls-Royce late in October (Bastow, above), Frere did not know of this. Indeed, Frere's affidavit filed on the 9th November 1931 included the following "...there has been no necessity to advertise the sale of the assets of the Company. My appointment as Receiver on the 10th July last was the subject of the greatest possible publicity in the newspapers and in the motor technical journals and from time to time there have been newspaper notices with regard to the sale of the business, the last being as recently as about a fortnight ago. There has been ample opportunity during the prolonged negotiations with Messrs. Napier for anyone else to have made definite counter-offers for the purchase and it will be observed that no counter-offer has been produced by the Defendant Company [Bentley Motors] and the only alternative is a sale piecemeal over an extended period..." This simultaneously disposes of the notion that Frere was in on the Rolls-Royce deal, and also that the deal with Napiers was fixed at a low price, as Napiers were the only serious public bidders.

To get the best deal for the creditors Frere pushed Napier's offer up to £103,675, and took the staff out to the theatre the night before the court hearing to seal the arrangement.

The court hearing to finalise the sale, held on the 17th November, went anything but according to plan. During the proceedings, a representative of the British Equitable Central Trust rose to his feet and announced that he was empowered to offer a sum for the assets of Bentley Motors, a figure higher than that proposed by Napiers. After a brief adjournment, Napier's counsel came back with a higher bid, but the Judge, announcing that he was not an auctioneer, told them to come back that afternoon at 4.30 with sealed bids. It would seem that he should have given the whole matter back to Frere to sort out, as Frere's responsibility was to get as much for the Company as possible.

Quite what happened outside the court is not

clear. Champneys of Claremont Haynes: "I accordingly was present in court when on application to approve the sale to the Napier Company Counsel on behalf of Rolls-Royce made an overbid, and the matter was adjourned till the afternoon to be decided on sealed tenders. It was my duty in the corridor of the court to read the Contract of purchase and other documents and advise as to their general effect." By the time they came back into court, a Conditional Agreement of Sale had been drawn up between the BECT and Frere dated 17th November 1931, to the effect that the BECT would pay £125,275 for Bentley Motors. Mr. Justice Maugham instructed that that sale agreement be ratified, at which Frere paid £11,500 into court as the BECT's deposit on the sale. Napier raised their bid to £104,775, which was of course greatly exceeded by the BECT's offer of £125,275.

As Sidgreaves put it: "Where as a result of competition between certain parties and (with the concurrence of the Chairman) that it had been necessary to increase our offer to £125,275 at which price the deal was completed." Sidgreaves also noted "that it was necessary to take over the agreement made between Bentley Motors Ltd. and Mr. W.O. Bentley" (referring to the 10th March, 1926, agreement between WO and Bentley Motors).

WO said in his autobiography "I don't know by how much precisely Napier were outbid, but the margin was very small, a matter of a few hundred pounds." WO's memory must have been at fault over this, because while the BECT's initial bid was indeed for a "few hundred pounds" more than Napiers their final bid was £20,500 greater. It seems surprising that Napier only raised their bid by £1,100 while Rolls-Royce raised theirs by so much, and is perhaps indicative of a degree of reluctance on their part. It was certainly a dejected quartet (WO, HM, Frere and Vane) who returned to Hanover Court after the hearing, having left that morning in high spirits to that which they thought would be a mere formality – apparently Vane had no authority to increase Napier's bid.

Almost immediately, Rolls-Royce filed for the Bentley trade marks, and a new limited liability company was formed. Claremont Haynes, Rolls-Royce's solicitors, filed a "Consent to take the Name" to register a new company, using the name Bentley Motors (1931) Ltd.

Sidgreaves had three immediate objectives:

"1. To see that we had a proper assurance of the name.

"2. To secure Trade Mark protection for the name.

"3. To see that the goodwill was not damaged by the activities of anyone connected with the old Company and that really meant two people...Capt. Woolf Barnato who had financed it, and who subsequently became a Director of our New Company, and W.O. Bentley."

WO said that it was to be some time before he discovered that the BECT were acting for Rolls-Royce. In fact, WO went to see Sidgreaves on the 20th November, three days after the court hearing on the 17th. WO had one meeting with Royce on the 11th December. Royce at that time was in failing health which apparently affected his decision-making capacity. (*Rolls-Royce – the Derby Bentleys*, A. Harvey-Bailey.)

According to WO, Royce's opening gambit was "I believe you're a commercial man, Mr. Bentley?" which seems rather surprising. One would think that Royce was well aware that as a "commercial man" WO had been somewhat less than successful. WO's retort was to point out that he had been a premium apprentice at Doncaster sometime after Royce had been a boy in the running sheds. It is likely that WO was still mentally very shaken by the turn of events and unlikely to be much more than polite with Royce. The two men should have got on, because of the similarities in their engineering backgrounds and Royce's noted liking for sporting cars. It is also possible that Royce felt a degree of envy of the tremendous successes that WO and Bentley Motors had amassed in just twelve years compared to the 30 odd years that it had taken Royce to achieve his position of pre-eminence in the manufacturing of luxury cars. WO himself does not seem to have had a very high opinion of Royce, feeling that Hives was probably more talented. This opinion was based on a great deal of respect for Royce's attention to detail, but a feeling that his designs themselves considered *in toto* were not in any way outstanding.

Royce offered WO a position, and to cancel the 1926 life agreement, but it was made clear that WO would not be employed in a design capacity. Perhaps understandably, Royce formed a poor impression of WO, writing to Sidgreaves: "If we were to let him [WO] have the run of the Derby designs, experiments and reputation, Rolls-Royce would teach him infinitely more than he would help us, and we should be making him more powerful to do us harm by perhaps in a year or two going to Napier or elsewhere." Frostick is particularly penetrating

on this point: "If only Royce could have seen that he might have been talking to his successor, what a future might have been planned, but such thinking demanded a mind of Churchillian proportions, which neither of them possessed."

WO wrote to Sidgreaves on the 12th December, declining the offer. Sidgreaves was concerned that WO would use the Bentley name to undermine the goodwill that Rolls-Royce had bought, and while they had no great wish to employ WO, Rolls-Royce nevertheless enforced the contract to stop the design work WO was doing for Napiers. Sidgreaves wrote to WO again on the 15th December, setting out the position they were prepared to offer. WO then rang up Sidgreaves and asked if he was free to work for Napiers. WO said he would relinquish all claims to royalties if Rolls-Royce would release him from a debt of £2480 to Bentley Motors, and cancel the life agreement. Sidgreaves declined this offer, so WO issued proceedings on the 14th January 1932, supported by Napiers, the case coming before Mr. Justice Eve on the 9th March.

WO argued that the agreement had been cancelled either by the appointment of the receiver or the liquidation of the Company. Rolls-Royce argued that WO could not take this line because of his conduct since the appointment of Frere on the 10th July. (That is, because WO had continued to accept payment from Frere under the terms of the agreement.) The Judge returned the matter to the lawyers on technical grounds, that he would have to adjourn the summons to allow the plaintiff (WO) to issue a writ. WO's counsel admitted in court that Bentley Motors could go on without WO being a part of it.

It is perhaps important to look at Napier's role. Vane was against going back into the motor trade, but had no alternatives to offer. At the turn of 1931–32, Vane put it to his co-directors that he saw "very great difficulties ahead of the company in designing and manufacturing a motor car and in arranging the necessary sales organisation to dispose of it in sufficient quantities to avoid a heavy loss, let alone making a profit." This was after Bentley Motors had gone to Rolls-Royce, and Napiers were thinking of an entirely new company with WO as the designer. Vane wanted the idea dropped, and put it to a general meeting, observing that "Should my views on this subject not prevail, I shall have carefully to consider my position as chairman and managing director of the company." Vane duly resigned on 18th February 1932, and was immediately replaced by Sir Harry Brittain on a temporary basis.

How are the mighty fallen! This is EX19 Phantom II, on test in France, sometime between April 1932 and February 1933. The photo is from WO's own album; he was one of the test drivers at the time.

Brittain soon announced his plans (unfortunately, the cutting is undated, but it is from the period 18/2/32 – 19/4/32): "One of the most famous motor engineering firms in the world – Messrs. D. Napier & Son, Ltd. – is to re-enter the motor car business after an interval of ten years. This announcement was made yesterday by Sir Harry Brittain when presiding at a meeting of the company's shareholders at Acton, W. He stated that: "Steps are being taken to enable the Napier Company to produce a motor car of first class performance, reliability and engineering merit." No details are yet available regarding the new Napier car, but I understand that there will be the minimum of delay in getting the new model into production. It is, I learn, to be the work of one of the most celebrated designers of motor cars and aero engines in the country. When his name is divulged, it will cause a sensation in motoring circles."

At this point, the interests of Rolls-Royce and WO started to converge. The registrar for trade marks expressed difficulties over registering "Bentley" as it was WO's surname, and Rolls-Royce were forced to obtain declarations from agents and prominent motorists to the effect that "Bentley" meant a car made by Bentley Motors, not one designed by W.O. Bentley. Sidgreaves wrote to WO after the court case on the 9th March, making WO an offer such that he would come to an agreement with Rolls-Royce over his debt to them in exchange for a declaration supporting their appli-

cation for the trade mark. At this point, Rolls-Royce still thought that WO was going to Napiers, and wished to preserve the value of the name they had bought.

Napiers by this stage were losing interest with the whole scheme, and their minutes record this. WO himself was in an increasingly precarious financial position. The upshot of this was that WO went to see Sidgreaves on the 15th April, and told him that he was upset with Napiers because they had reduced the royalties that had been agreed. WO went on that he felt he would never be happy with Napiers and asked Sidgreaves for a job. This was offered on the same basis as before, namely as an understudy to Northey. Northey was technical adviser to the managing director and sales staff, with responsibilities for liaison between Derby and the sales staff and activities connected with sales. (The author has heard a tape made by WO stating that he asked Sidgreaves for a job.)

Sidgreaves and WO came to an agreement, set out in a letter to WO on the 19th April. WO signed over to Rolls-Royce all rights to the use of his name connected with car manufacturing except in very limited circumstances for a period of ten years, and made a declaration that the right to use the name "Bentley" and the associated trade marks had passed to Bentley Motors (1931) Ltd. In return, the life contract was superseded by a contract with Rolls-Royce for three years, renewable on either side after that period. Despite the rather stiff terms drawn up by Rolls-Royce, WO had little choice but to agree to them. Apart from any other considerations, WO still owed Bentley Motors £2,480.10.11 and the interest due on that sum. Rolls-Royce wrote off that debt and paid him a further £1,210.

Napiers themselves gained a new chairman in the form of Sir Harold Snagge, and at his very first board meeting on 10th May 1932 it was agreed that no further moves would be made towards bringing WO into the company, and indeed all Napier's aspirations towards motor cars were abruptly terminated.

Leaving WO's problems and reverting to the business of Bentley Motors, on the 14th December 1931 F.W. Turner, Rolls-Royce's financial secretary, revealed to the court that Rolls-Royce were backing the BECT. Subsequently, the identity of the buyers was revealed publicly on the 21st December 1931, when title passed from the BECT to Bentley Motors (1931) Ltd., the latter concern paying the final balance due on the sale of £14,090.8.11 into the court. As both Rolls-Royce and Napiers knew the true value of Bentley Motors it seems doubly suprising that Napiers raised their bid by so little – Sidgreaves valued Bentleys at £148,015, so even for nothing more than asset-stripping the Company was still cheap. In many ways, that is of course what Rolls-Royce did.

In January 1932 Rolls-Royce revealed their plans for their new aquisition. The announcement read: "The directors of the company recently formed to take over the assets of Bentley Motors announced last night that in regard to future production of Bentley cars, the manufacture of the 8 Litre and 4 Litre models will be discontinued. The new company, however, propose to develop and produce a smaller model of the sporting type having engine capacity of approximately 2½ litres and operable with and without a supercharger. The company are unable to give any information in regard to specifications and price, an announcement in regard to which will be made as soon as is practicable. Meantime, special arrangements are being made concerning guarantees to existing owners of Bentley cars, full particulars of which can be obtained from the Kingsbury Works, The Hyde, Hendon N.W."

Although it has been widely denied, there seems little doubt but that Rolls-Royce were indeed motivated by a desire to remove the proposed Napier-Bentley as a competitor to the Phantom II in their take-over of Bentley Motors. Robotham, who in 1931 was the Chief Development Engineer for Motor Cars at Rolls-Royce in Derby, observed in 1964 that: "Round about 1930 it became apparent that sales of the large six-cylinder Bentley were beginning to make appreciable inroads into the traditional Rolls-Royce market of luxurious chauffeur-driven limousines." "In the higher-priced class nothing seems able to stop Bentleys. The Bentley Six is steadily making people in Derby thoughtful." (from *World Today*, August 1929). Rolls-Royce's concerns were obviously greatly enhanced by Napier's moves and the rather under-hand way in which Rolls-Royce secured the assets and goodwill of Bentleys while taking great care not to reveal their identity until after the event, does little for their reputation. The author was recently told by Warren Allport that "Doc" Llewellyn Smith, who knew Sidgreaves well, said that at least part of Sidgreaves' motivation in buying Bentley Motors was that Sidgreaves was ex-Napiers, and hated his former employers. Hence Rolls-Royce's purchase of Bentley was a double strike – the elimination of their closest rivals, and an opportunity for Sidgreaves' to do Napiers down. Sidgreaves later commented (of WO's difficulties with

Napiers): "I had myself worked for some years for the Napier Company and I had some fellow feeling for him." It was also, as Lloyd put it, "...an astute move on Sidgreaves' part. Even had Rolls-Royce made no use of the name to re-enter the sports car field the venture would have been both financially and commercially profitable."

In *Thoroughbred and Classic Cars*, September 1974, Graham Robson observed that "the inference of dirty dealing is made in the Napier book [*Men and Machines*, Wilson and Reader, Weidenfeld & Nicolson 1958], and on hindsight there seems to be reasonable evidence that someone at Napiers let it be known what Napier's final bid was to be so that Rolls' nominees could jump in with a better offer." Again, Bastow comments that: "[W.O.] Bentley was convinced that Rolls-Royce knew the amount of the final Napier bid and he was sure that he knew the source of that information. British Equitable must have known: they received £2,500 for their services in this matter." This puts WO's comment on the BECT's offer – "a figure that was, by an extraordinary coincidence, a fraction more than that offered by Napiers" – dramatically into context. The BECT/Rolls-Royce partnership must have been thrown by the sealed bids scenario, which would have negated their intelligence service and forced them into increasing their bid excessively, as it turned out.

Rolls-Royce's source of information could have been any of Carruth, Montgomerie, or Barnato. Had this information come from Barnato and had WO known, it would make sense of WO's comment: "I saw him [Barnato] often; we were still good friends, if there was a trace of reserve in a relationship which was now on an entirely social level. Business was never discussed between us, and when we talked of the old days it was of our cars and our racing."

Once the financial affairs had been sorted out, the new company Bentley Motors (1931) Ltd. was formalised with a capital of £20,000. The first annual general meeting of the new company was held at Vernon House on 23rd December 1931, attended by G.H.R. Tildesley of Claremont Haynes and F.P. Daniell, at which the Certification of Incorporation of Bentley Motors (1931) Ltd., dated 17th December 1931, was produced. The first meeting of the board of directors was held on the 30th December, attended by Sidgreaves, Wormald, F.W. Turner and G.H.R. Tildesley. Claremont Haynes reported the completion of the purchase of the assets of Bentley Motors from the receiver, covered by £20,000 from Rolls-Royce

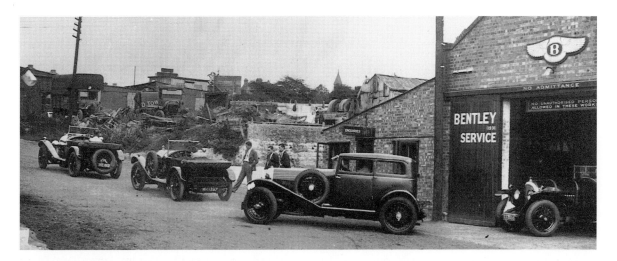

After the main Works at Cricklewood had been closed down, the service department at Kingsbury was kept open. In the early 1930s, under the direction of JC Neville, a number of second-hand Bentleys were bought back by the company and reconditioned for resale. This photo shows a number of such cars at the Kingsbury Service; 3 Litres YE812, MH1397, AX6758 and XX6856. Hubert Pike remained in charge of service under Rolls-Royce management. Note the 'Bentley (1931) Service' on the door – over both doors it read 'Bentley Motors (1931) Ltd. Service Department'.

Ltd. for 20,000 shares in the new company and a loan of £105,275 for the balance. Shares 1 and 2 were allotted to G.H.R. Tildesley and F.P. Daniell, the rest going to Rolls-Royce Ltd. Nos. 1 and 2 were then immediately transferred to Sidgreaves and Wormald respectively. The formalities had been completed and Rolls-Royce took full control of Bentleys.

The Service Department at Kingsbury was kept open under Hubert Pike and Nobby Clarke and the repair depot at 112/118 North St., Glasgows also continued trading. The rest of the firm was shut down. The remaining 8 Litre, 4 Litre and Super-charged chassis were sold off to Jack Barclay and Jack Olding "for a very satisfactory price", and all the spares moved to Kingsbury. Many of these spares went into the six RC (Reconstructed) series 4½ Litres (RC41-46) built under the direction of Nobby Clarke in 1935/36. Four RC series 3 Litres (RC 31-34) were also built, but these were based on secondhand cars bought in and completely worked over. Everything made of Elektron was defaced on the machined faces with sledgehammers and care-fully accounted for before being taken away by lorry, reputedly to be sold to the German aircraft industry! Certainly Conway, the storeman, recalled keeping meticulous records and returns to Rolls-Royce of Elektron shipped. Conway also burnt all the drawings and contents of WO's desk, on WO's instructions. Fortunately an incomplete set of blueprints survives. It has been suggested that spares for the 8 Litre were deliberately destroyed.

The new Machine Shop was disbanded, the tools going either to Derby or for sale. Pollen House was retained by Barnato until 1941, when the lease expired. It is still possible to see the holes in the frontage of the building for the "Bentley Motors" lettering. The Cricklewood Works were sold to Addressograph Multigraph in October 1932, later passing to Messrs. Smiths Instruments Ltd., who continued to manufacture in the old buildings up to the early 1980s, when they were demolished to make way for a new trading estate. Robert Montgomerie, the Company Secretary, completed the winding up by December 1932. The remaining £11,724.6.11 after the debentures had been paid was divided at the rate of 6d in the pound for the first dividend and 5d in the pound for the second. Barnato received some £42,000 back out of an investment of the order of £238,000: a loss of nearly £200,000. As Barclays Bank also appears in the list of creditors, it would seem that Barnato did not have to meet that liability. Barnato's later com-ments that he only lost £90,000 was a face-saving estimate, that no-one could challenge without ac-cess to all the Company's financial records.

Barnato retained Wally Hassan and "Old Num-ber One" Speed Six, which won the 1931 500 Miles Race but crashed in the 1932 500 Miles Race killing Clive Dunfee. This unhappy event finally put Barnato off motor racing completely. He had

After the Company had gone under, a large number of ex-works and demonstration cars, and finished chassis, were disposed of to Jack Barclay and Jack Olding. Seven of the last eight Supercharged chassis were fitted with this drop-head coupé style by Vanden Plas for Jack Barclay.

other matters on his mind as well in late 1931. In December he laid a wager of £500 to £100 that he would reduce his golf handicap from 13 to scratch in a year, (uncharacteristically, he failed) and then in March 1932 he married Miss Jacqueline Claridge Quealy in America. It is possible that the rumour that Barnato was in America and remote from the events in London started from the fact that Barnato took a short holiday in the South of France in late July/early August and then went to America at the end of August. The *Daily Telegraph*'s correspondent said on 26th August – "Rumour in America has been busy with the real purpose of Capt. Barnato's trip. All sorts of fantastic stories have apparently been in circulation over there, but I am in a position to state on his own authority that there is no vestige of truth in any of these. His trip, following on a prolonged period of negotiations relating to the future of the motor manufacturing concern with which he is identified, has been planned essentially as a holiday one." The newly-weds spent their honeymoon on a 10,000 mile tour around America in Barnato's favourite Bentley, "Old Number One" by now fitted with a coupé body.

Sidgreaves had identified two people to muzzle, WO and Barnato. The former had already been dealt with, and on the 26th February 1934 Sidgreaves wrote to Barnato offering him a seat on the board of Bentley Motors (1931) Ltd. By then, enough time had passed that no-one would question Barnato's role in the events of 1931/32, and he accepted the offer on the 5th March. It has to be questioned whether the fact that on 16th September 1948 the Board of Bentley Motors (1931) Ltd. minuted his death saying "a director of this company since its formation" was a purely accidental error.

Barnato withdrew his patronage completely in

1936 when he parted company from both Wally Hassan and the Barnato-Hassan. The Barnato-Hassan was the second fastest car ever at Brooklands, lapping at 142.60 mph to John Cobb's Napier-Railton's 143 mph. However, a full set of tyres for every race on an unbeatable handicap was too much of an indulgence compared to the negligible return, and even Barnato became fed up.

WO stayed at Rolls-Royce until June 1935 as a development engineer, first as an understudy to Percy Northey, and then as "Technical Assistant to the Managing Director" (Sidgreaves) before going to Alan Good's Lagonda concern shortly after their victory in the Le Mans race of that year. It was many years before WO revealed how difficult Sidgreaves made life for him. WO was involved in developing the Lagonda M45 into the LG45, and then designing the LG6 and the V12-engined car which finished 3rd and 4th at Le Mans in 1939.

Although he might have thought it was all over, WO found himself back in court early in 1945. In 1944 Lagonda ran a series of adverts, using the Bentley name in a manner that met the letter, if not the spirit, of WO's ten year agreement with Rolls-Royce, but broke the laws concerning trade marks. WO, joined by Lagonda, instituted proceedings against Rolls-Royce, on the grounds that Rolls-Royce had registered the Bentley trade mark by concealment and misrepresentation, because Rolls-Royce had not informed the registrar of the terms of the ten year agreement between WO and Rolls-Royce. WO tried to have an injunction taken out preventing Rolls-Royce using the Bentley name, and to have it removed from the Register of Trade Marks. The whole case was a sorry affair, the only redeeming feature being a decision made by Rolls-Royce not to follow a line of attack suggested by their legal people to attack WO's reputation as a designer. Needless to say, WO and Lagonda lost.

After the war WO designed the 2.6 litre Lagonda engine, which led to the acquisition of Lagonda by David Brown in order to use this engine in the Aston Martin DB2. He later designed the prototype 3 litre Armstrong-Siddeley. In 1950, WO found himself somewhat prematurely retired, to Shamley Green, where he died in 1971. In later years the Bentley Drivers Club and the reverence paid to the Cricklewood Bentleys was a considerable comfort to a man whose dreams and endeavours were four times shattered by world events.

Bentley Motors (1931) Ltd. soon dropped their plans to produce a supercharged car, producing instead the 3½ Litre and 4¼ Litre cars at Derby – the "Silent Sports Car" - in many ways a worthy successor to the Cricklewood cars. The Mark V with independent front suspension was killed by the Second World War, during which the Car Division of Rolls-Royce moved to Crewe, to be succeeded in 1946 by the Mk VI with (a first for Rolls-Royce) standard steel coachwork by Pressed Steel Ltd. The Mk VI was followed by the *R* type, in Continental form a superb motor car. However, badge engineering was on its way through the *S* and *T* series and it was not until the Mulsanne Turbo and the Turbo *R* that Crewe were to produce a different and faster Bentley than the contemporary Rolls-Royce product. The old name of just Bentley Motors Ltd. was re-instated in 1971 after the collapse of Rolls-Royce over the RB 211 aero-engine. In the introduction to *Bentley – Cricklewood to Crewe*, Michael Frostick laments "the dreadful dichotomy of enthusiasts". The author is unashamedly a Cricklewood enthusiast, and those wishing to read about the Derby and Crewe Bentleys will have to search elsewhere!

The vintage Bentley cars themselves hit an all-time low in the late 30s, but rose in value considerably after the Second World War in similar boom conditions to that following the First. The 4½ Litre particularly remained a competitive car for some years, particularly in club racing, and many were extensively rebuilt and modified. During the fifties and sixties many were driven daily by enthusiasts and maintained by like-minded people, until the increase in values (and hence desirability, for whatever reasons) that has at least ensured that many Bentleys have been preserved. It is surely impossible that a single car will now be lost to posterity, or even suffer from neglect. Of the 3024 cars built between 1921 and 1931 some 1400 survive, of which maybe 1000 are roadworthy, and it is to be hoped that these living creations of a previous golden age will be seen on the highways for many years to come.

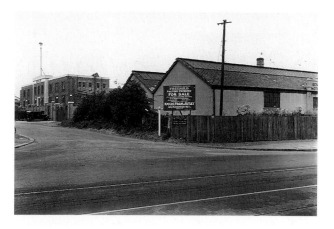

View of Oxgate Lane and the office block – note the Bentley weathervane, clearly visible (compare with p.292).

The interior of the erecting shop (compare with p.281).

The inside of the machine shop, after the machines had been stripped out.

20 Speculations and Afterthoughts

So why did Bentley Motors go bankrupt? Carruth attributed it to the delivery of new and untried models: "Broadly speaking, our main troubles in the past can be traced to the delivery of new models to the public before adequate tests of production have been made. . .it is an error of policy, and may even be a fatal error of policy." Ian Lloyd points out that "This interesting statement indicates the strong propensity to search for a single reason for failure in a complex situation where it may in fact be quite beyond the control of the management, as was undoubtedly the case with Bentley Motors at this time." There is no doubt but that it was indeed a very complex situation, and a quite insoluble one. The diversity of models produced and the lack of interchangeable components, the need to run an expensive service operation to maintain a small number of cars, and the weight of indirect expenditure in the form of design and experimental work, sales, racing and other related activities, were too high for the manufacturing profit to sustain.

Carruth's primary solution was a radical move into a new market and it is conceivable that the problems might have been overcome had the 4 Litre been successful. It is interesting to note the steps they took – first of all tackling the difficult sales position with "Bentley Weeks" at dealers up and down the country, then later in 1930 the establishment of the Société Européenne Bentley Automobiles SA to promote sales abroad, and then finally the introduction of a new, cheaper model – none of which measures had any impact on the gathering crisis. What is rather remarkable in view of all that WO tells us about Barnato is the latter's role in the whole affair. After WO, Barnato is the most important player in the Bentley scenario, and

yet he remains a rather nebulous figure. We know WO quite well, not least because he wrote much of the literature on the Bentley marque, but we have no such insight into Woolf Barnato.

We know quite a bit about Barnato the sportsman, in terms of motor racing, cricket, golf and boxing, and we are familiar with Barnato the socialite, with the parties at Ardenrun, Grosvenor Square and the South of France. However, of the other side, that of the business man who was "uncommonly perceptive, quick on the uptake, quick to act and quite ruthless" (*WO – An Autobiography*), there is remarkably little evidence. First of all, it has to be remembered that Barnato inherited a vast amount of money, the first £250,000 of it when he was 18. Barnato could have maintained his lifestyle and supported Bentley Motors virtually by living off the money he already had.

WO tells us that his business affairs were run by his advisers, and indeed Barnato himself listed his occupation as "gentleman" in the Bentley Motors company returns. The only company listed in the Public Records Offices' files under the name Barnato is Barnato Brothers, but Woolf was never listed as a director. Indeed, Woolf steered clear of the family firms, Barnato Brothers and the Johannesburg Consolidated Investment Co. Ltd. With the exception of Bentleys and Baromans, we do not know of any other company with which Woolf Barnato was involved. It is likely that most of his dealings were in commodities, as Barnato is quoted (by WO) as referring to a specific diamond deal made in the twenties out of which he made some £120,000. According to Stanley Jackson (*The Great Barnato*, Heinemann, 1970) "He [Woolf Barnato] made shrewd investments in the commodities markets and in a variety of paying companies."

Barnato's lifestyle is hardly that to make one think of a high powered entrepreneur. Hillstead refers to taking an open four seater 6½ Litre to Ardenrun at the time that Barnato was being persuaded to join the Company (late 1925/early 1926), rousing Barnato out of bed before a short run followed by lunch. Rivers Fletcher also refers to delivering a 4½ Litre to Grosvenor Square, obviously fairly well into the morning, and being ushered in to find Barnato having breakfast in bed. Admittedly business in this day and age is far removed from the mores of the 1920s, but Barnato's lifestyle and his seeming ability to be always available for motor racing and other sporting activities does seem hard to reconcile with the hard driving business man.

Barnato's business affairs were later conducted from a small first-floor office in Park Lane, shared with his cousin, Stanhope Joel. These two dealt in the commodities market. (The Park Lane offices were taken after the lease on Pollen House had expired.) After his death it was recorded that Barnato did not make much money because he enjoyed life too much, and he was quoted as saying "I never speculate". Estimates of the value of his estate ranged from £1.5 million to £5 million, neither of which lend much credibility to Barnato's reputation as a money-maker in view of his inheritance of £1.4 million by 1925.

The truth of the matter seems to be that WO's impressions of Barnato are largely misleading. Barnato had a shrewd business brain, but chose not to use it to any great degree. He employed capable advisers to look after his affairs, and spent the money that he earned. He enjoyed a very good life, and had a wide circle of friends who shared in his good fortune. On his death, he left his inheritance largely intact. He was certainly not the ruthless businessman of WO's descriptions. He was a man of integrity and honesty, and the gesture of paying a month's salary to the staff and a week's wages to the shop floor employees at Bentleys when the Company went into receivership in July 1931 was a typical gesture.

His decision to invest in Bentley Motors is an interesting study in contradictions. First of all, why did he do it? Bentleys had been selling cars for over four years, and had yet to show any evidence of profit – indeed the finances were distinctly hand to mouth, and at the time of the Barnato reconstruction the Company's trade debts alone amounted to £75,000. On top of this, of course, was the out-

standing mortgage for £40,000. Barnato put up roughly £115,000 in the first place, a sum which he must have been advised he was highly likely to lose. WO: "He conducted his business largely through his ambassadors – his financial advisers. These gentlemen, who devoted themselves to preserving and expanding Barnato's fortune, heartily disapproved of their master's rash venture into the motor car business, which they viewed with the gravest suspicion and of which they were totally ignorant." It is quite likely that they were totally ignorant, but they were accountants and more than capable of reading a balance sheet.

One of Barnato's advisers, J.H.C. Faire, said in a letter to WO in 1958: "Obviously I am not one of his cronies that you have mentioned, as I flatly refused his very kind offer to go into Bentley Motors, having begged of him not to go into the Company – leaving my colleagues Ramsey Manners and Jack Carruth to help him in Bentleys." Faire comments that "there are quite a number of things which you could not possibly know about Woolf Barnato just before he bought his first Bentley [3 Litre chassis no. 1106 in May 1925] and one or two things that you obviously do not know about him at the tail end of Bentley Motors." It was said of one tycoon that he regarded his companies merely as profit or loss generating entities with no consideration of what they did, and if they were loss making they shut down pretty quickly. One is left with the conclusion that Barnato's financial advisers were definitely against his acquisition of Bentley Motors on commercial grounds. Hence, as Barnato must have over-ruled them, one has to ask why.

It has always been looked at from the view point of what Barnato did for Bentleys, and not the other way round. Without Barnato, Bentleys would have been in receivership by early 1926. The 3 Litre would still be a highly regarded vintage sports car, but much less so than it is now. The 6½ Litre would be one of those "might have beens" with two prototypes built, and possibly one or two production cars. But what did Bentleys do for Barnato? In much the same way, Barnato might well have been a forgotten figure without the association with Bentleys. The cars provided Barnato with racing successes that made him internationally famous, and his position as Company Chairman invested him with an enormous amount of acquired prestige. WO put it: "He gained the distinction of being associated with the marque Bentley in its years of greatest prestige; and he was not entirely above such nourishment to his vanity." Barnato's East

End background was very humble, and as a Jewish financier, and very nouveau riche at that, at a time when society was much more closed and formalised than now, the association with such an ostensibly British undertaking must have been very attractive. Hillstead said (in the early years) that "almost the entire executive consisted of public school men", and clearly disapproved of the new regime, who were not. If one is seeking status, being Chairman of a Company with a customer list reading like Debrett's would have been an added attraction. In view of Barnato's over-ruling of his advisers and the manner in which they were proven correct, he must have had reasons other than purely financial considerations.

This is further supported by the manner in which Barnato clung on despite the odds being against him – supporting the Company virtually entirely out of his own pocket for the last year, and then stating (via Carruth) that he was "prepared to go a long way to maintain the name of Bentley." The financial figures make it abundantly clear that it was no longer realistically possible for even a man of Barnato's prodigious wealth to continue to plug the gaps. Then, having lost the Company, Barnato became a director of Bentley Motors (1931) Ltd., and was on sales drives for the new Mk VI Bentley in North America after the Second World War. This is, of course, wholly incompatible with WO's image of the ruthless, hard driving business man. There is also the question – why did Barnato buy a large quantity of shares in Rolls-Royce shortly before the liquidation? We know that Carruth's brief was to divest Barnato of his financial liabilities as far as Bentley Motors were concerned, and Carruth's approach to Rolls-Royce was obviously made with Barnato's knowledge, and presumably, approval. It is not inconceivable that Barnato's acquisition of Rolls-Royce shares was all part of the political wrangling to ensure the continuance of Bentley Motors in one form or another, as well as Barnato's continued association with the Company. It is clear that while Bentleys would have ceased to exist without Barnato's philanthropy, he too gained enormously from the deal. In his case, in the acquisition of things that money alone could never buy, a commodity of which he had sufficient to speculate. It is almost certainly fair to say that Barnato would have been consigned to obscurity without Bentleys, and perhaps (allowing for divergent opinions upon immortality) the £200,000 that bought that fame was a bargain.

Appendices

Appendix I. The Motor Cars.

This appendix is a form of directory to the cars referred to in the text. In all cases, further information will be found in the author's *Bentley – the Vintage Years* either in the main tables, or under the Non-Production Chassis tables.

3 LITRE

Chassis	Reg No	Details
EXP 1	BM8287	1st prototype
EXP 2	BM8752	2nd prototype
EXP 3	BM9771	3rd prototype
EXP 4	ME2431	4th prototype
EXP 5	YW3774	Team spares car 1928/29/30
7	XH9047	Team car Pike
	BM8287	"The Fire Engine"
42, 72, 74	ME1884, ME3115, ME3494	1922 TT cars
94	ME4976	Indy car, became No. 1 TT ME 1884
141	XM6761	Team car Duff
164	MF1632	Hewitt's saloon
221	FR5189	Porter's racer
238	PD40	Pike's saloon
246		Team car Works
No. 2 2 seater	MF330	Clement/Benjafield racer 1926 Montlhéry
582	XT1606	Team car Duff
976	OM6832	Team spares car 1927
1040	YM7646	Team car Duff 1925 Montlhéry
1106	PE3200	Barnato Brooklands racer
1138	MH7580	Team car Works
1179	KM4250	Team car Thistlethwayte
RT1541	YO3595	Team car Callingham
LM1344	MK5206	"Old No. 7" Team car Benjafield
LM1345	MK5205	Team car Kidston
BL1601	YE6029	Team car Birkin
ML1501	YF2503	Team car Scott
ML1513	YV8585	Team car Cook
SCRAP 1, 2, and 3.	MH1757?	BM lorries

6½ LITRE

EX1 – "The Sun"	MF7584	1st prototype
EX2	MH1030	2nd prototype, open tourer later saloon "The Box"/"The Coffin" used as 8 Litre prototype
WB2569		Driven by Rolls-Royce (pp. 161–4)
KD2112	YT56	Driven by Rolls-Royce (pp. 213–5)
LB2332	MT3464	"Old No. 1" team car Barnato
HM2855	GJ3811	"Blue Train" Barnato
HM2868	GF8507	Team car Works
HM2869	GF8511	Team car Works

4½ LITRE

EXP 6		No details known
EXP 7		Experimental engine – no details known
ST3001	YH3196	Effectively 4½ Litre prototype Team car, known as "Old Mother Gun"
HF3187	YU3250	Team car Rubin later 41/2S/C
KM3077	YV7263	Team car Birkin
KM3088	YW2557	Team car "Bobtail" Gunter
TX3246	YW5758	Team car Cook
FT3209	YW8936	Team car Durand
FB3301	UL4471	Team car Holder
NX3451	UU5580	Team car Scott
DS3568	UV6088	Hanley/Dutton car pp. 250, 253.

4½ LITRE SUPERCHARGED

SM3902	PO3265	1929 Show car
SM3903	UW3761	1929 Show car
HB3402	UU5871	No. 1 Birkin Blower single-seater
HB3403	UU5872	No. 2 Birkin Blower
HB3404/R	YU3250	No. 3 Birkin Blower
HR3976	UR6571	No. 4 Birkin Blower

8 LITRE

YF5002	GK706	WO's car

Appendix II. Resumé of Policy.

The Resumé of Policy is taken from "Plus Four", the Le Mans victory booklet produced by Bentleys to celebrate their victory in the 1930 Le Mans race.

A RESUMÉ OF POLICY

By A. Longman

" HE IS THE GREATEST ARTIST WHO HAS EMBODIED IN THE SUM
OF HIS WORKS THE GREATEST NUMBER OF THE GREATEST IDEAS."
—*Ruskin*

*I*T is with considerable gratification that we are able to prepare this the fourth description of a Bentley victory in the most famous of all European Road Races—the 24 hours Grand Prix d'Endurance at Le Mans, France.

It was in 1923 that the Automobile Club de l'Ouest conceived the excellent idea of organising a race under actual touring conditions for a period sufficiently long to ensure the most exacting test for any car. Only minor modifications from standard practice were permitted and a fully equipped touring body was insisted upon with weight equivalent to three passengers in addition to the driver.

We were then a very young Company, the first Bentley only having been produced about eighteen months before the first Le Mans Race. Our designer, Captain W. O. Bentley, was from the beginning a very strong believer in racing standard production cars. As most motorists remember we entered three such models in the R.A.C. Tourist Trophy Race in the Isle of Man, on the 21st June, 1922, and finished 2nd, 4th, and 5th, being the only team to complete the course while all our competitors were specially built racing cars.

The Company were very busily engaged in meeting the public demand for Bentley cars so that time could not be given to consideration of the 1923 Le Mans Race. However, an enthusiastic owner, Captain John F. Duff, persuaded us to allow him to enter a 3-litre, his own property, in the 1924 competition. Having enlisted Mr. F. C. Clement, as second driver, these two optimists went over to Le Mans, taking just a few spares and two mechanics, all travelling in the actual car which was entered ! With no pit organisation, this combination astounded the motoring world by winning easily at an average speed of 57.5 m.p.h., and set all the inhabitants of that ancient city of Le Mans talking about " le beau Bentley." Monsieur le Capitaine, as Duff will always be known to Le Mans enthusiasts, left motor racing for other fields shortly afterwards and interest in the 24 hours race rather waned in Bentley circles for a time.

By 1927, however, the Company were in a position to give serious attention to testing their models in public competitions, and now having as Chairman Captain Woolf Barnato, who has always been keenly interested in motor racing, we commenced to tackle the Le Mans Race seriously.

The result was a victory for a 3-litre model in the hands of Mr. S. C. H. Davis and Dr. J. D. Benjafield. Their average speed was 61.3 m.p.h. This race will always go down in history as one of the most exciting owing to the famous crash which put the two other Bentley entries out of action and damaged the winner.

It was also memorable because for the first time the British Daily Press published excellent and very full reports on the race, which were naturally deeply appreciated by us as victors and which undoubtedly were of very great benefit to the prestige of the British Motor Industry as a whole. Since that time the Press have consistently supported us with splendid reports and our grateful thanks are due to them.

The lessons learned at Le Mans in 1927, were taken full advantage of in 1928, when a 4½-litre driven by Captain Barnato and Mr. B. Rubin, was successful at an average speed of 69.1 m.p.h.

It will be noted that each year the average speed increased and the Company were fully alive to the fact that to be continuously victorious more and more efficiency was necessary, not to obtain extra speed alone but to ensure the required margin of safety with the higher average.

In 1929, therefore, the 6½-litre was entered and brought victory to England for the fourth time and the third successive occasion. This time the successful drivers were Captain Barnato and Captain Birkin and the average speed 73.6 m.p.h.

And so to this year—when once again—our Chairman, Captain Barnato, with Lt.-Commander Glen Kidston, R.N., drove a 6½-litre Bentley to victory at the even higher average of 75.8 m.p.h., for 24 continuous hours it should be remembered.

And now for the reasons for this extensive racing programme—is it for the honour of winning ?—for the very great publicity attached to victory ? We can assure readers that this alone would not warrant the very great expense, the lengthy preparations and the extremely detailed organisation necessary to enter these competitions. Of paramount importance are the lessons learned from racing, and these have been of the utmost value to the Company. Experience teaches—we now have a very considerable store of knowledge which is being applied to the production of even more perfect cars—data which could only be obtained, if at all, under ordinary testing conditions over a much longer period.

It is inevitable that participation in races should stamp the Bentley in the eyes of a certain section of the public as a racing car. Nothing is further from the truth. On the contrary as our racing successes have increased our cars have become more silent, more docile, more refined.

The Company have after serious consideration, decided to refrain from racing for a time, as we are of opinion that Bentley cars have reached the stage when they have more than sufficient speed for present day road conditions. In addition they possess outstanding reliability as results have shewn, and in our opinion our education is complete for the time being. The whole forces of the Company will therefore be applied to the production of cars built throughout with the help of their valuable knowledge gained in racing—cars which in reliability—silence, power, safety, and every other respect will, thanks chiefly to the Grand Prix d'Endurance at Le Mans, be the finest in the world.

To all those drivers, mechanics, members of the Works and our suppliers, our very best thanks are tendered for their invaluable services in enabling Bentley cars to thus end an unprecedented sequence of meritorious racing successes.

Appendix III. Financial and Sales Figures

These figures are taken from WO's own personal note-book, preserved at Long Crendon. The note-book starts in April 1926, and was then kept meticulously every week until early 1930. The figures for 1930 (and particularly those for the Supercharged 4½ Litres) need to be regarded with caution, as they are uncaptioned. It is clear from the way the figures peter out and then stop for no apparent reason that WO was well aware of the situation and the likely outcome.

Looking at trends, 3 Litre sales consistently outstrip production with the exception of the second quarter of 1927. This verifies assumptions that the 3 Litre was not selling as quickly as was needed in 1925 and early 1926, resulting in the NR series of 3 Litres (being unsold 1925 cars updated and sold as 1926 cars). That lack of sales was, of course, a factor leading to the Barnato takeover. Sales of the 3 Litre fell away to almost nothing after the introduction of the 4½ Litre.

The 6½ Litres maintained a fairly steady sales rate, apart from a spurt coincident with the introduction of the Speed Six at the 1928 Motor Show.

If the figures for the Supercharged cars are indeed correct, it is clear right from the start that they were difficult to sell. The figures show an initial demand for both the Speed Six and the 4½ Litre that was not matched by the Supercharged cars.

The dramatic fall-off in sales in the latter half of 1929 and into 1930 is all too evident.

ANNUAL SALES

Year	3 Litre	6½ Litre	4½ Litre	Total
1927	238	105		343
1928	63	90	219	372
1929	16	134	244	394
1930		116	157	273
Total		445	620	

MONTHLY PRODUCTION AND SALES FIGURES

1926	3 Litre		6½ Litre	
	Sales	Prod	Sales	Prod
Jan				
Feb				
Mar				
Apr	24	14	2	8
May	27	23	2	8
Jun	16	12	4	8
Jul	24	9	8	15
Aug	21	3	4	9
Sep	8	9	11	13
Oct	22	15	13	17
Nov	28	10	17	15
Dec	18	11	13	15
1927				
Jan	14	15	7	17
Feb	13	14	16	12
Mar	23	12	8	11
Apr	8	11	12	11
May	13	15	11	10
Jun	7	12	4	9

1927 (contd)	3 Litre		6½ Litre		4½ Litre	
	Sales	Prod	Sales	Prod	Sales	Prod
Jul	7	12	12	8	43	6
Aug	4	0	2	8	6	9
Sep	3	0	0	16	23	20
Oct	3	0	9	13	24	26
Nov	3	0	2	10	33	20
Dec	3	0	13	9	31	22
1928						
Jan	1	0	12	5	21	25
Feb	6	0	4	7	11	28
Mar	5	0	9	6	27	36
Apr	1	3	5	4	11	24
May	2	3	10	4	23	24
Jun	4	3	9	7	18	29
Jul	1	5	7	11	28	5
Aug	0	3	6	10	21	12
Sep	5	0	13	12	15	30
Oct	1	0	18	9	33	24
Nov	2	0	29	8	39	24
Dec			15	13	12	25
1929						
Jan			3	14	8	20
Feb			11	11	14	21
Mar			8	15	22	28
Apr			6	12	19	19
May			7	13	12	20
Jun			20	14	14	25
Jul			25	11	22	18
Aug			12	12	20	18
Sep			4	12	10	14
Oct			8	14	7	8
Nov			7	21	15	10
Dec			12	10	7	7

1930	4½ Litre s/c					
	Sales	Prod				
Jan			4	11	5	0
Feb	0	1	4	11	11	4
Mar	0	2	5	6	15	19
Apr	5	4	7	0	9	16
May	2	7	1	2	7	16
Jun	0	6	2	0	4	8
Jul	0	1	0	12	15	0
Aug	0	4	0	5	6	0
Sep	0	2	0	1	1	0
Oct						
Nov						
Dec						

STOCK TURNOVER

Year		3 Litre	6½ Litre	4½ Litre	4½ Litre Supercharged
Mar 27	Sales	238	105		
	Prod	147	148		
	Balance	+ 91	− 43		
Mar 28	Sales	63	90	219	
	Prod	50	112	192	
	Balance	+ 13	− 22	+ 28	
Mar 29	Sales	16	134	244	
	Prod	17	118	266	
	Balance	− 1	+ 16	− 22	
Mar 30	Sales		116	157	5
	Prod		147	162	6
	Balance		− 31	− 5	− 1
Sep 30	Sales		10	42	0
	Prod		20	40	7
	Balance		− 10	+ 2	− 7

FINANCIAL FIGURES

Fortunately, some at least of the financial data survives, from two sources. One set of figures is reproduced from Ian Lloyd's "Rolls-Royce – The Years of Endeavour", these having been culled by Lloyd from Bentley Motors records preserved by Rolls-Royce. The other set comes from WO's own notebook, and the two show a high degree of correlation. They also (in both cases) substantiate the maligned figures for the cost of racing quoted by WO in his own books.

The direct expenditure figures quoted show the way in which the manufacturing profit was turned into a loss in almost every year. The small profit shown in 1930 was of course contrived. It is possible that shown in 1925 was similarly contrived, as they were at the time looking for a backer. The three years to March 1929, March 1930 and March 1931 show a steady fall-off in the number of chassis and income received, with no reduction in expenditure. The result of not laying off staff to match the fall-off in demand for the cars is shown up starkly by the increase in the percentage of wages to the chassis value, from 18.3% to 21.4% to a staggering 48.2% in 1931. It is little wonder that WO later admitted that it would have been better to have cut costs and laid off some of the staff. The invoices file from WO's notebook shows a steady fall-off in income from the first quarter of 1929 onwards.

From the indirect figures quoted by Lloyd, this lack of economy is shown by the fact that the indirect expenditure did not really change between 1928 and 1931. Indirect expenditure of course relates to all the activities of a company that are not directly related to the manufacturing, ie all staff not on the shop floor, design, development, sales, support activities, etc. In hard times it is almost invariably the indirect expenditure that is scrutinised in order to reduce costs, while not reducing the manufacturing side (which generates the income). That Bentleys failed to do this is indicative of wishful thinking at best, and at worst sheer incompetence.

Notes.
The indirect expenditure figures for 1931 of £85,071 is after deducting a manufacturing loss of £9291 and a loss on the machine shop of £7034.

The direct figures for chassis turnover excludes income from spares and repairs.

Although quite a lot of money must have been spent on the new machine shops and office block in 1929/30, there is no evidence of that money in the figures.

Concerning profit margins, the following (apparently hypothetical) extract from WO's notebook is revealing:

Materials and machining	550
Erecting and test	60
Scrap, modifications, petrol etc	30
	£640

Selling price 1200
Agent's discount (20%) 240
£960

Overheads £75000

Cars per year	Profit per year
250	5000
300	15000
350	37000
400	43200
450	69300
500	85000
550	101200
600	117000

Direct Expenditure (Lloyd)

Year	Number of Chassis	Value of Chassis	Discount	Discount %	Chassis cost	Gross manufacturing profit	Net profit
1919	0	0					− 5933
1920	0	0					− 10268
1921		26759					− 7405
1922							
1923							
1924	322	290000	42647	14.8	233824	13529	− 56700
1925						56484	19037
1926	306	280076	48299	17.1	198744	33033	− 19249
1927	330	380380	64534	17.1	258285	57561	− 2219
1928	337	393275	67803	17.3	238084	87388	1201
1929	414	498875	86315	17.2	312414	100146	28467
1930	301	407365	73378	17.9	249900	84087	1023
1931	147	198850	35640	18.0	145949	17261	− 84174

Year	Exp work	Exp %*	Sales	Sales %*	Wages	Wages %*	Depr	Depr %*
1919	7791		348		2425		78	
1920	8436		43		7197		137	
1921	8000	30.7	501	1.9	13310	51.1	265	0.1
1922								
1923								
1924	15284	5.1	6806	2.1	49315	17.0	1098	0.3
1925	6009		5188					
1926	8130	2.8	4949	1.7	63566	22.6	2190	0.7
1927	11026	2.9	19399	5.0	72592	19.1	2367	0.6
1928	10727	2.8	22322	5.5	84315	21.4	2580	0.6
1929	15600	3.0	26054	5.2	91542	18.3	3170	0.6
1930	23356	5.6	27609	6.7	98337	21.4	3540	0.8
1931	17505	8.7	21526	10.8	96539	48.2	8199	4.2

* As percentage of value of chassis

Indirect Expenditure (Lloyd)

Year	Total indirect	% of total direct	Experimental	Exp %	Sales & Racing	S & R %
1924	77968	33.4	15284	19.8	9553	12.4
1925			6009★			
1926	55359	27.9	8130	14.7	7361	13.3
1927	64724	25.1	11026	17.2	22768	35.5
1928	77212	32.4	10727	13.9	24938	32.3
1929	78667	25.1	13600	20.0	28541	36.5
1930	88253	35.4	23356	26.5	30979	35.2
1931	85071	63.6	17505	19.0	22080	25.9

Year	Admin	Admin %	Interest	Interest %	Loss on Service	Service %	Balance	Bal %
1924	11517	14.9	5379	6.9	9687	12.5	26548	35.7
1925	16360★		7122★		10050★			
1926	16657	30.2	7861	14.3	8557	15.5	6787	12.3
1927	15321	23.9	3862	6.0	11746	18.1		
1928	15960	20.7	4564	5.9	14642	19.0	6381	8.2
1929	16110	20.6	4622	5.9	11091	14.2	2703	3.4
1930	17096	19.4	5150	5.8	9030	10.2	2642	3.0
1931	20137	23.6	10586	13.6	9215	10.8	5548	6.5

Indirect Expenditure (W.O. Bentley)

Year	Racing	Admin	Interest	Service	Advertising	Works Overhead	Design	Sales
1925	833	16360	7122	10050	4811	13750	8220	9658
1926	2412	16657	7861	8558	4545	14902	8140	7186
1927	3369	15321	3862	11746	8613	14872	11026	10787
1928	2616	15960	4564	14642	9777	16567	14642	12546
1929	2487	16110	4662	11091	11060	17682	11091	14995

Invoices (W.O. Bentley)

Year	Quarter	6½	4½	3	Receipts
1928	2	17	74	7	137666
1928	3	24	56	10	121427
1928	4	32	77	0	149083
1929	1	41	68	2	153733
1929	2	44	58		147529
1929	3	36	48		125175
1929	4	31	30		95451
1930	1	23	33		97661
1930	2				91220

★Figures for 1925 are from W.O.

Appendix IV. Bentley Motors Policy

The following comes from W.O. Bentley's personal papers, held by the BDC. It is not dated, and there is no explanation attached to it, but nevertheless it makes interesting reading and is significant in that it comes from WO's own pen. It also raises again the old chestnut of the "bread and butter" car, which would suggest that it was written before *The Cars in My life*, which was published in 1961.

"There was a very definite policy behind the direction of Bentley Motors. These were;

1. To get a good name.
2. To build the best car in its class in the world and a cheaper smaller car when we had our own machine shops.
3. To make it a public company as soon as we were accepted as a good car with a good name and that we were on the road to make profits. Then we would have been able to build and equip a machine shop.

From 1922 until the 6½ and 8 Litres were brought out we made losses, not large ones, but we were always short of working capital which meant for one thing that we could never buy on favourable terms. During this time we were, however, building up a very good name, both in England and abroad. The introduction of the large six cylinder cars, owing to the fact that they could carry a much higher selling price, changed this loss into a good profit and in 1929 we made £28,467. In the first half of the next year the profit was higher.

Then two things happened; a financial slump started which cut off the sales of high priced cars everywhere, and secondly we brought out a bad car, bad as regards performance.

If the slump had not started we could have carried on, because the fault of the 4 Litre was its weight, and this was caused by our using the 8 Litre frame, gearbox, axles etc with a 4 Litre engine. Given the time we could have used up all the 8 Litre chassis parts on the 8 Litre and put the 4 Litre into a modified 4½ Litre frame using those axles, etc., which would have produced a good car.

We have been criticized for racing too much and spending and thinking only of racing. This is quite untrue, as the car we were out to produce wanted all the advertisement it could get and in no other way could we have achieved this. Our racing cost the firm:

	Racing	Other advertising
1925	£833	£4811
1926	£2412	£4545
1927	£3369	£8613
1928	£2616	£12546
1929	£2487	£14995

It will be seen from the above that racing cost considerably less than our other advertising and I don't think anyone will question from which we got the better value. We also improved the car very much from the lessons learnt in racing.

No-one was frightened off a 6½ Litre owing to its being the most successful car we had ever raced and we never intended to race the 8 Litre.

If the policy and achievements of Bentley Motors had not been eminently good, firms such as Napiers and Rolls-Royce would not have wanted to buy it when it went into liquidation, and it is known that this was not done with the idea of killing it. Napiers were going to produce a car called the Napier Bentley and Rolls-Royce had obviously realized the goodwill in the name and kept it very much alive".

Appendix V. Directors of Bentley Motors

First Company January 1919		*Other Directorships*
H.M. Bentley	15/1/19	Bentley & Bentley
W.O. Bentley	15/1/19	Bentley & Bentley
A.H.M.J. Ward	15/1/19	Bentley & Bentley
		The Aerolite Piston Co.

Second Company July 1919

H.M. Bentley	8/7/19	Bentley & Bentley
(Jt. Managing Director)		
W.O. Bentley	8/7/19	Bentley & Bentley
(Jt. Managing Director and Chairman)		
M.V. Roberts	8/7/19	Light Production Co.
(Resigned 9/6/20)		
A.H.M.J. Ward	17/6/20★	
(Resigned 26/10/20)		
W.S. Keigwin	17/6/20★	
(Replaced M.V. Roberts)		
C.F. Stead	26/10/20★	
(Replaced A.H.M.J. Ward)		
C.F. Boston	26/10/20★	
(Chairman after WO)		
H. Pike	14/3/22★	
G.A. Peck	14/3/22★	
C.L. Breeden	3/7/22★	
(Replaced Boston. Resigned 28/6/22)		
S.A. de la Rue	3/7/22★	Thos de la Rue & Co.
(Chairman after Boston)		
F.K. Prideaux Brune	26/6/23★	

Third Company March 1926

W.O. Bentley	2/3/26	
(Managing Director, later Jt. MD up to 10/30)		
H. Pike	2/3/26	
J.W. Barnato	2/3/26	
(Chairman)		
J.K. Carruth	2/3/26	
(Managing Director 10/30 on after WO)		
R. Manners	2/3/26	
(Deceased 26/1/30)		
Marquis of Casa Maury	4/5/27★	
(Jt. Managing Director to 10/30)		
Sir W. Sinclair	13/1/31★	British Goodrich Rubber Co
R.S. Witchell	13/1/31★	

★ Dates of returns, not dates of appointments.

Appendix VI. Director's Report

DIRECTORS' REPORT TO THE SHAREHOLDERS OF

BENTLEY MOTORS, LIMITED.

FINANCIAL YEAR ENDED 31ST MARCH, 1928.

→•←

In the course of the last financial year the most important change in the Company's programme has been the introduction of the 4½-litre model. This model has proved very popular, and has well maintained the Company's high reputation both for design and workmanship.

The audited accounts for the year ended 31st March, 1928, show a net profit of £1049 10s. 1d. During the first six months of the financial year there was a lull in sales which adversely affected trading results, and the Company incurred a loss throughout that period. There was a marked revival on the introduction of the 4½-litre model, and the profit earned in the latter half of the year was more than sufficient to counterbalance the earlier loss, notwithstanding the fact that during the latter period an amount of over £6000 was debited to the accounts in respect of the cost of modifications to existing 6-cylinder models, to which more detailed allusion is made below. Although general conditions throughout the motor trade for a considerable time are understood to have been very bad the Company has, ever since the introduction of the 4½-litre model, experienced a steady demand for its products which has enabled it to work both economically and profitably.

The sales of the 6-cylinder model have for some time shown a steady upward tendency. Although the reputation of the Bentley car for speed and reliability is universally established, the Company found at first a certain difficulty in entering the very conservative market for which the 6-cylinder model was designed. Any initial prejudice has now been overcome, and every fresh 6-cylinder Bentley car sold adds to the growing reputation of this model.

Great attention has been given to the record of all 6-cylinder Bentley cars sold and of their performance in the hands of very diverse types of drivers under every conceivable kind of traffic and road conditions. The testimony of all owners was such that it was clear that no fundamental change of design was called for, and this model remains, and is likely to remain, substantially the same as when first produced. Certain improvements, however, were made to the chassis at the date of the last Olympia Show. The most important of these improvements consisted of a greater cooling capacity and of a dynamo with a higher output driven directly from the crankshaft, and as a result of these improvements this model can now, under whatever conditions it may be used, challenge comparison with the most luxurious products of the oldest established manufacturers of all countries.

With a view to fostering the reputation of this model, your Directors decided that the improvements above cited should be made in all existing cars free of charge to their owners. The expense of doing so has necessarily been considerable. Approximately three quarters of the cars in question have been converted in the year under review and the cost of doing so amounted to £6378, the whole of which has been written off. This expenditure has reduced the net profit in the year to a very small margin but the advertising value of this move has been very great, and it is confidently hoped that this expenditure will lead to greatly increased profits in the future and to the firm establishment of the model in its class. The upward tendency of the sales of this model to which reference has been made above tends to confirm this view.

There have been no important changes in the specification of the 3-litre model. This model retains its popularity and it has a steady demand.

The Company has continued its policy of entering its cars for certain important speed and endurance trials, and it has always been your Directors' policy to enter for events where success is dependent on all round excellence of design rather than upon any one specific quality such as speed. In this context it might not be remembered that BENTLEY MOTORS, LIMITED, was the first Company of importance to anticipate the modern tendency to race cars of standard design, and the Company's success at the Tourist Trophy Race of 1922 in the Isle of Man deserves not to be forgotten. The success of the Company's entries at Le Mans in the year under review and in the current year is too well known for comment. It is not, however, so well known that at the Essex Six Hour Trial held at Brooklands on 12th May, where road conditions were simulated, the three 4½-litre Bentley cars entered were respectively 1st, 2nd, and 3rd.

It is gratifying to report that during the current year, owing to the generosity of certain of the drivers and to the kind support of various suppliers, the cost of competing at these events will not fall upon the Company.

During the year under review there has been no change in the composition of the Board of Directors. In accordance with Article 95 of the Company's Articles of Association, Captain WOOLF BARNATO retires by rotation and offers himself for re-election.

Shareholders are invited to write to the Company for further information on any specific point which may be of interest to them.

A copy of the audited accounts will be forwarded to any Shareholder on application to the Secretary.

Index